MEASUREMENT IN FLUID MECHANICS

Measurement in Fluid Mechanics is an introductory, up-to-date, general reference in experimental fluid mechanics, describing both classical and state-of-the-art methods for flow visualization and for measuring flow rate, pressure, velocity, temperature, concentration, and wall shear stress. Particularly suitable as a textbook for graduate and advanced undergraduate courses, *Measurement in Fluid Mechanics* is also a valuable tool for practicing engineers and applied scientists. This book is written by a single author, in a consistent and straightforward style, with plenty of clear illustrations, an extensive bibliography, and over 100 suggested exercises. *Measurement in Fluid Mechanics* also features extensive background materials in system response, measurement uncertainty, signal analysis, optics, fluid mechanical apparatus, and laboratory practices, which shield the reader from having to consult with a large number of primary references. Whether for instructional or reference purposes, this book is a valuable tool for the study of fluid mechanics.

Stavros Tavoularis is a Professor of Mechanical Engineering at the University of Ottawa and has served as Department Chair and Director of the Ottawa-Carleton Institute for Mechanical and Aerospace Engineering. His research interests include turbulence structure and diffusion, vortical flows, aerodynamics, biofluid dynamics, nuclear reactor thermal hydraulics, and experimental methods. Professor Tavoularis is a Fellow of the American Physical Society, the Canadian Academy of Engineering, and the Engineering Institute of Canada, as well as the Canadian Society for Mechanical Engineering. He is a recipient of the George S. Glinski Award for Excellence in Research.

Measurement in Fluid Mechanics

STAVROS TAVOULARIS

University of Ottawa

CAMBRIDGE UNIVERSITY PRESS
Cambridge, New York, Melbourne, Madrid, Cape Town, Singapore,
São Paulo, Delhi, Dubai, Tokyo, Mexico City

Cambridge University Press
32 Avenue of the Americas, New York NY 10013-2473, USA

Published in the United States of America by Cambridge University Press, New York

www.cambridge.org
Information on this title: www.cambridge.org/9780521138390

First published 2005
Reprinted 2006
First paperback edition 2009

A catalogue record for this publication is available from the British Library

Library of Congress Cataloging in Publication Data

Tavoularis, Stavros.
Measurement in fluid mechanics / Stavros Tavoularis.
p. cm.
Includes bibliographical references and index.
ISBN-13: 978-0-521-81518-5 (hardback)
ISBN-10: 0-521-81518-5 (hardback)
1. Fluid mechanics – Measurement. I. Title.
QA901.T38 2005
532 – dc22 2 005014865

ISBN 978-0-521-81518-5 Hardback
ISBN 978-0-521-13839-0 Paperback

To Sofia, Christina and Jason

Contents

Preface

The purpose of experimental fluid mechanics is to measure the properties of a flowing fluid. Combined with theoretical analysis, measurements are used for understanding the operation of a fluid-containing system and then applying this knowledge towards designing improved systems and predicting their future operation. One may also use measurement to monitor and control a physical process, thus ensuring efficient and safe operation of a system. Performing a fluid mechanics experiment requires theoretical and practical knowledge and skills from a variety of fields of science and engineering. The experimental fluid mechanicist will likely need, in addition to a solid education in fluid mechanics, an advanced background in material properties, physics, mathematics, statistics, and electronics, with the list often expanding to include computer science, chemistry, biology, physiology, and environmental sciences. Much of the necessary background is covered in typical engineering education curricula, although segmented and presented in ways that are not focussed on the needs of experimental fluid mechanics. The diversity of background information, combined with the need for in-depth understanding of many different topics, can be intimidating to the novice in this field. Conducting an apprenticeship of substantial length under the guidance of an experienced experimentalist would certainly be the most sensible approach, but not one that is always available or compatible with time constraints. The next option is to learn through published literature. A literature search in even a narrow aspect of experimental fluid mechanics will most likely reveal an overwhelmingly lengthy list of related sources, widely uneven in scope, objectives, and styles. One would have to steer judiciously among these sources in order to identify and extract the truly needed material. This is by no means a negative reflection on the fluid mechanics community, which has put extraordinary efforts in disseminating the available knowledge in hundreds of books, review articles, and reports, both at introductory and advanced levels. It reflects on the understandable frustration of the non-expert when dealing with expert-written material. Some sources are very specialized and advanced, presuming that the reader is already familiar with the topic and has readily available all required background. Many available sources of broad scope constitute collections of separate articles, with little or no connecting material among the different topics. In other cases, the information presented is practical and targeted towards a specific audience, such as process engineers or technologists.

The present book is an attempt to fill in the observed need for a consistently written, introductory-level, up-to-date, general reference in experimental fluid mechanics. Its main intended use is as a textbook in an introductory graduate course, and, in fact, the material is based on a set of notes I developed over several years for such a course at the University of Ottawa. Selected sections may also serve as a textbook in an advanced undergraduate or a combined undergraduate–graduate course on this topic. The book contains extensive background material to shield the reader from having to consult a large number of primary references in diverse areas of science and engineering. The book may also be of interest to practicing engineers and applied scientists in many areas of application, as much of the instrumentation and methods described here are used not only in fluid mechanics research but also in many other fundamental and applied fields.

Like all areas of engineering and science, experimental fluid mechanics has been profoundly influenced by recent advances in electronics, optics, computers, and information technology. Yet most experimental methods are based on classical scientific principles, which must be understood well for their correct application. The emphasis in selecting and presenting the material was on time-resisting fundamentals, rather than on giving a detailed description of the latest technologies, which, in any case, would likely be of ephemeral duration. A main strength of any educational material is its use of illustrations. I have tried to supplement the text with simple, consistent sketches and plots, the great majority of which are original, although often based on previously published illustrations. Considering the breadth of the included topics and the diversity, in quality and style, of the available information, this represents an option for uniformity and clarity, rather than exactness in scale and completeness. To restrict the length of the exposition and the cost of publication, a significant number of methods discussed briefly in the text have not been accompanied by illustrations, and the reader is referred to the cited references for further details. Fluid mechanics is a field that is distinguished by the ample availability of images, often spectacular ones, illuminating the physical phenomena under study and suitable for both qualitative and quantitative purposes. Once more, restrictions on the length and cost of the present book have dictated the inclusion of only a small number of such images, mostly obtained by relatively modest means. The reader and instructors who consider using this book as a textbook are encouraged to augment the material with images easily accessible in collective works. Examples of suitable sources include *An Album of Fluid Motion* (assembled by M. Van Dyke, Parabolic Press, Stanford, California, 1982), *A Gallery of Fluid Motion* (edited by M. Samimy et al., Cambridge University Press, Cambridge, UK, 2003), *Visualized Flow* (compiled by the Japan Society of Mechanical Engineers, Pergamon, Oxford, UK, 1988), and the website *www.efluids.com.*

It is impossible to acknowledge all persons who provided ideas, specific material, or criticisms on the different topics discussed in this book. The long-lasting influence of my mentor, the late Stan Corrsin, has unquestionably affected the style and organization of the material, particularly the urge for clarity of presentation, whether it was actually achieved or not. The input and feedback of the many students who attended my classes have had a strong effect on the selection of topics and the level and scope of the presentation. During recent years, while the book was getting formalized, I gratefully acknowledged the valuable suggestions of the following individuals, in alphabetical order: Yiannis Andreopoulos, Sean Bailey, Warren Dunn, Mohamed Gad-el-Hak, Gordon Holloway, Jacques Lewalle, Martin Maxey, Cliff Weissman, and Phil Zwart. Conscious of possible limitations in the present edition, I welcome any feedback and suggestions of all readers, which I would gladly consider in future amendments or revisions.

Stavros Tavoularis
Ottawa, 2005

PART ONE

GENERAL CONCEPTS

1 Flow properties and basic principles

Before being able to measure a flow property, it is necessary to understand its nature and its relationship to other properties. Furthermore, for a proper use of a measuring instrument, one must be thoroughly familiar with the principles of its operation, which usually involve concepts and relationships from several different fields. In this chapter, the basic principles of fluid mechanics are reviewed and the properties of interest and their groupings in dimensionless form are identified. This is meant to be a refresher of familiar concepts, as well as to identify a possible need for more in-depth reviews of fluid mechanics [1–3], thermodynamics [4], and heat transfer [5]. Background material from system dynamics, signal analysis, and optics is reviewed separately in later chapters.

1.1 Forces, stresses, and the continuum hypothesis

All material objects are subjected to external forces, which are of two types, body forces and surface forces. *Body forces* act on the bulk of the object from a distance and are proportional to its mass; the most common examples are gravitational and electromagnetic forces. *Surface forces* are exerted on the surface of the object by other objects in contact with it; they generally increase with increasing contact area.

Any surface force acting on an elementary surface section of an object can be decomposed into a *normal* component, with a direction normal to the local tangent plane, and a *tangential* or *shear* component, with a direction parallel to the local tangent plane (see Fig. 1.1).

The *stress* at a point of an object is defined as the corresponding surface force per unit area; consequently there are two types of stresses, *normal stresses* and *shear stresses*. With respect to a Cartesian coordinate system, all stresses acting on three planes normal to the three axes form a second-order Cartesian tensor, which has nine components, only six of which (three normal stresses and three shear stresses) are independent. In classical fluid mechanics, the (static) *pressure* is defined as the average normal stress along any three orthogonal directions.

According to the classical definition, a *fluid* is a material that cannot withstand a shear stress when at rest; in other words, a fluid subjected to a shear stress will always be in motion or deformation. Fluids are easily deformable materials and take the shape of

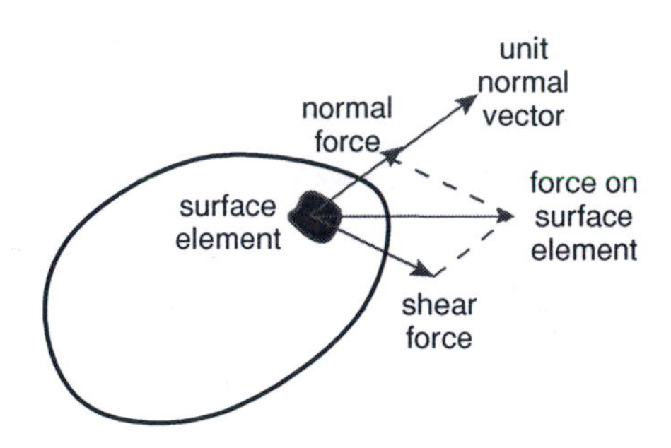

Figure 1.1. Force on a surface element and its decomposition to normal and shear components.

any container in which they are contained. They are further distinguished into *liquids*, which have a relatively high density and require an extremely large change of normal stresses for a change of their volume, and *gases*, which have a relatively low density and can easily change their volume. Unlike liquids, gases tend to occupy the entire available volume of their container. Besides 'simple' liquids and gases, there are a number of materials that, although not satisfying all properties of classical fluids, exhibit fluid-like properties. Examples include viscoelastic materials, which may sustain a certain amount of shear stress without being set in motion but behave like fluids when the shear stress exceeds a certain level, and plasmas, which form when gases are exposed to extremely high temperatures, in which case their molecules are dissociated into free atoms.

In most applications within the scope of conventional fluid mechanics, the phenomena of interest are characterized by scales that are far larger than the distances between molecules. Then the flow properties are defined as statistical averages over a volume that contains a very large number of molecules. In such cases, one need not be concerned about individual molecular or atomic motions and masses; instead, one should invoke the *continuum hypothesis*, by which any property of the fluid is assumed to have a continuous distribution within the volume of the fluid. Thus one may define the *local* value of the property as the limit of the volume-averaged value of this property as the volume collapses towards a mathematical point. One may refer to a *fluid element*, or *fluid particle*, as a material entity that has an infinitesimal volume, in which case its properties are uniform within this volume. In multiphase flows of immiscible fluids, the continuum hypothesis applies within each individual fluid, whereas some properties may be considered as discontinuous at the interface. Obviously there are also situations in which the continuum hypothesis does not apply at all; for example, in rarefied gases, in which the distances between gas molecules are relatively large, one must account for individual molecules and their motions.

1.2 Measurable properties

A property of a fluid element can be measured directly or estimated from measurements of other properties only if it has a precise and unambiguous scientific definition, associated with a measurement procedure. The following list identifies measurable properties of common interest, classified into four general classes.

- *Material properties:*

  mass | density | specific volume
  viscosity | thermal conductivity | molecular diffusivity
  specific heat under constant pressure | specific heat under constant volume | gas constant
  bulk modulus of elasticity | coefficient of thermal expansion | electric conductivity
  surface tension | index of refraction | fluorescence

- *Kinematic properties*, namely properties that describe the motion of a fluid without consideration of applied forces:

  position | displacement | velocity
  volume flow rate | mass flow rate | acceleration
  vorticity | strain rate | angular position
  angular displacement | angular velocity | angular acceleration
  momentum | angular momentum

- *Dynamic properties*, namely properties related to the applied forces:

  force | stress | torque
  pressure (mechanical definition)

- *Thermodynamic properties*, namely properties related to heat and work:

  temperature | internal energy | enthalpy
  entropy | heat flux | work
  energy | pressure (thermodynamic definition)

Material properties are usually not the subject of experimental fluid mechanics, as their values can be found in handbooks [6] or other sources. However, if a material property is unknown or overly sensitive to the particular experimental conditions, its value may have to be determined either as part of the overall experiment or by a specific experimental investigation.

1.3 Flow velocity and velocity fields

A position in space is specified in terms of its coordinates x_i, $i = 1, 2, 3$, with respect to a Cartesian coordinate system. At any time t, this position is occupied by some fluid element, assumed to maintain its mass within an infinitesimally small volume. With the fluid considered as a continuum, we may define the flow velocity at a given position and a given time as the velocity of a fluid element that occupies that position at that time. We are also interested in defining the *velocity field*, which consists of the velocities of all fluid elements that comprise a material system. Thus it becomes necessary to distinguish the fluid element in question from any other fluid element. For clarity, we specify as X_i the coordinates of the fluid element that occupies position x_i at time t. This element moves along its trajectory, indicated by the dotted curve in Fig. 1.2. At time $t + \delta t$, the

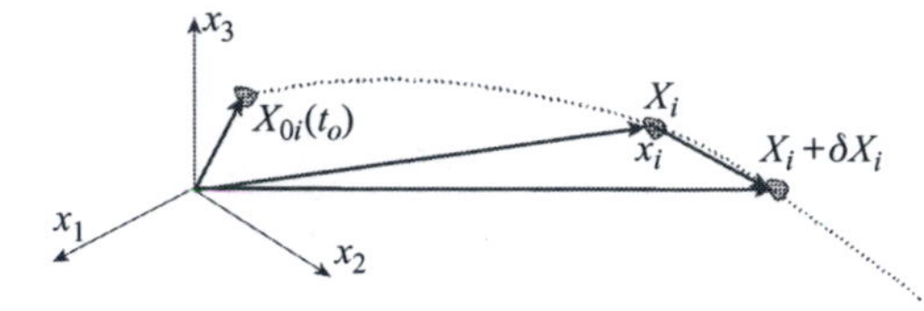

Figure 1.2. Fluid element positions.

same element will have the coordinates $X_i + \delta X_i$. Then the *flow velocity* is defined as

$$V_i = \lim_{\delta t \to 0} \frac{\delta X_i}{\delta t} = \frac{\mathrm{d}X_i}{\mathrm{d}t}. \tag{1.1}$$

One approach to identifying the fluid element is to specify its initial coordinates X_{0i}, namely the coordinates of the position that the element occupied at the origin of time t_o. Then its coordinates at any time t are functions of only the initial coordinates and t, namely $X_i = X_i(X_{0i}, t)$. The velocity field may also be specified as a function of these two variables, as $V_i = V_i(X_{0i}, t)$. This approach is known as the *material* or *Lagrangian* description of flow motion. Because identifying ('tagging') individual fluid elements is not usually practical, it is also customary to express the velocity field in terms of a fixed position with respect to the coordinate system and, of course, time; this approach is known as the *spatial* or *Eulerian* description. To avoid confusion when differentiating, the velocity field according to the Eulerian description is denoted by a different symbol, as $U_i(x_i, t)$; it is understood nevertheless that, at all positions and for all times, the definition of flow velocity is unique and that

$$V_i(X_{0i}, t) = U_i(x_i, t). \tag{1.2}$$

Following the Lagrangian description, the fluid element *acceleration* is defined as

$$a_i = \frac{\mathrm{d}V_i}{\mathrm{d}t}. \tag{1.3}$$

However, if we follow the Eulerian description, we must account for changes of the velocity from one location of the particle to the next, and the fluid acceleration becomes

$$a_i = \frac{\partial U_i}{\partial t} + U_1 \frac{\partial U_i}{\partial x_1} + U_2 \frac{\partial U_i}{\partial x_2} + U_3 \frac{\partial U_i}{\partial x_3}, \tag{1.4}$$

where the first term on the right-hand side is called the *local acceleration* and the remaining terms are called the *convective acceleration*. The right-hand side of Eq. (1.4) is called the *material* or *substantial derivative* of U_i and is usually denoted as $\mathrm{D}U_i/\mathrm{D}t$.

Fluid deformation under the influence of stresses is described by the *rate of strain*, or *rate of deformation, tensor*:

$$e_{ij} = \frac{1}{2}\left(\frac{\partial U_i}{\partial x_j} + \frac{\partial U_j}{\partial x_i}\right), \quad i, j = 1, 2, 3. \tag{1.5}$$

1.4 Analytical description of flows

The physical relationships among various flow properties are represented by analytical expressions, which constitute algebraic, differential, integral, or integrodifferential equations. Such relationships include *axiomatic principles*, which are generally accepted as natural laws subjected to experimental verification, and *empirical relationships*, which range from semi-theoretical concepts, based on physical arguments with some experimental input, to purely empirical expressions, obtained by statistical curve fitting to experimental results.

Conventional fluid mechanics and thermodynamics are based on four basic axiomatic principles: the conservation of mass, the momentum principle, and the first and second laws of thermodynamics. When applied to a *closed fluid system*, these principles result in a set of integral relationships that describe the fluid motion. However, it is usually more convenient to apply these principles to the contents of a fixed *control volume*, in which case one may derive the following four basic relationships.

- The *conservation of mass* or *continuity* equation:

$$\frac{\partial}{\partial t}\int_{\mathcal{V}} \rho \mathrm{d}\mathcal{V} + \int_{\mathcal{A}} \rho \vec{V} \cdot \mathrm{d}\vec{A} = 0, \tag{1.6}$$

  where ρ is the fluid density, $\vec{V}$ is the velocity vector, $\mathcal{V}$ is the control volume, and $\mathcal{A}$ is the area of its surface. For non-reacting multicomponent flows, one may formulate similar equations expressing the *conservation of species*.

- The *momentum* equation (Newton's second law), which, for an inertial coordinate system, can be written as

$$\vec{F} = \frac{\partial}{\partial t}\int_{\mathcal{V}} \vec{V} \rho \mathrm{d}\mathcal{V} + \int_{\mathcal{A}} \vec{V} \rho \vec{V} \cdot \mathrm{d}\vec{A}, \tag{1.7}$$

  where $\vec{F}$ is the net external force acting on the control volume.

- The *energy* equation (first law of thermodynamics):

$$\dot{Q} - \dot{W} = \frac{\partial}{\partial t}\int_{\mathcal{V}} \left(u + \frac{1}{2}V^2 + gz\right) \rho \mathrm{d}\mathcal{V} + \int_{\mathcal{A}} \left(u + \frac{p}{\rho} + \frac{1}{2}V^2 + gz\right) \rho \vec{V} \cdot \mathrm{d}\vec{A}, \tag{1.8}$$

  where $\dot{Q}$ is the rate of heat transfer from the surroundings to the control volume, $\dot{W}$ is the mechanical power produced by moving solid components, shear stresses acting on the boundary or electromagnetic forces, but not normal stresses acting on the control volume (the latter have been included on the right-hand side), u is the specific internal energy, p is the pressure, z is an upwards vertical axis, and g is the gravitational acceleration.

- The *second law* of thermodynamics:

$$\frac{\partial}{\partial t}\int_{\mathcal{V}} s \rho \mathrm{d}\mathcal{V} + \int_{\mathcal{A}} s \rho \vec{V} \cdot \mathrm{d}\vec{A} \geqslant \int_{\mathcal{V}} \frac{1}{T}\left(\frac{\dot{Q}}{A}\right) \mathrm{d}A, \tag{1.9}$$

  where s is the specific entropy and T is the absolute temperature.

By letting the control volume vanish towards a point, one may convert the previous equations into differential forms (see examples in the next section). These must be complemented by initial and boundary conditions. The usual conditions employed at a solid boundary in conventional fluid mechanics are the *no-penetration* condition, which specifies that the relative velocity normal to the contact surface vanishes, and the *no-slip* condition, which specifies that the relative velocity tangential to the contact surface vanishes. In two-phase flows, the pressure difference Δp across the interface is related to the surface tension σ by the expression

$$\Delta p = \sigma \left(\frac{1}{R_1} + \frac{1}{R_2} \right), \tag{1.10}$$

where R_1 and R_2 are the principal radii of curvature; obviously, $\Delta p = 0$ across plane interfaces. In liquid–gas flows, it is customary to neglect the shear stress on the liquid side of the interface (*stress-free boundary*).

Certain types of flow require the use of additional axiomatic principles, such as the following ones:

- The *laws of chemical reaction* for reactive flows.
- *Magnetohydrodynamic laws* for electrically conductive fluids in magnetic fields.

Among the common types of empirical relationships, one may mention the following ones:

- *Equations of state* or *constitutive relationships*, relating various thermodynamic properties of a material. The simplest example of an equation of state is the perfect-gas law,

$$\frac{p}{\rho} = RT \tag{1.11}$$

  where R is a gas constant.
- *Stress–strain rate* (also called 'constitutive') *relationships*. Besides the linear stress–strain rate relationship that describes *Newtonian fluids*, a variety of other such relationships have been proposed to describe deformation of *non-Newtonian fluids*.
- *Turbulence models*, which are relationships among various statistical properties of turbulent flows. A widely used turbulence model is the gradient transport model, by which the turbulent stresses are linearly related to the mean strain.

1.5 The choice of analytical approach

Although accurate mathematical models of fluid flow are available, it is generally advisable, and even necessary, to employ simplifications whenever possible and to the greatest possible extent. This strategy has been applied extensively to the analysis of measuring instrument operation. As a rule, one should strive to use the simplest possible analytical model that permits the desired measurement with an acceptable uncertainty. Of course, the differences between an approximate and a more 'exact' method must be analysed, and, if they are found to be excessive, one must either apply appropriate corrections or abandon the approximate method in favour of a more exact one.

Among the important effects, which may or may not be accounted for in a particular type of analysis, are deformation, friction, compressibility, and turbulence; additional effects complicating the analysis may be present under specific circumstances. Accordingly, one may distinguish the following theoretical approaches used commonly to describe fluid motion:

1. **Fluid statics.** When a fluid is at rest, or in rigid-body motion, it does not deform, and therefore it cannot sustain shear stresses. In fluid statics, the three normal stresses are equal in magnitude to each other and to the pressure. Static fluids are subjected to gravity, which causes the development of hydrostatic pressure and buoyancy force. Static fluid analysis is useful in analyzing the performance of many instruments, notably those of manometers, barometers, and certain types of pressure transducers. On the other hand, the use of static fluid analysis in situations in which there are fluids in motion could lead to substantial errors. For example, a static analysis of a liquid manometer subjected to pressure fluctuations would result in erroneously low readings of pressure differences.
2. **Inviscid incompressible flows.** The simplest mathematical model of fluid flow is one that neglects the effects of friction and compressibility. Continuity imposes the requirement of conservation of volume of an incompressible fluid. In differential form, this can be expressed as

$$\frac{\partial U_1}{\partial x_1} + \frac{\partial U_2}{\partial x_2} + \frac{\partial U_3}{\partial x_3} = 0. \tag{1.12}$$

When friction is neglected, turbulence must also be disregarded. The differential momentum equation for an *inviscid, incompressible* fluid is known as the *Euler* equation:

$$\frac{\mathrm{D}U_i}{\mathrm{D}t} = g_i - \frac{1}{\rho}\frac{\partial p}{\partial x_i}. \tag{1.13}$$

Together, the continuity equation and the Euler equation form a closed system, which is sufficient for the determination of the fluid velocity and pressure. Integration of Euler's equation along a streamline leads to the simple and frequently used (as well as misused), steady-flow *Bernoulli*'s equation, which is an algebraic expression relating velocity magnitude and pressure as

$$p + \frac{1}{2}\rho U^2 + \rho g z = \text{const.} \tag{1.14}$$

The flow analysis can be further simplified by the assumption of *irrotationality*, namely the vanishing of *vorticity*,

$$\vec{\zeta} = \operatorname{curl} \vec{U}, \tag{1.15}$$

everywhere in the flow domain, with the possible exception of isolated singularities. Such a flow is called *potential* and is described by the *velocity potential*, which satisfies Laplace's equation. In potential flow, Bernoulli's equation can be applied not only along a streamline but also from one streamline to another. Potential flow analysis is a common approach in aerodynamics. It is also used to explain the operation of several simple instruments. For example, the measurement of velocity with the use of immersed pressure tubes routinely employs the steady-flow Bernoulli's equation; this approach is acceptable in many wind-tunnel applications, in which the effects of friction are known to be below the uncertainty level, but it introduces large errors in measurements near walls, where friction effects are important, or in high-speed flows, for which compressibility must be accounted.

3. **Viscous incompressible flows.** Application of Newton's second law to an incompressible, Newtonian fluid with constant material properties leads to the *Navier–Stokes* equations:

$$\frac{\mathrm{D}U_i}{\mathrm{D}t} = g_i - \frac{1}{\rho}\frac{\partial p}{\partial x_i} + \nu\left(\frac{\partial^2 U_i}{\partial x_1^2} + \frac{\partial^2 U_i}{\partial x_2^2} + \frac{\partial^2 U_i}{\partial x_3^2}\right), \tag{1.16}$$

where ν is the kinematic viscosity. The main parameter characterizing such flows is the *Reynolds number* Re, which represents the ratio of inertia forces and viscous forces. In the limiting case of vanishing Re, inertial effects become negligible and the Navier–Stokes equation is reduced to the *Stokes* equation [3,7], which contains no non-linear terms. The form of applicable solutions and the measuring instrument response in low-Re flows are radically different from those at higher Re. On the other extreme, as Re $\to \infty$, one would expect that viscous effects should become negligible. However, this limit contradicts the physical fact that, as the Reynolds number increases, a flow would become increasingly unstable and eventually turbulent. As a rule, at high Re, one must consider friction as well as account for turbulence effects. The infinite-Re limit also leads to an analytical singularity because the order of the Navier–Stokes equation changes from second to first.

4. **Compressible flows.** In compressible flow, density variation is significant. Density changes are related to pressure changes and are also accompanied by temperature changes. Unlike incompressible flow, compressible flow is not divergence free, and the corresponding differential continuity equation includes the unknown variable density. Compressible flow models contain, besides velocity and pressure, two additional unknowns, density and temperature, and therefore require two additional equations. One independent differential equation (*energy* equation) is provided by the first law of thermodynamics, and a second relationship is provided by an equation of state. To simplify the analysis of compressible flow, friction is often neglected. *Isentropic* flow, namely a reversible and adiabatic change of state, is commonly used as an approximation in compressible flow instrumentation. Allowing for density changes permits the modelling of propagation of weak disturbances with a finite speed, which, for isentropic flows, is called the *speed of sound*. The ratio M of fluid velocity to the speed of sound, called the *Mach number*, is a dimensionless parameter describing the effects of compressibility. It can be viewed as representing the ratio of inertia forces to elastic forces. When $\mathrm{M} > 1$ (supersonic flow), dramatic and

sudden changes of fluid velocity and thermodynamic properties may occur across a *shock wave*, which is an irreversible, and therefore non-isentropic, process.

5. **Turbulent flows.** Turbulence is a state of motion in which flow properties, including velocity, pressure, and vorticity, vary rapidly and randomly in space and time. Randomness requires a statistical description, which is customarily obtained by the decomposition of each variable property into a *mean* and a *fluctuation* (*Reynolds decomposition*). Statistical equations for the means and fluctuations have been formulated but lead to systems containing unknowns whose number exceeds the number of available equations. Turbulence affects the operation of even the simplest instruments. In some cases, relatively simple corrections for turbulence effects have been devised. An important parameter is the *turbulence intensity*, i.e., the ratio of the root-mean-square velocity fluctuation and the corresponding mean velocity; however, information on the time and length scales of turbulent motions is often required in the correction method. The measurement of turbulent properties themselves can be achieved with the use of special instrumentation that has particularly refined spatial, temporal, and amplitude resolutions. Special methods, including *phase averaging* and *conditional sampling*, have been developed to resolve quasi-deterministic turbulence patterns, known as *coherent structures*.
6. **'Complex' flows.** Fluids and flows are sometimes complicated by particular conditions that necessitate the use of specialized analytical models and methods as well as special instrumentation, calibration, and measuring procedures. Of particular interest are the following cases:

   - *Multiphase* flows, including liquid–solid, gas–solid, gas–liquid, and liquid–liquid flows, and flows with phase change.
   - Flows with *chemical reaction*, including combustion. These involve changes of composition, accompanied by release or absorption of chemical energy and changes of temperature.
   - *Low-density* (*rarefied*) gas flows, for which the continuum hypothesis is inappropriate and whose measurement relies on molecular and atomic phenomena.
   - Flows of *non-Newtonian* fluids having nonlinear stress–strain rate relationships (e.g., polymer flows).
   - *Magnetohydrodynamic* flows, namely flows of electrically conductive fluids in magnetic fields.

1.6 Similarity

Similarity and non-dimensionalization: Measurable properties may be either *primary*, namely independent of each other, or *secondary*, namely related to other properties through their definition or a basic principle. International conventions regulate the choice of a sufficient set or primary properties (called *dimensions*) and appropriate standard amounts (called *units*) that may be used as scales for expressing magnitudes. The SI (Système International d'Unités) System of Units has adopted the following properties as primary (the corresponding units and their abbreviations are given in the second and third columns, respectively):

Property	*Unit name*	*Abbreviation*
length	metre	m
mass	kilogram	kg
time	second	s
temperature	degrees Kelvin	K
electric current	ampere	A
amount of substance	mole	mol
luminous intensity	candela	cd
plane angle	radian	rad
solid angle	steradian	sr

Although measurement may occasionally be performed directly on the actual system of interest under actual operating conditions, more often than not the measurement is performed either on a model of the system or on the actual system but under modified and controlled conditions, which may differ from the actual operating conditions. The concept of *similarity* (also known as *similitude*) permits the application of information obtained with models to the actual systems they represent. An essential requirement is that an actual system and its experimental model be *geometrically similar*; this means that they should have the same shape, so that the ratios of all corresponding dimensions are the same. In addition, the flows in the model and in the actual system must be *kinematically similar*; this requires that the velocities at corresponding positions have the same directions and a constant ratio of their magnitudes. For complete similarity, the flows must also be *dynamically similar*; this requires that the corresponding forces have the same directions and a constant ratio of their magnitudes. When two flows are similar, results collected in one may be used for the prediction of the other, although usually not directly but following proper *scaling*, i.e., resizing of the values of the different properties. Measurable flow properties are normally *dimensional,* having their magnitudes expressed as multiples of the corresponding units. *Non-dimensionalization* of the results is the process of converting measured properties into dimensionless numbers by dividing their magnitudes by appropriate scales. Besides the economy in presentation, non-dimensionalization produces results that are independent of the unit system, and it serves as a guide for the selection of optimal geometrical and operating conditions. Similarity and non-dimensionalization permit the derivation of scaling laws and the design of models.

Common dimensionless parameters: A large number of dimensionless groups of various flow properties have already been identified in previous analyses of problems in fluid mechanics, heat and mass transfer, and related fields. In many cases, these groups appear naturally, when all terms in the governing equations are converted to dimensionless forms by dividing all properties by appropriate scales, a process often called *similarity analysis*. To illustrate similarity analysis, consider the Navier–Stokes equations, equation (1.16), for incompressible flow with constant viscosity in a gravitational field. Assume that a length scale L and a velocity scale V_0 can be identified, for example as a characteristic dimension of the boundary and an average or free-stream velocity. Then one may non-dimensionalize all distances and velocity components by dividing them by the corresponding scales, and time is made dimensionless by the time scale L/V_0. Assume further that pressure variation is non-dimensionalized by a reference value p_0, whereas the gravitational acceleration vector is non-dimensionalized by its magnitude. Denoting dimensionless variables by asterisks, one may then express the Navier–Stokes equations in dimensionless form as

$$\frac{\mathrm{D}^* U_i^*}{\mathrm{D}^* t^*} = \frac{gL}{V_0^2} g_i^* - \frac{p_0}{\rho V_0^2} \frac{\partial p^*}{\partial x_i^*} + \frac{\nu}{V_0 L} \left(\frac{\partial^2 U_i^*}{\partial x_1^{*2}} + \frac{\partial^2 U_i^*}{\partial x_2^{*2}} + \frac{\partial^2 U_i^*}{\partial x_3^{*2}} \right), \quad i = 1, 2, 3. \tag{1.17}$$

The dimensionless coefficients of the three terms on the right-hand side have been identified as the squared reciprocal of the *Froude number*, the *Euler number*, and the reciprocal of the *Reynolds number* for the system. The left-hand side of Eq. (1.17) is the dimensionless acceleration of a fluid element; brought to the right-hand side, it may be considered as representing a dimensionless 'inertia force', which balances the external forces. The three terms on the right-hand side represent, respectively, the dimensionless gravitational, pressure, and viscous forces. Thus the dimensionelss groups just identified may be considered as representing the relative magnitudes of external forces, compared with the inertia force.

Among the most common dimensionless groups that are encountered in experimental fluid mechanics are the following.

- *Reynolds* number:

$$\mathrm{Re} = \frac{\rho V d}{\mu}, \tag{1.18}$$

where V is a characteristic velocity, d is a characteristic length, and μ is the viscosity; it represents the ratio of the inertia forces to the viscous forces.

- *Mach* number:

$$\mathrm{M} = \frac{V}{c}, \tag{1.19}$$

where c is the speed of sound; it represents the ratio of the inertia and elastic forces (compressibility).

- *Pressure* coefficient (or *Euler* number):

$$C_{\mathrm{P}}(\equiv \mathrm{Eu}) = \frac{p - p_{\mathrm{ref}}}{\frac{1}{2}\rho V^2}, \tag{1.20}$$

where p_{ref} is a reference pressure; it represents the ratio of pressure and inertia forces. Some sources define Eu as one half of the preceding pressure coefficient.

- *Drag* coefficient:

$$C_{\mathrm{D}} = \frac{F_{\mathrm{D}}}{\frac{1}{2}\rho A V^2}, \tag{1.21}$$

where F_{D} is the drag force and A is the frontal (i.e., normal to the free stream) area of the immersed object for bluff objects or the planform (i.e., projected on a plane parallel to the free stream) area for streamlined objects; it represents the ratio of the drag force and the inertia force.

- *Lift* coefficient:

$$C_{\mathrm{L}} = \frac{F_{\mathrm{L}}}{\frac{1}{2}\rho A V^2}, \tag{1.22}$$

where F_{L} is the lift force and A is the planform area; it represents the ratio of the lift force and the inertia force.

- *Prandtl* number:

$$\mathrm{Pr} = \frac{\nu}{\gamma} = \frac{c_p \mu}{k}, \tag{1.23}$$

where γ is the thermal diffusivity, c_p is the specific heat under constant pressure, and k is the thermal conductivity of the fluid $[\gamma = k/(\rho c_p)]$; it represents the ratio of the rates of diffusion of momentum and heat that are due to molecular motions.

- *Schmidt* number:

$$\mathrm{Sc} = \frac{\nu}{\gamma_c}, \tag{1.24}$$

where γ_c is the molecular diffusivity of a species in a fluid mixture; it represents the ratio of the rates of diffusion of momentum and mass in the fluid.

- *Froude* number:

$$\mathrm{Fr} = \frac{V}{\sqrt{gL}}, \tag{1.25}$$

where L is a characteristic length; Fr^2 represents the ratio of inertia to gravitational forces and applies mainly to liquid flows with a free surface.

- *Weber* number (for liquids):

$$\mathrm{We} = \frac{\rho V^2 L}{\sigma}, \tag{1.26}$$

where σ is the surface tension; We represents the ratio of the inertia to the surface-tension forces.

- *Capillary* number (for two-phase flows):

$$Ca = \frac{\mu V}{\sigma} = \frac{\mathrm{We}}{\mathrm{Re}}; \tag{1.27}$$

it represents the ratio of viscous forces to surface-tension forces.

- *Cavitation* number (for liquids):

$$\sigma_c = \frac{p - p_v}{\frac{1}{2}\rho V^2}, \tag{1.28}$$

where p_v is the vapour pressure.
- *Nusselt* number:

$$\mathrm{Nu} = \frac{hL}{k}, \tag{1.29}$$

where h is the overall heat transfer coefficient and k is the thermal conductivity of the fluid; it represents the ratio of total and conductive heat transfer rates in a fluid.
- *Biot* number:

$$\mathrm{Bi} = \frac{hL}{k}, \tag{1.30}$$

where h is the overall heat transfer coefficient from a solid surface to a fluid and k is the thermal conductivity of the solid; it represents the ratio of heat transfer rates to the surrounding fluid and the solid interior.
- *Péclet* number:

$$\mathrm{Pe} = \frac{VL}{\gamma} = \mathrm{Re}\,\mathrm{Pr}; \tag{1.31}$$

it represents the ratio of heat convection and heat conduction.
- *Grashof* number (for free thermal convection):

$$\mathrm{Gr} = \frac{\alpha g L^3 \Delta T}{\nu^2}, \tag{1.32}$$

where α is the thermal expansion coefficient and ΔT is a temperature difference; it represents the ratio of buoyancy forces and viscous forces.
- *Rayleigh* number (for free thermal convection):

$$\mathrm{Ra} = \frac{\alpha g L^3 \Delta T}{\nu\gamma} = \mathrm{Gr}\,\mathrm{Pr}. \tag{1.33}$$

- *Marangoni* number (for convection induced by surface-tension gradients):

$$\mathrm{Ma} = \frac{\frac{\partial\sigma}{\partial c}\frac{\partial c}{\partial x}L^2}{\mu\gamma_c} \quad \text{(for concentration gradients)}, \tag{1.34}$$

$$\mathrm{Ma} = \frac{\frac{\partial\sigma}{\partial T}\frac{\partial T}{\partial x}L^2}{\mu\gamma} \quad \text{(for temperature gradients)}. \tag{1.35}$$

- *Richardson* number (for density-stratified flows):

$$\mathrm{Ri} = -\frac{g\rho/z}{\rho(V/z)^2} = -\frac{gz}{V^2} = -\mathrm{Fr}^{-\frac{1}{2}}; \tag{1.36}$$

it represents the ratio of potential energy associated with gravity and kinetic energy.

- *Taylor* number (for rotating flows):

$$\mathrm{Ta} = \frac{\Omega^2 L^4}{\nu^2}, \tag{1.37}$$

where Ω is the rotation rate.
- *Rossby* number (for rotating flows):

$$\mathrm{Ro} = \frac{V}{\Omega L}; \tag{1.38}$$

it represents the ratio of inertia and Coriolis forces.
- *Strouhal* number (for periodic vortex shedding from bluff objects):

$$S = \frac{fL}{V}, \tag{1.39}$$

where f is the frequency of vortex shedding.
- *Knudsen* number for gases:

$$\mathrm{Kn} = \frac{\lambda}{L}, \tag{1.40}$$

where λ is the mean free path.

A large number of additional dimensionless parameters have been identified in specialized areas [8].

1.7 Patterns of fluid motion

Many flow visualization and measurement methods provide images or other records of the fluid or of transported admixtures, which contain information on the fluid motion or the variation of some flow property. For a correct interpretation of the observed patterns, it is necessary to understand their relationship to the actual flow characteristics. Among the simplest types of visual patterns that can be obtained are these:

The *instantaneous position* of a visible fluid particle or marker as, for example, provided by a still camera with a short exposure time.

A *pathline*, namely the locus of positions of a fluid particle during a time interval; this would, for example, be recorded by a still camera with a relatively long exposure time; short pathlines provide a measure of the local velocity direction and magnitude.

A *streakline*, namely the locus of all fluid particles that have passed through a fixed position in the fluid during some time interval; one can achieve this by introducing continuously fluid markers at some point and taking a short-exposure photograph of the flow.

A *timeline*, namely the locus, at a given time, of all fluid particles that formed a continuous line in the fluid at some previous time; one can achieve this by introducing the markers through a line source operating over short time intervals.

Combinations of the preceding patterns, for example, time–streaklines or photographic exposures of markers over intermediate time intervals.

It is essential to understand that all the preceding patterns are distinct, in general, from the instantaneous *streamlines*, namely lines in the fluid, which are tangent at all their points to the local velocity vector. In a Cartesian coordinate system, a streamline is defined by the following pair of equations:

$$\frac{dx_1}{U_1} = \frac{dx_2}{U_2} = \frac{dx_3}{U_3}. \tag{1.41}$$

In steady flows, pathlines, streaklines, and streamlines coincide, whereas, in unsteady flows, they may be vastly different. In stationary turbulent flows of relatively low intensity, these lines may approximately coincide on the average, but their instantaneous features are distinct. One may reconstruct a set of streamlines in unsteady flows by fitting a family of tangent curves to an image of many short pathlines generated by adjacent particles and thus representing the velocity vector field.

QUESTIONS AND PROBLEMS

1. Provide definitions for the following measurable flow properties: angular momentum, entropy, thermal conductivity, molecular diffusivity, and surface tension.
2. List the established names for the SI units of force, pressure, energy, and power and their relationships to primary units. Also list the conversion factors of these units to corresponding units in the British gravitational system.
3. Provide expressions for the components of the rate of strain tensor and the velocity vector in cylindrical coordinates.
4. Consider Couette flow, in which the velocity vector is parallel to the x_1 axis and its magnitude changes linearly along the x_2 axis. Determine the components of the rate of strain tensor and the vorticity vector and discuss them comparatively.
5. Write the van der Waals equation of state for a gas and compare it with the perfect-gas law.
6. Define a viscoelastic material and give an example of a constitutive equation for such a material.
7. Define and discuss briefly the concept of turbulent, or ‘eddy’, viscosity.
8. Explain whether Bernoulli’s equation may be used (a) in air flow entering the engine of a cruising jet airplane, (b) in flow around swimming bacteria, and (c) in the gas plume rising from a smokestack.
9. Assume that gas bubbles rise very slowly in a glass of beer. Explain how you would estimate the pressure inside the bubbles by measuring their size.
10. Provide a physical example of a supersonic flow with a very small Reynolds number (i.e., less than 1), also providing numerical values for the appropriate length and velocity scales.
11. How is the surface-pressure coefficient around a circular cylinder in uniform flow related to its drag coefficient? What are the assumptions involved? Is this approach appropriate when the Reynolds number is (a) 1, (b) 10^4, or (c) 10^6?
12. From tables, find the values of the Prandtl number for (a) air at standard temperature and pressure, (b) water at standard atmospheric pressure and a temperature of (i) 4 °C, (ii) 20 °C, and (iii) 80 °C, and (c) mercury at 20 °C. Comment on these numbers.

13. Give typical values of the Strouhal number for vortex shedding from (a) circular and (b) square cylinders at a Reynolds number of 5×10^4.
14. The wings of butterflies are covered with scales, the height of which is typically 0.1 mm. Based on simple analysis, it is estimated that the relative air velocity just over the scales is about 15 mm/s. For a visual–optical study of their aerodynamic characteristics, it is advisable to use a large-scale model, having scales at least 10 mm high. Is it possible to do these tests in a wind tunnel? A water tunnel? If so, describe and discuss the conditions and possible problems arising. Can you describe alternative types of experiment that would be more suitable for this research? Hints: Consider using different fluids and/or moving the model.
15. A very large flat plate, immersed in a fluid, oscillates in the x direction with a velocity $V_0 \cos \omega t$, while remaining parallel to the x–y plane. Assume that the surrounding fluid is induced in motion with a velocity also in the x direction and satisfying the following theoretical solution:

$$V = V_0 e^{-z/\delta} \cos(\omega t - z/\delta),$$

where z is the direction normal to the plate and $\delta = \sqrt{2\nu/\omega}$ is the *Stokes length*. Sketch typical streamlines for this flow. Then sketch pathlines for particles at distances 1, 2, and 5 times the Stokes length. Further sketch the time evolution of timelines generated by a stationary wire stretched in the z direction, while neglecting gravitational effects. Describe how you can use this method to measure the viscosity of the fluid.

REFERENCES

[1] G. K. Batchelor. *An Introduction to Fluid Dynamics*. Cambridge University Press, Cambridge, UK, 1970.

[2] R. W. Fox and A. T. McDonald. *Introduction to Fluid Mechanics* (6th Ed.). Wiley, New York, 2004.

[3] F. M. White. *Viscous Fluid Flow* (2nd Ed.). McGraw-Hill, New York, 1991.

[4] R. E. Sonntag, C. Borgnakke, and G. J. Van Wylen. *Fundamentals of Thermodynamics* (5th Ed.). Wiley, New York, 1998.

[5] J. P. Holman. *Heat Transfer* (8th Ed.). McGraw-Hill, New York, 1997.

[6] D. R. Lide (Editor). *CRC Handbook of Chemistry and Physics* (78th Ed.). Chemical Rubber Co., Cleveland, OH, 1997.

[7] L. Rosenhead (Editor). *Laminar Boundary Layers*. Oxford University Press, Oxford, UK, 1963.

[8] S. P. Parker (Editor-in-Chief). *Fluid Mechanics Source Book*. McGraw-Hill, New York, 1987.

2 Measuring systems

This chapter contains a review of definitions and concepts that apply to all types of measuring systems. This is meant to assist in the understanding of the generic operations and steps that are involved in any measurement process. Some general discussion of the response of common types of measuring systems is also provided. More thorough coverage of these topics may be found in texts dealing with measuring systems [1–3].

2.1 Measuring systems and their components

Inputs and outputs: A fluid mechanics experiment is conducted by an experimenter, who measures properties of the flow. Thus, in any fluid mechanics experiment, one may identify three essential and distinct systems:

- The *physical system*, consisting of one or more flowing fluids, the flow-producing apparatus, and, possibly, test models and other related objects.
- The *measuring system*, consisting of a number of interconnected components, such as sensors, electric and electronic circuits, data acquisition, and processing devices and software.
- The *experimenter(s)*, namely one (or more) person(s) who plans, executes, and interprets the measurements.

It is obvious that, for a successful experiment, all three systems must be compatible with each other and functioning properly.

A measuring system, and each of its components, has one or more *input* and one or more *output* (Fig. 2.1). The output of each component may represent an input to another component. Each input or output corresponds to a physical property, as, for example, a displacement or an electric voltage. The relationship between the values of an input and an output is called the *response* of that component to the particular input.

The flow properties that the measuring system is intended to respond to and measure are the *desired* inputs. In addition, a measuring system is also subjected to *undesirable* inputs, which can be further classified into *interfering* and *modifying* inputs. An interfering input is a property to which the system is unintentionally sensitive, and a modifying input is a property that modifies the response to a particular desired input.

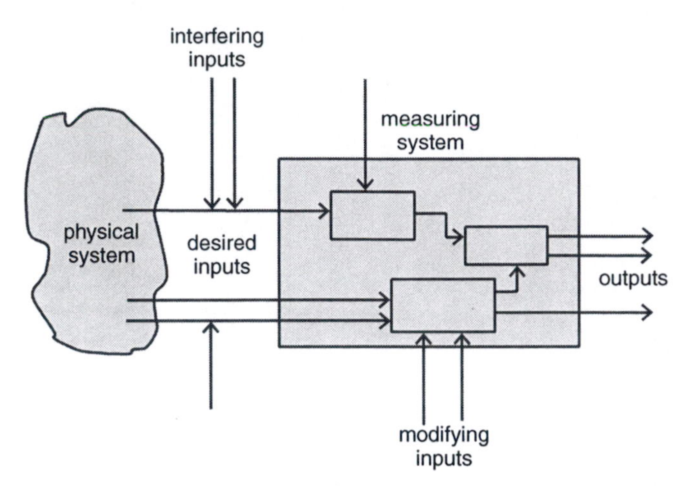

Figure 2.1. Inputs and outputs of a measuring system.

As an example, consider an experiment by which a velocity transducer (e.g., a hot-wire anemometer) is used to measure the local air velocity in a jet issuing through a nozzle into the laboratory. The desired input is the flow velocity. The draft of air produced by the ventilation system in the laboratory acts as an interfering input because it distorts the jet's velocity field. A change in room temperature, which may be due to the same ventilation system, might also have an interfering effect; however, its main undesirable effect is that it modifies the response of the electronic circuit supplying the transducer as well as causes deviations from the transducer's calibrated response; thus the room-temperature change acts as a modifying input for the velocity measurement.

Filtering, compensation, and output correction: Undesirable inputs cause measurement errors and may produce an output even in the absence of desired inputs. For accurate measurement, the experimenter must identify all undesirable inputs and estimate their effects, which, if found significant, must be accounted for. It goes without saying that the preferred approach would be to redesign a component that is sensitive to undesirable inputs in a way that it becomes insensitive to them. For example, the use of materials with a near-zero thermal expansion coefficient (within the range of interest) may eliminate the temperature sensitivity of a mechanical flow meter that has been calibrated to measure flow rate in terms of displacement of a certain component.

Inputs and outputs may be constant within a certain time interval or fluctuating (i.e., time dependent). A fluctuating property that depends on time (*time domain*) may be transformed (e.g., through a Fourier series, Fourier transform, Laplace transform, or wavelet transform) into an equivalent property that depends on frequency (*frequency domain*). Any operation on the original property will also affect its transform. A common way to reduce or eliminate undesirable input effects is the use of *filters*, which may be applied directly to an input or an output of the measuring system or one of its components. Filters are normally classified according to the frequency range of the fluctuations that they remove, as follows:

- *no-pass* filters remove all fluctuations, permitting only a steady component, if at all;
- *low-pass* filters remove fluctuations with frequencies above a cutoff value;

- *high-pass* filters remove fluctuations with frequencies below a cutoff value;
- *band-pass* filters remove all fluctuations except those with frequencies within a certain band;
- *band-reject* filters remove all fluctuations with frequencies within a certain band.

In terms of physical operation, filters may be of several different types:

- *electrical–electronic* filters, applied to electric signals;
- *mechanical* filters, designed to filter motion or forces (an example is the use of shock absorbers to reduce vibrations of an apparatus);
- *thermal* filters, designed to remove temperature fluctuations (the simplest type is thermal insulation);
- *electromagnetic* filters, designed to remove the interfering effects of electric and magnetic fields (the containment of instrumentation within a grounded metallic shield is an effective method of such filtering);
- *digital filters*, applied to recorded signals; these are very versatile and could be applied in multiple passes.

An alternative approach to reduce undesirable effects is the use of *compensation*. By this term, we understand the deliberate introduction of additional interfering or modifying inputs, which may partly or entirely cancel the original undesirable effects. For example consider the case of a strain gauge whose output is sensitive to ambient-temperature fluctuations. We may eliminate this effect by connecting this strain gauge into one leg of a Wheatstone bridge and connecting an identical strain gauge into another leg of the same bridge. Another form of compensation is to restore the high-frequency content of a fluctuating property that is measured with an instrument of relatively low-frequency response. This may be achieved either with the use of an electric–electronic circuit designed to boost the high frequencies of its input or digitally by the application of digital filters that enhance the high-frequency content of a recorded output.

Besides filtering and compensation, analytical corrections could be used to remove undesirable effects and errors in the measurement from the output. This approach clearly requires knowledge of both the values of undesirable inputs and the dependence of the measuring system response to such inputs.

Modes and functions of measuring system components: Each component of a measuring system may operate in an analogue, discrete (digital, binary, etc.), or hybrid mode. It may be *passive*, if the energy necessary for producing its output is supplied by the input, or *active*, if this energy is supplied mainly by an external excitation source (Fig. 2.2). To understand the detailed operation of a measuring system, it is instructional to identify the function of each component. The main functions that take place during measurement are usually the following:

1. *Sensing:* This is the step during that the measured property excites a sensor (or transducer) that produces an output that depends on the magnitude of this property.

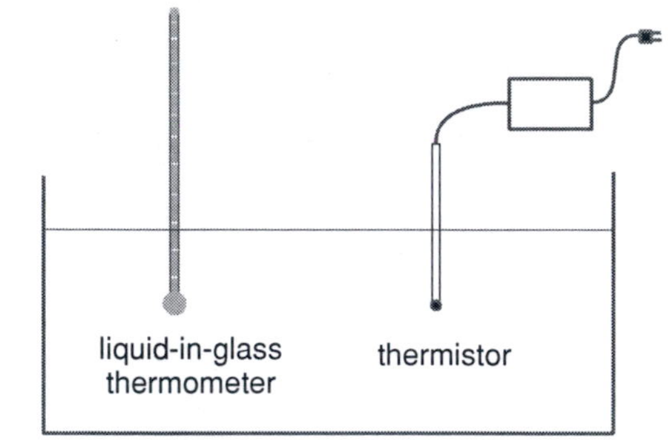

Figure 2.2. Measurement of temperature with a passive sensor (thermometer) and an active sensor (thermistor).

2. *Conversion* and *conditioning:* This is the process by which one or more components transform the sensor's output to a form, amplitude, or both, that is more suitable for observation or further processing.
3. *Transmission:* This includes the step(s) of transferring signals or other information from one component to another.
4. *Presentation* and *storage:* This is the final act of measurement, during which the output is displayed or stored.

One may note that each component of a measuring system may contribute to more than one of the preceding functions, and each function may be performed by more than one component.

An example of the functions effected by a measuring system is shown in Fig. 2.3. A strain gauge is mounted on a shaft that is subjected to tension. The measured property (desired input) is the elongation of the shaft. The strain gauge is the sensor, whose electric resistance is sensitive to elongation. The strain gauge is wired to a Wheatstone bridge, which converts the change of resistance to an electric voltage. Further use of an electronic amplifier and a low-pass filter conditions this voltage to a measurable,

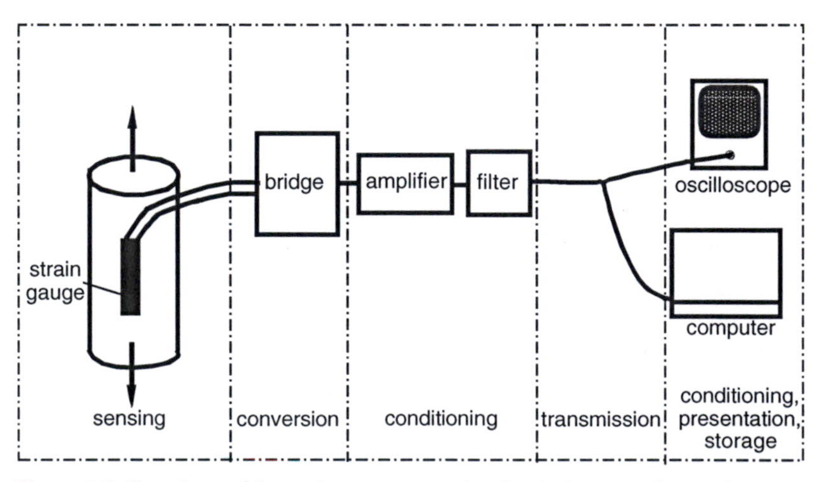

Figure 2.3. Functions of the various components of a strain measuring system.

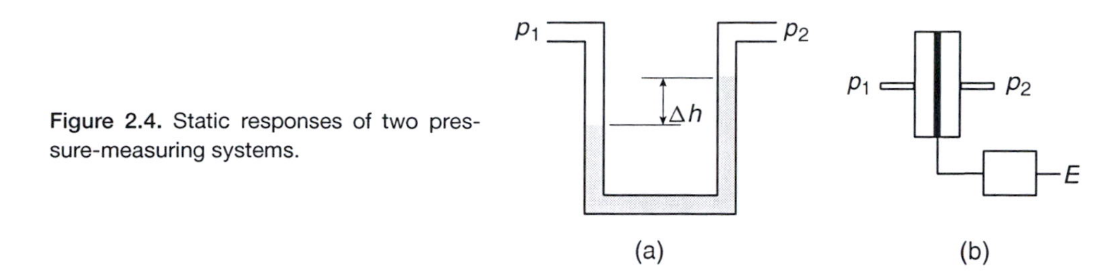

Figure 2.4. Static responses of two pressure-measuring systems.

low-noise electric signal. Cables are used to connect the various components and to transmit the amplifier output to an oscilloscope where the electric signal is displayed, and to a computer, where the signal is digitized, further conditioned by software and stored on magnetic storage devices.

2.2 Static response of measuring systems

Static response and static calibration: The operation of a measuring system is called *static* when all its inputs and outputs are either constant or varying very slowly with time. The relationship that provides the value of the input as a function of the system's output during static operation is called the *static response* of the measuring system. If there is a physical law that connects the input and output of the measuring system, the static response may be entirely theoretical. More commonly, however, the input–output relationship is based on static calibration and contains adjustable coefficients, the optimal values of which are determined by curve fitting.

Static calibration is preferably performed separately for each desired input. In case of multiple inputs, all inputs should be kept fixed, except for one, which is given successively different values within a range, causing the output to reach corresponding values. The values of the input are measured independently by some other instrument whose response is known and that serves as a standard; therefore the accuracy of a calibration depends on the accuracy of the instruments used as standards. Sometimes it is impossible or difficult to maintain all but one input constant, because changes in the desired input may trigger changes in other desired or undesirable inputs; in such cases, all changing inputs must be measured during calibration and the different effects must be identified and separated, with the use of, for example, available theoretical relationships among the various inputs. If possible, all interfering and modifying inputs during calibration should be maintained at the same levels as those during the actual measurement. Undesirable inputs may distort the static response by introducing a *zero drift*, namely a parallel shift of the primary calibration curve, or a *sensitivity drift*, namely a change in the slope of the primary calibration curve.

As an example of the two types of static response, consider the measurement of the pressure difference $p_1 - p_2$ in a gas (input) by using a liquid manometer and a variable-reluctance pressure transducer (see Fig. 2.4). In the first case, the input is related to the difference in heights of the two liquid columns (output) by the hydrostatic law $p_1 - p_2 = \rho_l g \Delta h$, where ρ_l is the liquid density. In the second case, the same input has been correlated to the electric voltage output E by the approximate expression $p_1 - p_2 = \alpha E + \beta$, where α and β are constants determined by calibration of the transducer vs. a liquid manometer.

Static performance characteristics: The static performance of a measuring system is characterized by several parameters:

- The *static sensitivity* is the slope of an input–output relationship. Obviously the static sensitivity of a linear system is constant over its operating range, whereas non-linear systems have a *local* sensitivity, which varies over their input range. In the case of multiple inputs, it is understood that the static sensitivity with respect to a particular input is specified for particular, fixed values of the other desired inputs.
- The *scale readability*, exclusively referring to analogue instruments, is the minimum change in the output that can be recognized by an observer. Obviously this is a subjective property and should not be confused with the accuracy of the instrument.
- The *span*, or *input full-scale*, is the range of an input that the measuring system is designed to measure with an acceptable accuracy.
- The *full-scale output* is the algebraic difference between the output values measured with maximum and minimum input values applied.
- The *dynamic range* is the ratio of largest to smallest values of the input that the system can measure.
- The *linearity* (actually a measure of non-linearity) is the maximum deviation of the actual response from the linear least-squares straight line fitted to the calibration measurements; manufacturers specify linearity as a percentage of the instrument's reading, a percentage of full scale, or a combination of the two.
- The *threshold* is the smallest input level that will produce a detectable output.
- The *resolution* is the smallest input change that will produce a detectable output change.
- The *hysteresis* is the difference between the output value corresponding to an input value that was reached from below and the output value corresponding to the same input value reached from above.

Normality tests and removal of outliers: During static calibration, as well as during static operation, one may discover that the values of repeat measurements are not identical, even if all inputs are seemingly constant. The reasons for such a variation could be *systematic*, such as a fluctuation in the powerline voltage that affects the operation of the flow apparatus or a measuring instrument, or *random*, namely attributable to several unspecifiable and, presumably, small effects. Although random effects cannot be easily eliminated, appreciable systematic effects can be detected and removed. This requires one to assess the 'randomness' of repeat measurements. A standard procedure of assessing randomness is the *normality test*, which entails comparing the statistical properties of the repeat measurement population with those of a *normal* or *Gaussian* random variable (see Section 4.4). Justification for the use of the Gaussian probability

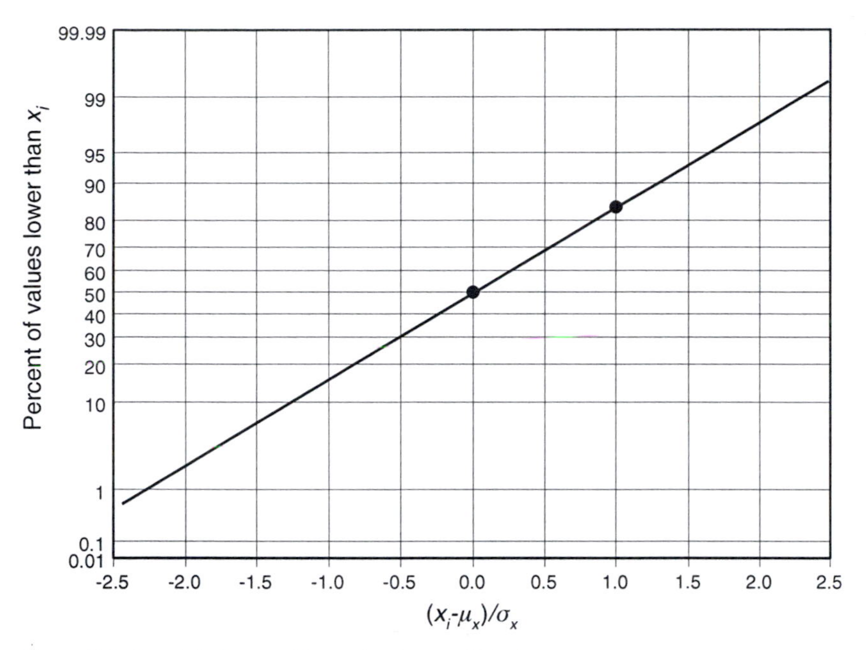

Figure 2.5. Graph to be used for a simple normality test.

density function (pdf) in many scientific and engineering tests is provided by the *central limit theorem* [4], which states that the pdf of the sum of a very large number of independent random variables will approach the Gaussian pdf as the number of variables increases. Thus the pdf of a large number of repeat measurements is likely to be close to the Gaussian pdf if the experiment is under *statistical control*, which indicates that it is likely affected by a large number of different undesirable inputs, each of which contributes to the output variation by a small amount. Thus a *normality test* provides some reassurance, though not conclusive evidence, that an experiment is under statistical control and not subjected to strong systematic influences (such as a voltage drift that is due to temperature increase).

A simple normality test would be to plot the histogram of the repeat measurements on the same graph as the Gaussian pdf and visually compare the two. This task may be assisted by use of probability graph paper, on which the distribution function of Gaussian random processes appears as a straight line (see Fig. 2.5). For this purpose, one may proceed as follows:

- Let x_i, $i = 1, 2, \ldots, N$, be a number of repeat values. Rearrange them by magnitude such that $x_i \leqslant x_{i+1}$.
- Compute the percentage of repeat values

$$y_i = 100\frac{i-1}{N}\% \tag{2.1}$$

that are lower than or equal to each value x_i.

- Compute the mean of the sample as

$$\mu_x = \frac{1}{N} \sum_{i=1}^{N} x_i \tag{2.2}$$

and the variance of the sample as

$$\sigma_x^2 = \frac{1}{N-1} \sum_{i=1}^{N} (x_i - \mu_x)^2. \tag{2.3}$$

- Normalize the repeat values as

$$x_i^* = \frac{x_i - \mu_x}{\sigma_x}. \tag{2.4}$$

- On probability graph paper (Fig. 2.5), plot y_i vs. x_i^*.
- By inspection, assess whether the plotted points deviate significantly or not from the corresponding Gaussian line [a straight line that passes through the points (0,50%) and (1,84.1%)].

A more sophisticated, quantitative, normality test is the χ^2 (*chi-square*) goodness-of-fit test. Here are the steps to be followed for this test:

- Let x_i, $i = 1, 2, \ldots, N$, be a number of repeat values, ordered such that $x_i \leqslant x_{i+1}$. For a proper test, $20 \leqslant N$. Compute the mean and the variance of the sample as previously.
- Group the ordered values into m groups. If $20 \leqslant N \leqslant 40$, select m such that each group contains at least five values. If $40 < N$, select m to be as closely as possible to the value $1.87(N-1)^{0.4}$ (Kendal–Stuart grouping guide), and assign at least five points to each group. Let n_j, $j = 1, 2, \ldots, m$, be the number of values in each group. Calculate the number of degrees of freedom as $m - 3$.
- Assign boundary values y_{j1} and y_{j2} to each group, such that they are halfway between the largest value of a group and the smallest value of the next group. Obviously, y_{11} is $-\infty$ and y_{m2} is ∞. Using probability tables or a scientific calculator, compute the probability $p_j = F(y_{j2}) - F(y_{j1})$, $j = 1, 2, \ldots, m$, that the value of a Gaussian distribution with mean μ_x and variance σ_x^2 lies between the boundaries of each group.
- Compute the value of chi square as

$$\chi^2 = \sum_{j=1}^{m} \frac{(n_j/N - p_j)^2}{p_j}. \tag{2.5}$$

- Using the computed number of degrees of freedom (d.o.f.) and value of χ^2 in Fig. 2.6, determine the probability that the sample of repeat values has a normal pdf. For example, for 2 d.o.f. ($m = 5$ groups) and $\chi^2 = 0.20$, the probability of normality is about 90%, or, in other words, the level of confidence that the sample has a normal distribution is about 90%. Which confidence level may be acceptable would depend on many factors; in general, a 95% level would likely be more than

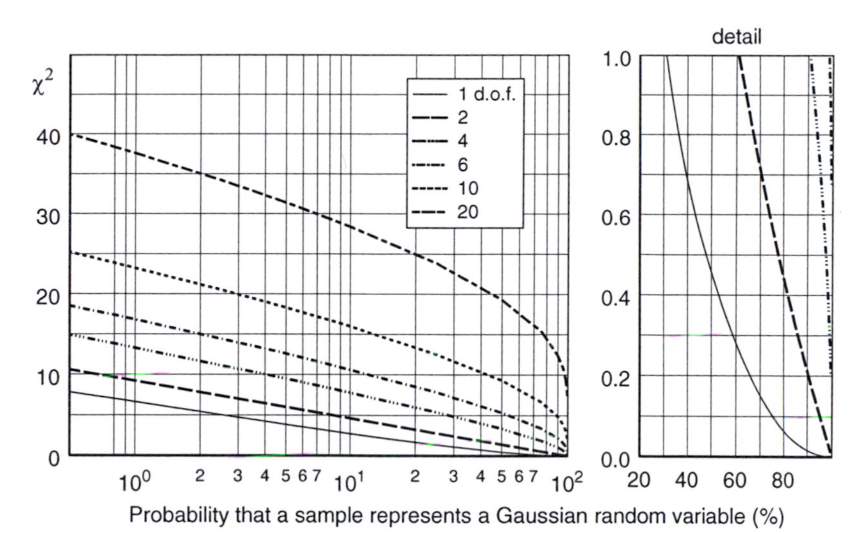

Figure 2.6. Graphs for normality test based on the chi-square method.

adequate for most engineering measurements. If a set of repeat measurements fails the normality test, it is advisable to trace the causes of systematic variation and eliminate them or apply appropriate corrections.

Occasionally a set of repeat measurements may contain *spurious values*, namely values containing errors far greater than those attributable to usual uncertainty. These may be due to gross human error (e.g., misreading of an instrument's output) or to a temporary or intermittent undesirable input (e.g., a power surge caused by the turning-on of a nearby, large electric motor). Such values are called *outliers* and are best discarded. Outliers may be identified by application of *Chauvenet's criterion* [5], which states that a value x_i is an outlier if

$$\tau\sigma \leqslant |x_i - \mu_x|, \tag{2.6}$$

where the parameter τ is given in the following table:

N	3	4	5	6	8	10	15	20	25	50	100
τ	1.38	1.54	1.65	1.73	1.87	1.96	2.13	2.24	2.44	2.57	2.81

Obviously, the smaller the number of repeat measurements, the larger the effect of an outlier on the statistical analysis. For this reason, if outliers are detected in a small population, it would be advisable to repeat the test and collect additional data.

Trend identification: Although repeat measurements of a process under statistical control should fluctuate about a constant average, it often happens that the results contain a *trend*, namely a monotonic increase or decrease in value that due to a systematic effect, while also subjected to random fluctuations, which are expressed as *scatter*. Trends may be revealed by inspection of the values plotted vs. time, but, especially when scatter is appreciable, it would be advisable to perform a quantitative statistical test to determine whether a trend is significant. The *reverse arrangement test* [6] is capable of detecting monotonic trends. Let $x_i, i = 1, 2 \ldots N$ be a set of repeat measurements arranged in chronological sequence. The *total number of reverse arrangements* R is defined as

$$R = \sum_{i=1}^{N-1} \sum_{j=i+1}^{N} R_{ij}, \tag{2.7}$$

where

$$R_{ij} = \begin{cases} 1, & \text{if } x_i > x_j \\ 0, & \text{if } x_i \leq x_j \end{cases}. \tag{2.8}$$

If the process were under statistical control and N were sufficiently large, R would be a Gaussian random variable with a mean $\mu_R = N(N-1)/4$ and a variance $\sigma_R^2 = N(2N+5)(N-1)/72$. Thus one may test the hypothesis that there is no trend in the data within a *level of significance* by evaluating the probability that R would be within a certain interval of values. For example, if R were a Gausssian random variable, the probability that $\mu_R - 2\sigma_R < R < \mu_R + 2\sigma_R$ would be approximately 0.95, the probability that $\mu_R + 2\sigma_R < R$ would be 0.025, and the probability that $R < \mu_R - 2\sigma_R$ would also be 0.025. To illustrate this process, consider a set of repeat values for which $N = 30$. The mean and standard deviation of a Gaussian reverse arrangement set of 30 values are, respectively, $\mu_R = 217.5$ and $\sigma_R = 28.0$. Assume that the actual number of reverse arrangements, calculated according to Eq. (2.7) is R. If $161.45 < R < 273.55$, then one may say that, at a level of significance of 5%, there is no trend in the data. If $R < 161.45$, then one may say that the data have a downward trend, whereas, if $273.55 < R$, one may say that the data have an upward trend. One could remove a trend by first fitting a linear function (see next subsection) to the complete set of data, subtracting this function from the original data and then adding the data average. Trend removal should be done with caution to avoid eliminating true data features. Similarly, the fitting of non-linear expressions to data for the purposes of non-linear trend removal should be avoided, unless there is sound theoretical reason to anticipate such trends.

Curve fitting and linearization: A measuring system or component is called *linear*, within its range of operation, if its input–output relationship can be represented by a straight line. In determining this relationship by direct calibration, it is important to remove random errors, as much as possible, by use of statistical methods. On many occasions, it is sufficient to fit a straight line to the plot of input–output data pairs such that the mean-square departure of all output measurements from that line is minimal. This is called a *linear least-squares fit* (llsf). Given a set of calibration measurements, $(x_i, y_i), i = 1, 2, \ldots, N$, a llsf provides the line

$$y(x) = ax + b, \tag{2.9}$$

where

$$a = \frac{N\sum x_i y_i - \sum x_i \sum y_i}{N\sum x_i^2 - \left(\sum x_i\right)^2}, \quad b = \frac{\sum y_i \sum x_i^2 - \sum x_i y_i \sum x_i}{N\sum x_i^2 - \left(\sum x_i\right)^2}. \tag{2.10}$$

If there are theoretical reasons to believe that the response of a system is non-linear and further suggesting a more suitable expression, one may use similar statistical methods to fit calibration data with non-linear curves, such as polynomials or exponentials. Most scientific graphics software packages contain curve-fitting capabilities and usually provide an indication of the *goodness of fit*, namely how much the measurements depart statistically from the fitted line, as well as of the uncertainty in the estimated coefficients of the fitted expressions. In selecting the type of expression to be fitted to a set of data that contain scatter, one must judiciously compromise between the smoothness of a fitted curve and the goodness of fit to a specific sample of measurements. In particular, it is advisable to fit only expressions having a significantly smaller number of adjustable coefficients than the number of calibration points available. Otherwise, one may obtain an expression with an excellent goodness of fit to a specific data sample but with wild fluctuations in the intervals between consecutive points, which is obviously an unacceptable situation.

In practice, all measuring systems will become non-linear if the input magnitude becomes sufficiently large; for example, a metallic spring that behaves linearly for small extensions will become non-linear when its extension exceeds the elastic limit. Conversely, any continuous non-linear response may be linearized within a limited range of input variation. One may achieve this by expanding the response equation into a Taylor series about a reference value (for example, the midpoint of the input range) and retaining only the lowest-order terms, up to first order. Thus a non-linear expression

$$y = f(x) \tag{2.11}$$

may be replaced with the linear expression

$$y \simeq f(x_0) + \frac{\mathrm{d}f(x_0)}{\mathrm{d}x}(x - x_0) \tag{2.12}$$

within a range of x around x_0. For example, the cubic function $y(x) = ax^3 + bx + c$ may be approximated by the linear function $y(x) \simeq (3a + b)x - 2a + c$ near $x = 1$, and by simply $y(x) \simeq bx + c$ near $x = 0$.

In the case in which an output depends on the multiple inputs $x_1, x_2, \ldots$, one may linearize the response

$$y = f(x_1, x_2, \ldots) \tag{2.13}$$

about a *reference point* $(x_{10}, x_{20}, \ldots)$, with respect to one or more input, by expanding it into a multi-variable Taylor series expansion and retaining the terms up to first order, as

$$y = f(x_{10}, x_{20}, \ldots) + \sum \frac{\partial f(x_{10}, x_{20}, \ldots)}{\partial x_i}(x_i - x_{i0}). \tag{2.14}$$

Although this procedure is of general applicability, in practice, linearization becomes a futile exercise when the operation of an instrument involves input changes that are sufficiently large for non-linear effects to become significant. The maximum error that is due to linearization can be computed by a comparison of the values of the linearized expression with those of the original expression at the boundaries of the linearization range or at other appropriate points, if the response is not a monotonic function.

Effects of fluctuations on mean measurements: In many experimental settings, the input $x(t)$ of a measuring system may be fluctuating periodically or randomly in time such that an average value can be defined by integration over a sufficiently long window T, as

$$\overline{x} = \lim_{T\to\infty} \frac{1}{T} \int_0^T x(t)\mathrm{d}t. \tag{2.15}$$

Then the instantaneous value of the input can be considered as consisting of a mean component $\overline{x}$ and a superimposed, mean-free, fluctuation x', such that $x = \overline{x} + x'$. The output y of the measuring system would also be, in general, time dependent, and one could similarly define a mean $\overline{y}$ and a fluctuation y' such that $y = \overline{y} + y'$.

Let us assume that we are using a measuring system to measure $\overline{x}$ by reading a mean output $\overline{y}$. If the system is linear, linearity of the integration operation ensures that the static response expression can be equally used to connect the means of inputs and outputs. For example, let $y = \alpha x + \beta$, where α and β are calibration constants; then it is obvious that $\overline{y} = \alpha\overline{x} + \beta$. However, if the measuring system is non-linear, the static relationship could be quite different from the relationship between the two means. For example, let us assume that $y = \alpha x^n$, where α and n are constants fitted to static calibration data. Then the mean $\overline{y}$ can be computed as

$$\begin{aligned}\overline{y} = \overline{\alpha x^n} &= \alpha \lim_{T\to\infty} \frac{1}{T} \int_0^T x^n(t)\mathrm{d}t = \alpha \lim_{T\to\infty} \frac{1}{T} \int_0^T \left(\overline{x} + x'\right)^n \mathrm{d}t \\ &= \alpha\overline{x}^n \lim_{T\to\infty} \frac{1}{T} \int_0^T \left(1 + \frac{x'}{\overline{x}}\right)^n \mathrm{d}t. \end{aligned} \tag{2.16}$$

Assuming that the fluctuations are small, one may use the binomial expansion to derive

$$\overline{y} = \alpha\overline{x}^n \left[1 + \frac{n(n-1)}{2} \frac{\overline{x'^2}}{\overline{x}^2}\right]. \tag{2.17}$$

Several popular instruments used in fluid mechanics research (Pitot-static tubes, hot-wire anemometers, and thermistors, among others) are distinctly non-linear and subjected to fluctuation-induced errors. A consequence of the preceding analysis is that time averaging of the output of non-linear measuring systems (either crude mental averaging of a fluctuating reading, use of an averaging voltmeter or integrator, or even discretization and digital averaging) would produce a result that would not correspond to the average input. Whether the 'apparent' average input would be higher or lower than the actual average input would depend on the response of the measuring system. For the power-law response just presented, $\overline{y} < \alpha\overline{x}^n$ if $0 < n < 1$, and $\overline{y} > \alpha\overline{x}^n$ if $n < 0$ or $n > 1$; if $n = 1$, $\overline{y} = \alpha\overline{x}^n$, and the case $n = 1$ is trivial. In the case of Pitot-static tube, whose output is the dynamic pressure $\frac{1}{2}\rho V^2$, the apparent average input, i.e., flow velocity, would always be higher than the actual average velocity.

2.3 Dynamic response of measuring systems

Models of dynamic response: The operation of a measuring system is called *dynamic* if at least one of its inputs is time dependent. Unlike the static response, which is usually specified by an algebraic relationship, the *dynamic response* of a system is generally specified by a differential equation that contains time derivatives. A measuring system, or one of its components, is called *linear* if its input–output relationship is a linear algebraic or differential equation; otherwise it is called *non-linear*. Because the solution of non-linear differential equations is much more difficult than that of linear ones, it is customary to linearize, whenever possible, these systems. Also for simplicity, it is common to examine certain non-linear phenomena, such as hysteresis, as part of the static response of the system and to neglect their effects in the determination of the dynamic characteristics.

The dynamic response of measuring systems is usually modelled by mathematical equations describing physical principles (e.g. electric circuit laws, momentum balance, or energy balance). When an output depends on several inputs, the full mathematical model of the system would likely be a set of non-linear, coupled, partial differential equations. One may simplify this by considering the time variation of the output, y, when only one input, x, changes and under conditions such that the input–output relationship can be approximated by a single, linear, ordinary differential equation with constant coefficients, such as

$$a_n \frac{\mathrm{d}^n y}{\mathrm{d}t^n} + a_{n-1} \frac{\mathrm{d}^{n-1} y}{\mathrm{d}t^{n-1}} + \cdots + a_1 \frac{\mathrm{d}y}{\mathrm{d}t} + a_0 y = b_m \frac{\mathrm{d}^m x}{\mathrm{d}t^n} + b_{m-1} \frac{\mathrm{d}^{m-1} x}{\mathrm{d}t^{m-1}} + \cdots + b_1 \frac{\mathrm{d}x}{\mathrm{d}t} + b_0 x, \quad n \geq m. \tag{2.18}$$

Such a model is called *time invariant*. In the following, we discuss the response of the simplest and most common systems, which are the zero-, first-, and second-order systems.

Zero-order systems:

$$y = Kx. \tag{2.19}$$

A zero-order system is characterized by a single parameter, the *static sensitivity* K, which has the same dimensions as the ratio y/x. The dynamic response of zero-order systems is independent of time, as the output remains proportional to the input at all times. Examples of zero-order systems are an electric resistor [Fig. 2.7(a)] and an elastic spring. Most physical systems, however, even those normally described by zero-order models, will show some time dependence when subjected to fast-changing inputs. For

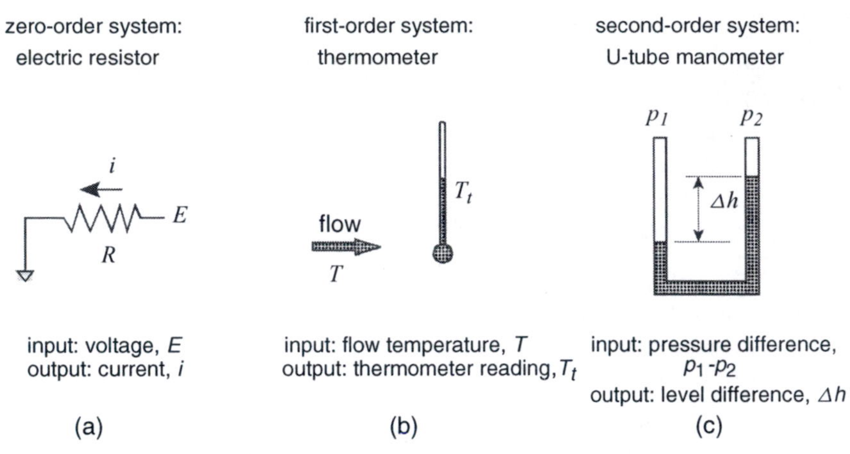

Figure 2.7. Examples of measuring systems whose dynamic response may be approximated by (a) zero-order, (b) first-order, and (c) second-order models.

example, an elastic spring, normally considered to have a negligible mass, will show effects of inertia when the rate of change of the applied force exceeds some limit; a more accurate representation of this spring would be by a second-order system.

First-order systems:

$$\tau \frac{\mathrm{d}y}{\mathrm{d}t} + y = Kx. \tag{2.20}$$

A first-order system is characterized by two parameters, the static sensitivity K and the *time constant* τ, which has dimensions of time. Examples are the common thermometer [Fig. 2.7(b)] and a resistor–capacitor (*RC*; note that *RC* can also stand for resistance–capacitance) electric circuit.

Second-order systems:

$$\frac{\mathrm{d}^2 y}{\mathrm{d}t^2} + 2\zeta\omega_n \frac{\mathrm{d}y}{\mathrm{d}t} + \omega_n^2 y = K\omega_n^2 x. \tag{2.21}$$

A second-order system is characterized by three parameters, the static sensitivity K, the *undamped natural frequency* ω_n (with dimensions of inverse time), and the *damping ratio* ζ (dimensionless). A second-order system is called *undamped* when $\zeta = 0$, *underdamped* when $0 < \zeta < 1$, *critically damped* when $\zeta = 1$, and *overdamped* when $1 < \zeta$. The parameter

$$\omega_d = \omega_n \sqrt{1 - \zeta^2}, \tag{2.22}$$

defined for underdamped systems only, is called the *damped natural frequency*. Damping is generally due to a mechanism that dissipates energy and opposes some action; mechanical friction is the most common damping mechanism. Because energy dissipation is always present in real systems, perfectly undamped systems do not exist; the closest approximation to an undamped system is a *lightly damped* system, having $0 < \zeta \ll 1$.

An example of a measuring system that can be represented approximately by a second-order model is the liquid manometer [Fig. 2.7(c)], in which damping is caused by friction between the liquid and the glass tube wall.

Types of input: With the exception of zero-order systems, the output of all other systems exposed to a time-dependent input depends not only on the current value of the input but also on its time history. For the description of dynamic response, it is customary to consider a few idealized types of input as approximate models of more realistic input types.

A relatively fast change of the input from one constant level to another may be idealized as proportional to the *unit-step* (or *Heaviside*) *function*,

$$U(t) = \begin{cases} 0, & \text{for } t < 0 \\ 1, & \text{for } 0 \leq t \end{cases}. \tag{2.23}$$

A sudden, impulsive application of a different value of the input, lasting only briefly before it returns to the original level, may be idealized as proportional to the *unit-impulse function* or *Dirac's delta function*, defined as

$$\begin{aligned} \delta(t) &= 0 \quad \text{for } t \neq 0, \\ \delta(t) &\to \infty \quad \text{for } t \to 0, \\ \int_{-\infty}^{\infty} &\delta(x)\mathrm{d}x = 1, \end{aligned} \tag{2.24}$$

and having the additional property, for any continuous function $f(x)$,

$$\int_{-\infty}^{\infty} \delta(x - x_0) f(x)\mathrm{d}x = f(x_0). \tag{2.25}$$

A gradual change of the input, starting from a constant level and persisting monotonically, may be modelled in terms of *the unit-slope ramp function*,

$$r(t) = \begin{cases} 0, & \text{for } t < 0 \\ t, & \text{for } 0 \leq t \end{cases}. \tag{2.26}$$

A function $f(t)$ is called *periodic* with a period T if $f(t) = f(t + nT)$, where n is any integer number. Periodic functions relevant to measuring systems can be decomposed into *Fourier series* [7], as

$$f(t) = \frac{a_0}{2} + \sum_{n=1}^{\infty} \left[a_n \cos\left(2\pi n \frac{t}{T}\right) + b_n \sin\left(2\pi n \frac{t}{T}\right) \right], \tag{2.27}$$

where

$$a_n = \frac{2}{T} \int_{-T/2}^{T/2} f(x) \cos\left(2\pi n \frac{t}{T}\right) \mathrm{d}t,\ n = 0, 1, 2, \ldots, \tag{2.28}$$

$$b_n = \frac{2}{T} \int_{-T/2}^{T/2} f(x) \sin\left(2\pi n \frac{t}{T}\right) \mathrm{d}t,\ n = 1, 2, 3 \ldots. \tag{2.29}$$

If the input of a linear system is the sum of different components, its output will be equal to the sum of outputs to the individual input components acting separately from each other. Therefore the response to periodic changes of the input may be determined as the sum of the responses to different *sinusoidal* inputs, $x(t) = A \sin \omega t$, where ω is the *frequency*. The output of a linear system that is subjected to a periodic input undergoes two states. When the input is first applied, there is a *transient state*, during which the output depends on the initial conditions. After a sufficiently long time, however, the output reaches a *steady state*, which is periodic with the same period as the input and independent of the initial conditions. It can be shown that the steady-state output of a linear system subjected to the sinusoidal input introduced previously is $y(t) = B \sin(\omega t + \varphi)$, where φ is the *phase*; thus, $y(t)$ is also sinusoidal, with the same frequency as $x(t)$ but out of phase with it. The *amplitude ratio* B/A and the phase φ depend, in general, on the frequency ω; their determination for different frequencies constitutes the *frequency response* of the system. If the frequency response is known, then the steady-state response of a linear system subjected to an arbitrary periodic input can be easily determined if both the input and the output are decomposed in Fourier series.

The objective of measurement is to determine the value of a physical property (input) based on a reading of the output of the measuring system. Under static operation of a linear system, the input is easily computed from the output as $y(t)/K$. Under dynamic conditions, however, the input value as indicated by a measuring system would differ from the actual input value. The *measurement error* $\Delta x(t) = x(t) - y(t)/K$, at a given time t, is defined as the difference between the value of the actual input and the value of the input estimated from the measured output of the measuring system by assuming a static response.

Dynamic response of first-order systems: The response of first-order systems to the idealized types of input mentioned in the previous subsection is as follows.

Step response (Fig. 2.8):

$$x(t) = AU(t), \quad \frac{y(t)}{KA} = 1 - e^{-t/\tau}, \quad \frac{\Delta x(t)}{A} = e^{-t/\tau}. \tag{2.30}$$

The accuracy of a first-order system improves as its time constant decreases. Following a step input change, the relative measurement error $\Delta x/A$ will be 37% after time τ, 13.5% after time 2τ, 5% after time 3τ, and negligible after a sufficiently long *settling time*.

Impulse response:

$$x(t) = A\delta(t), \quad \frac{y(t)}{KA} = \frac{1}{\tau} e^{-t/\tau}, \quad \frac{\Delta x(t)}{A} = -\frac{1}{\tau} e^{-t/\tau}. \tag{2.31}$$

The preceding relationships show that a system with a small time constant, although preferable in terms of its step response, would be subjected to large-amplitude spikes when exposed to impulsive inputs. As previously, the error decreases with time and eventually vanishes.

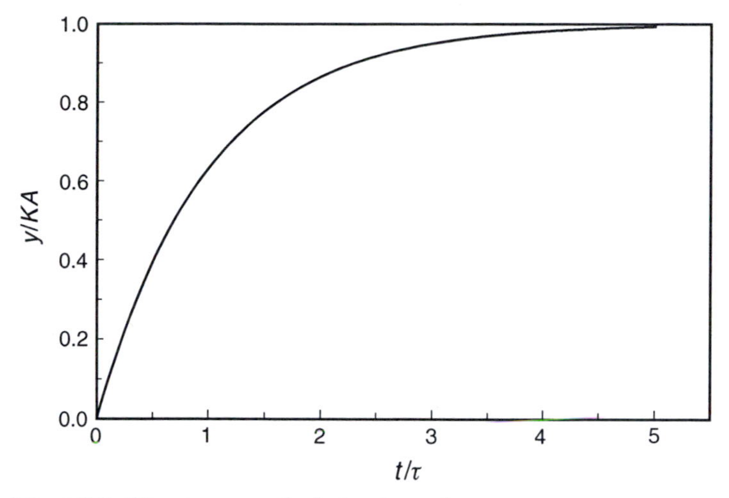

Figure 2.8. Step response of a first-order system.

Ramp response:

$$x(t) = Ar(t), \quad \frac{y(t)}{KA} = \tau e^{-t/\tau} + t - \tau, \quad \frac{\Delta x(t)}{A} = -\tau e^{-t/\tau} + \tau. \tag{2.32}$$

It is interesting to note that a first-order system will always be in error when subjected to ramp-type inputs. Even after a very long time, the error does not vanish but increases asymptotically to the constant value $A\tau$.

Frequency response (Fig. 2.9):

$$\frac{B}{KA} = \frac{1}{\sqrt{\omega^2\tau^2 + 1}}, \quad \varphi = -\arctan \omega\tau. \tag{2.33}$$

As $\omega \to \infty$, $B/A \to 0$ and $\varphi \to -\pi/2$. Thus a first-order system acts like a low-pass filter.

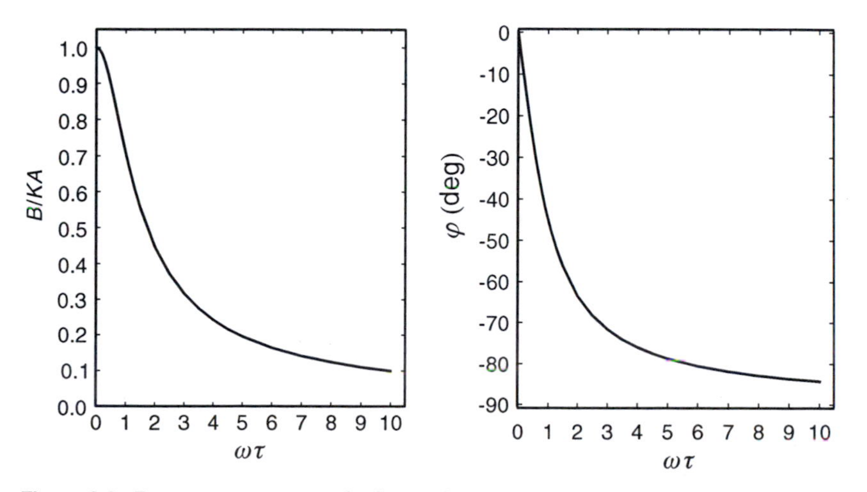

Figure 2.9. Frequency response of a first-order system.

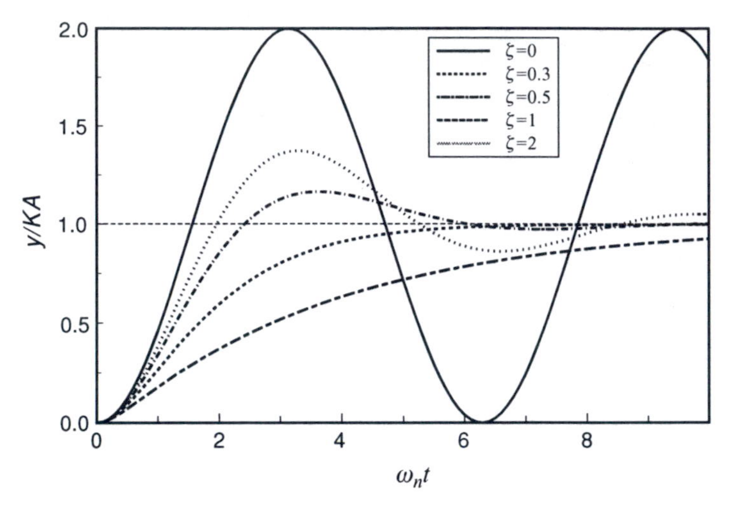

Figure 2.10. Step response of a second-order system.

Dynamic response of second-order systems: The response of second-order systems to the idealized types of input mentioned earlier is as follows [2, 3].

Step response (Fig. 2.10):

$$x(t) = AU(t); \tag{2.34}$$

$$\frac{y(t)}{KA} = 2\sin(\omega_n t), \quad \zeta = 0; \tag{2.35}$$

$$\frac{y(t)}{KA} = 1 - \frac{e^{-\zeta\omega_n t}}{\sqrt{1-\zeta^2}} \sin\left(\sqrt{1-\zeta^2}\omega_n t + \arcsin\sqrt{1-\zeta^2}\right), \quad 0 < \zeta < 1; \tag{2.36}$$

$$\frac{y(t)}{KA} = 1 - (1+\omega_n t)\, e^{-\omega_n t}, \quad \zeta = 1; \tag{2.37}$$

$$\frac{y(t)}{KA} = 1 - \frac{\zeta + \sqrt{\zeta^2 - 1}}{2\sqrt{\zeta^2 - 1}} e^{\left(-\zeta + \sqrt{\zeta^2-1}\right)\omega_n t} + \frac{\zeta - \sqrt{\zeta^2 - 1}}{2\sqrt{\zeta^2 - 1}} e^{\left(-\zeta - \sqrt{\zeta^2-1}\right)\omega_n t}, \quad 1 < \zeta. \tag{2.38}$$

Therefore the functional form of the step response of a second-order system changes depending on the value of the damping ratio. Whereas the output of critically damped and overdamped systems increases monotonically towards its static level, the output of underdamped systems oscillates about the static level with diminishing amplitude. Lightly damped systems ($\zeta \ll 1$) are subjected to large-amplitude oscillations that persist over a long time and obscure a measurement. All second-order systems, except for the unrealistic case of undamped systems ($\zeta = 0$), will eventually settle to their static output level, following a step change in input.

Impulse response:

$$x(t) = A\delta(t); \tag{2.39}$$

$$\frac{y(t)}{KA} = \omega_n \sin(\omega_n t), \quad \zeta = 0; \tag{2.40}$$

$$\frac{y(t)}{KA} = \frac{\omega_n e^{-\zeta\omega_n t}}{\sqrt{1-\zeta^2}} \sin\left(\sqrt{1-\zeta^2}\omega_n t\right), \quad 0 < \zeta < 1; \tag{2.41}$$

$$\frac{y(t)}{KA} = \omega_n^2 t e^{-\omega_n t}, \quad \zeta = 1; \tag{2.42}$$

$$\frac{y(t)}{KA} = \frac{\omega_n}{2\sqrt{\zeta^2-1}} \left[e^{\left(-\zeta+\sqrt{\zeta^2-1}\right)\omega_n t} - e^{\left(-\zeta-\sqrt{\zeta^2-1}\right)\omega_n t}\right], \quad 1 < \zeta. \tag{2.43}$$

Similar to their step response, critically damped and overdamped systems have a non-oscillatory impulse response, whereas underdamped systems have an oscillatory response. The long-time measurement error for an impulse-type input vanishes, except for $\zeta = 0$.

Ramp response:

$$x(t) = Ar(t); \tag{2.44}$$

$$\frac{y(t)}{KA} = t + \frac{2}{\omega_n} \sin(\omega_n t), \quad \zeta = 0; \tag{2.45}$$

$$\frac{y(t)}{KA} = t - \frac{2\zeta}{\omega_n} + \frac{e^{-\zeta\omega_n t}}{\omega_n\sqrt{1-\zeta^2}} \sin\left(\sqrt{1-\zeta^2}\omega_n t + \arctan\frac{2\zeta\sqrt{1-\zeta^2}}{2\zeta^2-1}\right),$$
$$0 < \zeta < 1; \tag{2.46}$$

$$\frac{y(t)}{KA} = t - \frac{2\zeta}{\omega_n} + \frac{2}{\omega_n}\left(1 + \frac{\omega_n t}{2}\right) e^{-\omega_n t}, \zeta = 1; \tag{2.47}$$

$$\frac{y(t)}{KA} = t - \frac{2\zeta}{\omega_n} - \frac{2\zeta^2 - 1 - 2\zeta\sqrt{\zeta^2-1}}{2\omega_n\sqrt{\zeta^2-1}} e^{\left(-\zeta+\sqrt{\zeta^2-1}\right)\omega_n t}$$
$$+ \frac{-2\zeta^2 + 1 - 2\zeta\sqrt{\zeta^2-1}}{2\omega_n\sqrt{\zeta^2-1}} e^{\left(-\zeta-\sqrt{\zeta^2-1}\right)\omega_n t}, \quad 1 < \zeta. \tag{2.48}$$

Similar to their step and impulse responses, critically damped and overdamped systems have an non-oscillatory ramp response, whereas underdamped systems have an oscillatory response. The long-time relative measurement error for a ramp-type input approaches the constant $2\zeta/\omega_n$, except for $\zeta = 0$.

Frequency response (Fig. 2.11):

$$\frac{B}{KA} = \frac{1}{\sqrt{[1-(\omega/\omega_n)^2]^2 + 4\zeta^2\omega^2/\omega_n^2}}, \quad \varphi = -\arctan\frac{2\zeta\omega/\omega_n}{1-(\omega/\omega_n)^2}. \tag{2.49}$$

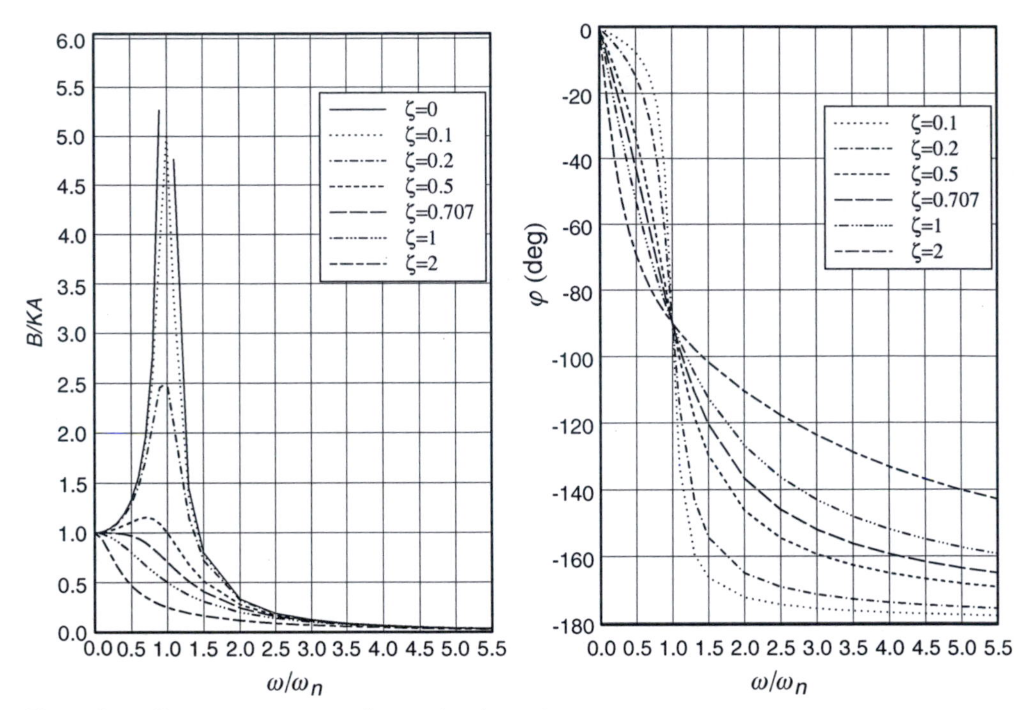

Figure 2.11. Frequency response of second-order systems.

As $\omega/\omega_n \to \infty$, $B/A \to 0$ and $\varphi \to -\pi$. Critically damped and overdamped systems act like low-pass filters and have diminishing output amplitudes when subjected to sinusoidal inputs of increasing frequency. Undamped systems have an infinite output amplitude when $\omega = \omega_n$ (*resonance*). Underdamped systems with $0 < \zeta < \sqrt{2}/2$ present a peak $B/KA = (2\zeta\sqrt{1-\zeta^2})^{-1}$ in the output amplitude at the *resonant frequency*

$$\omega_r = \omega_n\sqrt{1 - 2\zeta^2}. \tag{2.50}$$

Underdamped systems with $\sqrt{2}/2 < \zeta$ have no resonant peak and have an amplitude ratio that decreases monotonically with frequency.

Higher-order and non-linear systems: The dynamic response of linear systems of an order higher than 2 can be found with the use of analytical methods such as the Laplace transform (see next subsection). The results become, however, exceedingly complicated, as the order increases, because of the increasing number of parameters affecting not only the value but also the functional shape of the output. The dynamic response of non-linear systems can be found analytically for only a few special cases, which leaves numerical analysis as the only practical approach. For these reasons, the use of higher-order and non-linear models is rather rare in the measurement field.

Dynamic analysis by use of Laplace transform: Given any time-dependent property $f(t)$, such that $f(t) = 0$ for $t < 0$, one may define its *Laplace transform* [2, 3, 8] as

$$\mathcal{L}[f(t)] = \mathbf{F}(s) = \int_0^\infty f(t)e^{-st}\,\mathrm{d}t, \tag{2.51}$$

where s is a complex variable with a dimension of frequency and the transformed properties are denoted by boldfaced characters. The Laplace transform is unique and reversible. Given $\mathbf{F}(s)$, one may compute the original time-dependent property $f(t)$ by using the *inverse Laplace transform*

$$f(t) = \mathcal{L}^{-1}\left[\mathbf{F}(s)\right]. \tag{2.52}$$

An important property of Laplace transform that makes it very attractive for use with linear, time-invariant systems is the *differentiation property*:

$$\mathcal{L}\left[\frac{\mathrm{d}^n f(t)}{\mathrm{d}t^n}\right] = s^n \mathbf{F}(s) - \left[s^{n-1}\frac{\mathrm{d}^{n-1} f(0)}{\mathrm{d}t^{n-1}} + \cdots + s\frac{\mathrm{d}f(0)}{\mathrm{d}t} + f(0)\right]; \tag{2.53}$$

this property allows the Laplace transform of any derivative of $f(t)$ to be computed from the Laplace transform of $f(t)$ itself and a set of initial conditions.

Consider a linear time-invariant system with an input $x(t)$ and an output $y(t)$, and let $\mathbf{X}(s)$ and $\mathbf{Y}(s)$ be the corresponding Laplace transforms. The *Laplace transfer function* of this system is defined as

$$\mathbf{H}(s) = \frac{\mathbf{Y}(s)}{\mathbf{X}(s)} \tag{2.54}$$

under the assumption that the initial values of the input and all its derivatives are equal to zero. Then the differentiation property provides the expression

$$\mathbf{H}(s) = \frac{b_m s^m + b_{m-1}s^{m-1} + \cdots + b_1 s + b_0}{a_n s^n + a_{n-1}s^{n-1} + \cdots + a_1 s + a_0}. \tag{2.55}$$

In systems analysis, it is customary to use, instead of the complex variable s, the real frequency ω, defined as $s = j\omega$, where $j = \sqrt{-1}$. Then, instead of the term Laplace transfer function, one may use the term *sinusoidal transfer function* $\mathbf{H}(j\omega)$. As a complex function, $\mathbf{H}(j\omega)$ [or $\mathbf{H}(s)$] is composed of a real part and an imaginary part, namely

$$\mathbf{H}(j\omega) = \mathrm{Re}\,\mathbf{H}(j\omega) + j\,\mathrm{Im}\,\mathbf{H}(j\omega). \tag{2.56}$$

In polar form, $\mathbf{H}(j\omega)$ may be written as

$$\mathbf{H}(j\omega) = |\mathbf{H}(j\omega)|\, e^{j\varphi(j\omega)}, \tag{2.57}$$

where its *magnitude* and *argument* (*phase*) are, respectively, defined as

$$|\mathbf{H}(j\omega)| = \sqrt{[\mathrm{Re}\,\mathbf{H}(j\omega)]^2 + [\mathrm{Im}\,\mathbf{H}(j\omega)]^2}, \tag{2.58}$$

and

$$\varphi(j\omega) = \arctan\frac{\mathrm{Im}\,\mathbf{H}(j\omega)}{\mathrm{Re}\,\mathbf{H}(j\omega)}. \tag{2.59}$$

One of the useful properties of the sinusoidal transfer function is that its magnitude and phase are equal to the amplitude ratio and phase in the steady-state frequency response of a linear, time-invariant system of arbitrary order. Thus, given the differential equation that describes such a system, one may readily compute its steady-state frequency response.

Experimental determination of the dynamic response: When a measuring system is exposed to time-dependent inputs, it is advisable to determine its dynamic response by direct dynamic calibration, namely by monitoring the output under controlled and known changes of the input. The most common types of dynamic tests are the *square-wave test* and the *frequency test*. For the square-wave test, the input is made to switch periodically from one level to another. For electric signals, this is easily achieved with the use of a multi-function generator, and a binary field of velocity, temperature, or pressure may be approximately created with the use of a perforated disk rotating in front of a fluid jet. An essentially step-wise change in these parameters may also be achieved by the bursting of a diaphragm separating two chambers containing fluids under different pressures. Dynamic testing of fast-response transducers is sometimes done in shock tubes, which produce travelling shock waves across which there are drastic changes in flow properties. If it is known that the system under test is of first order, the square-wave test will provide the value of the time constant. The oscillatory output of an underdamped, second-order system may be analysed to provide the undamped natural frequency and damping ratio. Determination of these parameters for critically damped and underdamped systems is possible but more difficult because of the reduced sensitivity of this approach. The determination of the parameters of higher-order systems by use of the square-wave test is even more complicated or impossible. In contrast, the *frequency test* is relatively easy to interpret and applies equally to all-order systems. It consists of exposing the system to a sinusoidal input of a constant amplitude and varying frequency. Sinusoidal electric signals are conveniently produced by multi-function generators, whereas cam-driven pistons may be used to design devices producing sinusoidal-type flows. It is usually the amplitude response, B/A, that is measured and plotted vs. frequency to characterize the system.

2.4 Distortion, loading, and cross-talk

The physical presence or the operation of instruments often introduces undesirable influences to the physical system or to the measurement process. Among the most common adverse influences are *flow distortion*, *loading*, and *instrument cross-talk*.

Any instrument inserted in a flow causes a certain amount of distortion, such as blockage, streamline displacement, vortex shedding, instability, phase change, and turbulence or shock waves. Such effects are distributed over some part of the fluid and depend on the local flow conditions, so that they are not easy to correct for. It is advisable to minimize flow distortion by proper design of the apparatus.

Loading of a physical system occurs when a measuring component extracts significant power from it, thus altering the values of the measured properties. A common form of loading occurs in electric and electronic circuits. For example, the connection of an ammeter in series with some electric circuit would change the measured current, unless the ammeter resistance its negligible compared with the resistance of the circuit. Similarly, the use of a large turbine meter to measure flow rate through a pipe may introduce sufficient resistance to reduce the flow rate. Although, in some cases, it may be possible to compensate analytically for the effects of loading, it is obviously preferable to avoid or minimize such effects by proper design or choice of instrumentation.

Instrument cross-talk occurs when a measuring component is coupled in its operation with another measuring component, such that the output of one acts as an undesirable input to the other. Electric and electronic circuits are particularly prone to cross-talk, as one influences another through the generation of electric currents or electromagnetic fields. Instrumentation in fluid systems may also be subjected to thermal cross-talk, as for example in the mutual interference of the wakes of closely positioned heated thermal anemometers, and to mechanical cross-talk, commonly in the form of coupled vibrations. To ensure accurate measurement, any significant cross-talk must be identified and removed.

QUESTIONS AND PROBLEMS

1. Consider a U-tube manometer, containing water as the working liquid and used for measuring pressure differences in air flows. Identify all foreseeable interfering and modifying inputs to this measuring system, explaining their relationship, if any, to the desirable input. In particular, identify undesirable inputs that may occur if the manometer (a) is tilted with respect to the vertical direction, (b) is transported on a vehicle moving with variable speed, (c) is located at different elevations, (d) contains water with different amounts of salt, and (e) is used outdoors throughout the year.
2. A manometer is known to have a second-order response. Describe tests suitable for determining the values of the corresponding parameters.
3. Consider a pressure transducer that will be used for the measurement of wall pressure of an aircraft model in the wind tunnel. Explain how you would perform a static calibration and a dynamic calibration of this transducer. For both cases, sketch the calibration setup, identifying all instruments to be used and their functions.
4. Repeat measurements of the electric resistance of a resistor measured with a digital ohmmeter are as follows: 325, 315, 322, 309, 315, 318, 327, 319, 325, 313, 336, 310, 321, 316, 320, 319, 321, 302, 323, and 318 Ω. Perform normality tests with probability graph paper and the chi-square test. Identify and discard any possible outliers. Compute the possible effect of outliers on the mean and standard deviation of these measurements.
5. Consider the following sequence of readings of a fluctuating voltage: 1.861, 1.005, 1.655, 1.642, 1.442, 1.776, 1.505, 1.498, 1.608, 1.296, 1.947, 1.282, 1.857, 1.008, 1.297, 1.302 V. Would you discard any of these measurements as spurious? If so, which ones and why?
6. Turbulence can be generated in a wind tunnel when a uniform stream with velocity U is passed through a grid of parallel rods, spaced by a distance M ('mesh size') from axis to axis. The turbulence intensity u'/U (where u' is the standard deviation of the velocity fluctuations) can be represented by the empirical expression $u'^2/U^2 = a(x/M - x_0/M)^{-n}$, where x is the downstream distance from the grid

and x_0 ('effective origin') and n are empirical constants to be determined by curve fitting. The usual ranges are $-10 < x_0/M < 10$ and $-1.4 < n < -1.0$. During an experiment with $U = 10.0$ m/s and $M = 25.4$ mm, the following measurements have been taken with a hot-wire anemometer.

x(mm)	u'^2(m^2/s^2)
381	0.205
508	0.135
635	0.098
889	0.062
1270	0.039
1651	0.028
2159	0.020
2794	0.015
3556	0.011
4572	0.008

Plot these results in log–log axes to determine the optimal values of the coefficients x_0/M, a, and n.
Hint: Plot the preceding results by using different values of x_0/M. Select the value that produces the best straight line.

7. The output of a non-linear transducer subjected to the input pressure p is a displacement x given approximately by the equation

$$(Ax + B)\frac{\mathrm{d}x}{\mathrm{d}t} + \sqrt{x} = Cp,$$

where A, B, and C are constants. Linearize the preceding expression about the point (p_0, x_0) and provide expressions for the time constant and the static sensitivity of the linearized transducer.

8. A common expression for the response of a hot-wire anemometer is King's law, $E^2 = A + BU^n$, where E is the voltage output, U is the flow velocity, and A, B, and n are empirical constants determined by calibration. For a particular experiment, it was found that $A = 7.80$, $B = 3.12$ (both in SI units), and $n = 0.45$. Using Taylor's expansion, derive a linear relationship between E and U, valid in the vicinity of (a) $U = 10$ m/s and (b) $U = 15$ m/s. Determine the maximum linearization error for these cases, for velocity intervals of ± 0.05 and ± 0.50 m/s about the reference values. Then assume that, when the anemometer is inserted in a turbulent flow, the measured standard deviation of the anemometer voltage output is 0.033 V. Estimate the standard deviation of the velocity for the two preceding mean speeds.

9. Plot the impulse response and the ramp response, both as amplitude ratios and phases, of first-order and second-order systems, the latter ones for damping ratio values $\zeta = 0$, 0.2, 0.5, $\sqrt{2}/2$, 1, and 2. Discuss the appearance of these plots and

compare qualitatively the responses of first-order systems and those of second-order systems with different damping ratios.

10. A manometer at atmospheric pressure is suddenly connected to a high-pressure air cylinder. What can you conclude about the type of dynamic model that is appropriate for this manometer if its reading (a) monotonically increases to a value and (b) oscillates before it settles to a value. Which kind would you prefer to use and why?
11. To perform a dynamic calibration, a thermometer is attached to an oscillating arm such that it is immersed alternately in one or another of two air jets with different temperatures. The oscillation period is 8 s, and it may be assumed that the thermometer is exposed to a sinusoidal temperature field with a maximum of 330 K and a minimum of 300 K. The reading of the thermometer fluctuates between a maximum of 325 K and a minimum of 305 K. Assuming that the thermometer is a first-order system, determine its time constant. Also determine the difference between the time that the thermometer is exposed to the highest temperature and the time that the thermometer indicates the highest temperature. Hint: first find the phase shift.
12. A common type of active electronic filters is based on the Butterworth polynomials. The sinusoidal transfer function of a third-order low-pass Butterworth filter is

$$\mathbf{H}(j\omega) = \frac{K}{\left(\frac{j\omega}{\omega_0} + 1\right)\left(-\frac{\omega^2}{\omega_0^2} + \frac{j\omega}{\omega_0} + 1\right)},$$

where K is a numerical constant and ω_0 is a characteristic frequency, called the 3-dB high-frequency cutoff. (a) Write a differential equation that describes the response of this system. (b) Prove that, when the input frequency is $\omega = \omega_0$, the amplitude ratio of the filter is equal to $\sqrt{2}K/2$. (c) Plot the amplitude ratio (divided by K) and the phase of this filter vs. the frequency ratio ω/ω_0. Warning: Make sure that the phase plot is continuous and indicates a negative phase shift.

REFERENCES

[1] T. G. Beckwith, R. D. Marangoni, and J. H. Lienhard. *Mechanical Measurement* (5th Ed.). Addison-Wesley, Reading, MA, 1993.

[2] E. O. Doebelin. *Measurement Systems Application and Design* (5th Ed.). McGraw-Hill, New York, 2004.

[3] K. Ogata. *Modern Control Engineering* (3rd Ed.). Prentice-Hall, Englewood Cliffs, NJ, 1996.

[4] A. Papoulis. *Probability, Random Variables and Stochastic Processes* (3rd Ed.). McGraw-Hill, New York, 1991.

[5] Fluid Dynamics Panel Working Group 15, Advisory Group for Aerospace Research and Development. *Quality Assessment for Wind Tunnel Testing*. AGARD Adv. Rep. 304. NATO, Neuilly sur Seine, France, 1994.

[6] J. S. Bendat and A. G. Piersol. *Random Data Analysis and Measurement Procedures* (3rd Ed.). Wiley, Inc., New York, 2000.

[7] W. H. Beyer (Editor). *CRC Handbook of Mathematical Sciences* (6th Ed.). CRC Press, Boca Raton, FL, 1978.

[8] A. R. Cohen. *Linear Circuits and Systems*. Regents Publishing, New York, 1965.

3 Measurement uncertainty

This chapter is a brief review of definitions and estimation methods of measurement uncertainty and conventional practices for the rounding of reported values. Information on measurement uncertainty can be found in many elementary engineering textbooks, but some caution is necessary when older sources are used because uncertainty-related definitions and standards are set by international and national conventions, which are subject to occasional change. The present discussion is based on a simplified version of the 1993 standards set by the International Organization for Standardization (ISO) [1], which is deemed to be suitable for engineering experimentation [2–4].

3.1 Measurement errors

A measurable property can be assumed to have a *true value*, which ideally matches the magnitude of the physical property of interest and the determination of which is the objective of the measurement. The true value cannot be known; if it were known, there would not be a need for measurement. The *measured value* is an estimate of the true value provided by the measuring system; it is generally different from the true value, because no measuring system is perfect. The difference between the measured value and the true value, expressed in the same units as the measured property, is called the *absolute measurement error*. The ratio of the absolute measurement error and the true value, expressed as a percentage, in parts per thousand, or a fraction, is called the *relative measurement error*. Like the true value, the measurement error cannot be determined exactly.

Accuracy of a measurement is an indication of how close the measured value is to the true value. *Inaccuracy* is an indication of how imperfect the measurement is. The term 'accuracy' may be defined as the inverse of the relative error, expressed as a fraction of the measured property. Nevertheless, it is more common to use this term qualitatively and instead to quantify inaccuracy by providing values for the *measurement uncertainty* (see next section). Inaccuracy in a measurement may be caused by errors in calibration, data acquisition, and data reduction. In addition to errors that are due to imperfections of the measuring system and procedures, human errors may also be present in measurements, for example those caused by incorrect or inconsistent readings of analogue instruments

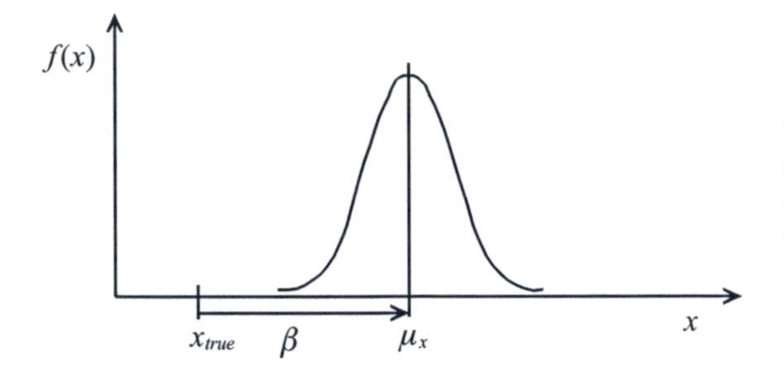

Figure 3.1. True value of a measured property x_{true}, bias error β, and probability density function $f(x)$ of precision error for a large sample of repeat measurements.

or by inaccurate eye averaging of fluctuating digital meters. There are two types or errors: *bias* (fixed or systematic) errors, and *precision* (random) errors.

Bias errors are constant throughout the experiment and could be positive or negative. If a bias error can be estimated (for example, by comparison to a standard, i.e., a more accurate instrument) and found to be non-negligible, it should be removed from the measured value. In most cases, however, the bias is unknown, and it contributes to the overall uncertainty of the measurement.

Precision errors are, presumably, the result of changes in several undesirable inputs, each of which can be assumed to have a relatively small effect on the output, so that the total effect is unpredictable. When this happens, the experiment is said to be *under statistical control*, in which case the *central limit theorem* [5] dictates that repeat measurements should have a *normal* (*Gaussian*) distribution. To verify 'randomness', one may employ a normality test (see Section 2.2); it is also advisable to detect possible outliers and remove them from the sample (also see Section 2.2).

Consider a measurable property that has a true value x_{true}. Now consider that this property is measured with a measuring system that provides repeat values x_i, $i = 1, 2, \ldots, N$, obtained under conditions of statistical control. Assuming that the sample population N is sufficiently large (typically, $N \geq 10$), the set of repeat values would approximately constitute a normal random variable with a mean μ_x and a variance σ_x^2 (Fig. 3.1). Using the mean μ_x as the estimate of x_{true} should eliminate the precision error, in which case the only error in the measurement would be the bias error β, assumed to be the same for all repeat measurements. Then the true value can be found as

$$x_{\text{true}} = \mu_x - \beta. \tag{3.1}$$

When, however, the measured value is based on a single measurement or a relatively small number of measurements, a precision error will also occur. A statistical measure of the likely magnitude of the precision error is the standard deviation σ_x, which is representative of the spread of the distribution of repeat values x about the mean μ_x.

A measurement is called *accurate* if both its bias and precision errors are small, i.e., when

$$|\beta|, \sigma \ll |x_{\text{true}}|; \tag{3.2}$$

otherwise it is considered to be *inaccurate*. Inaccurate measurements may be of three different kinds:

- precise but biased, when $\sigma \ll |x_{\text{true}}|$ but $|\beta|$ is not small compared with $|x_{\text{true}}|$,
- unbiased but imprecise, when $|\beta| \ll |x_{\text{true}}|$ but σ is not small compared with $|x_{\text{true}}|$,
- biased and imprecise, when neither σ or $|\beta|$ are small compared with $|x_{\text{true}}|$.

Repeatability of a measurement is the closeness of measurements of the same property repeated under the same conditions with the same measuring system and during a relatively short time interval. Good repeatability may be considered as a positive indication, but not conclusive evidence, that random errors are small. On the other hand, good repeatability cannot exclude the presence of bias and should not necessarily be interpreted as an indication of overall accuracy.

Reproducibility of a measurement is the closeness of repeat measurements of a property conducted in different laboratories and with different measuring systems. Good reproducibility is evidence that both the systematic and the random errors are small.

3.2 Measurement uncertainty

The *uncertainty* of a measurement represents the interval within which the true value of a measurable property lies with a certain probability. Let x represent a measured value of a measurable property. From recent international convention [2], the uncertainty u of this measurement is defined such that the experimenter is 95% confident that the true value of this property lies in the interval $[x - u, x + u]$. Uncertainty can be specified as *absolute*, expressed in the same units as the measured property, or *relative*, expressed as a percentage, in parts per thousand, or a fraction.

Both bias and precision errors contribute to the uncertainty, and their contributions have to be added appropriately. The measurement uncertainty is determined as

$$u = \sqrt{b^2 + p^2}, \tag{3.3}$$

where b and p are, respectively, the *bias limit* and *precision limit*, defined in the next two subsections.

Bias limit: The bias limit, or systematic uncertainty, b signifies that the experimenter is 95% confident that $|\beta| < b$, where β is the true (but unknown) bias error. When biases are introduced by K different components or steps of the measuring process (e.g., by calibration, data acquisition, or data processing), the total bias limit is determined as

$$b = \sqrt{\sum_{k=1}^{K} b_k^2}, \tag{3.4}$$

where b_k are the individual bias limits. Biases may be difficult to estimate, but this does not mean that they should be ignored. One may estimate a bias from the manufacturer's specifications (e.g., that a certain type of resistor is accurate within ±5% of the specified value), analytical considerations by using material properties, and the experimenter's personal experience with similar instruments.

Precision limit: Consider a random variable consisting of a very large number of measurements, repeated with the same measuring system and under the same conditions. Assume that this random variable has a mean μ_x and a standard deviation σ_x. The precision limit p signifies that, for every single repeat measurement x, μ_x would fall within the interval $[x - p, x + p]$ 95% of the time. Assuming that the random variable has a Gaussian distribution, one may calculate the precision limit as

$$p = 2\sigma_x. \tag{3.5}$$

The parameter σ_x can be estimated from a set of N repeat measurements, under the condition that $N > 10$. If a smaller population sample is used to compute σ_x, the preceding approximation is inaccurate and other methods become more appropriate [1]. Once σ_x is established for a measuring system that operates consistently, it can be used to estimate the precision limit in later experiments, without the need to obtain repeat measurements in each case.

The preceding discussion applies to cases in which the measured value is based on a single reading. As mentioned in the previous section, however, the precision error would vanish if the measured value were taken to be equal to the average μ_x of an infinitely large sample of repeat values, rather than a single reading. Now consider the case in which the measured value is taken to be equal to the average μ_{xN} of a finite sample of repeat measurements x_i, $i = 1, 2, \ldots, N$. Then, provided that $N > 10$, the variance of μ_{xN} would be

$$\sigma_{\mu xN}^2 \sim \frac{\sigma_x^2}{N}, \tag{3.6}$$

clearly lower than the variance σ_x^2 of single measurements x_i. In this case, the precision limit should be calculated as

$$p = 2\sigma_{\mu xN} \sim \frac{2\sigma_x}{\sqrt{N}}. \tag{3.7}$$

3.3 Uncertainty of derived properties

When a property is measured directly, its uncertainty is determined as indicated in the preceding section. Most commonly, however, a property y is computed from direct measurements of one or more *subsidiary properties*, x_m, $m = 1, 2, \ldots, M$, with the use of an analytical expression. Sometimes y is computed from x_m with the use of a graph or table; these may also be considered as equivalent to an analytical expression, to which they can be converted by curve fitting and interpolation. Thus we may assume that there exists an analytical expression $y = y(x_1, x_2, \ldots, x_M)$; we further assume that this expression is continuous in all variables, so that partial derivatives $\partial y/\partial x_m$ can be defined. Each of these derivatives represents the sensitivity of y to changes in the corresponding subsidiary property x_m. When a graph or table is involved in the determination of y, partial derivatives may be estimated as finite differences between neighbouring values.

As for directly measured properties, the uncertainty u_y of the *derived property* y can be computed from the corresponding bias limit b_y and precision limit p_y as

$$u_y = \sqrt{b_y^2 + p_y^2}. \tag{3.8}$$

These bias and precision limits are estimated from the corresponding bias limits b_{x_m} and precision limits p_{x_m} of the subsidiary properties as follows.

Bias limit: If each subsidiary property is measured independently of all others, so that the measurement uncertainties of all x_m are independent of each other, the bias limit b_y can be computed as

$$b_y = \sqrt{\sum_{m=1}^{M} \left(\frac{\partial y}{\partial x_m} b_{x_m} \right)^2}. \tag{3.9}$$

It may happen, however, that the biases of two or more subsidiary properties arise from the same source and are therefore correlated. For example, consider that the derived property y depends on two temperatures that appear only in the form of a temperature difference; if both temperatures are measured with the same thermometer having a bias b_T, it is clear that the temperature difference will be free of any bias that is due to the thermometer. In such cases, we say that the biases of the two subsidiary properties are perfectly correlated. Notice that if, instead of a temperature difference, the sum of the two temperatures was involved in the calculation of y, the bias error that is due to the thermometer would not vanish but would become twice as large as the bias of a single thermometer reading. When the biases of two subsidiary properties are due to more than one source, it is possible that only portions of the total biases are correlated. Following the previous example, let us consider the case of two temperatures measured with the same thermometer but by different persons, one of whom tends to read high and the other tends to read low. The portions of the biases corresponding to the thermometer are correlated, whereas the portions corresponding to the reading person are not. Now consider that, among the M subsidiary properties x_m, there are only two, say x_1 and x_2, whose biases are correlated, and let b'_{x_1} and b'_{x_2} be the portions of these biases that are perfectly correlated. Then the bias limit of y can be estimated from the bias limits of b_{x_m} as

$$b_y = \sqrt{\sum_{m=1}^{M} \left(\frac{\partial y}{\partial x_m} b_{x_m} \right)^2 + 2 \frac{\partial y}{\partial x_1} \frac{\partial y}{\partial x_2} b'_{x_1} b'_{x_2}}. \tag{3.10}$$

Terms similar to the second term on the right-hand side of Eq. (3.10) must be added for any additional pair of variables whose biases are correlated. Depending on the signs of the partial derivatives, these terms could be positive or negative and could therefore increase or decrease the overall bias limit.

Precision limit: When a derived property is estimated from a set of single subsidiary measurements x_m, its precision limit can be estimated from the precision limits of x_m as

$$p_y = \sqrt{\sum_{m=1}^{M} \left(\frac{\partial y}{\partial x_m} p_{x_m} \right)^2}. \tag{3.11}$$

As in the case of directly measured properties, however, we can substantially reduce the precision uncertainty of a derived property y by averaging a number of estimates of y based on repeat subsidiary measurements. Let us assume that $N > 10$ sets of subsidiary measurements $(x_1, x_2, \ldots, x_M)_i$, $i = 1, 2, \ldots, N$ are available, providing N estimates of y as $y_1, y_2, \ldots, y_N$. Then we may compute the average and variance of the sample as, respectively,

$$\mu_{yN} = \frac{1}{N} \sum_{i=1}^{N} y_i, \tag{3.12}$$

$$\sigma_{yN}^2 = \frac{1}{N-1} \sum_{i=1}^{N} \left(y_i - \mu_{yN} \right)^2. \tag{3.13}$$

If we use μ_{yN} as the estimate of y, rather than a single value y_i, we should estimate the precision limit as

$$p_y = 2\sigma_{\mu yN} = \frac{2\sigma_{yN}}{\sqrt{N}} \tag{3.14}$$

Propagation of uncertainty: From the previous discussion one may conclude that the uncertainty of any measured property generally contributes to the uncertainty of any other dependent property. This is commonly referred to as the *propagation of uncertainty*. The amount by which the uncertainty of each subsidiary property x_m affects the uncertainty of the dependent property y also depends on the sensitivity $\partial y/\partial x_m$. The bias limit of a derived property would clearly be unaffected by averaging of repeat values. When the biases of two subsidiary properties are correlated, their net effects on the overall bias could be cumulative or cancelling, depending on the signs of the corresponding sensitivities. The estimation of the uncertainty of a derived property from the precision limits of the subsidiary properties [Eq. (3.11)] is sometimes referred to as *general uncertainty analysis*, and the use of repeat sets of measurements and averaging is referred to as *detailed uncertainty analysis* [3,4].

Example: Let us estimate the uncertainty of convective heat transfer $\dot{q}$ in a pipe determined from the relationship

$$\dot{q} = c_p \rho Q \left(T_1 - T_2 \right), \tag{3.15}$$

where c_p is the specific heat of the fluid, ρ is its density, Q is the volume flow rate, measured with a flowmeter, A is the cross-sectional area of the pipe, and T_1 and T_2 are the inlet and outlet temperatures, measured with thermometers. For simplicity, let us assume that the uncertainty of the product $k = c_p \rho$ is negligible, a condition not likely to be met with in practice. We shall consider two cases: (1) the measurement of T_1 and T_2 by two separate thermometers of the same kind and (2) the sequential measurement of T_1 and T_2 by the same thermometer. We will assume that Q has a bias limit b_Q and a precision limit p_Q and T_1 and T_2 have a bias limit b_T and a precision limit p_T.

1. The bias and precision limits of $\dot{q}$ would be, respectively,

$$b_{\dot{q}} = \sqrt{\left[k(T_1 - T_2) b_Q\right]^2 + (kQb_T)^2 + (-kQb_T)^2}$$
$$= k\sqrt{\left[(T_1 - T_2) b_Q\right]^2 + 2(Qb_T)^2},$$

$$p_{\dot{q}} = \sqrt{\left[k(T_1 - T_2) p_Q\right]^2 + (kQp_T)^2 + (-kQp_T)^2}$$
$$= k\sqrt{\left[(T_1 - T_2) p_Q\right]^2 + 2(Qp_T)^2},$$

and the overall uncertainty of $\dot{q}$ would be

$$u_{\dot{q}} = k\sqrt{(T_1 - T_2)^2 \left(b_Q^2 + p_Q^2\right) + 2Q^2 \left(b_T^2 + p_T^2\right)}.$$

2. The bias limit of $\dot{q}$ would be

$$b_{\dot{q}} = \sqrt{\left[k(T_1 - T_2) b_Q\right]^2 + (kQb_T)^2 + (-kQb_T)^2 + 2(kQ)(-kQ) b_T b_T}$$
$$= k\,|T_1 - T_2|\, b_Q,$$

which is independent of b_T. The precision limit of $\dot{q}$ would be

$$p_{\dot{q}} = \sqrt{\left[k(T_1 - T_2) p_Q\right]^2 + (kQp_T)^2 + (-kQp_T)^2}$$
$$= k\sqrt{\left[(T_1 - T_2) p_Q\right]^2 + 2(Qp_T)^2}.$$

Finally, the overall uncertainty of $\dot{q}$ would be

$$u_{\dot{q}} = k\sqrt{(T_1 - T_2)^2 \left(b_Q^2 + p_Q^2\right) + 2Q^2 p_T^2}$$

which is lower than that in case 1.

3.4 Rounding of reported values

When presenting numerical results of an experiment, whether direct measurements or calculated parameters, one must report all values rounded to a *number of significant digits* or *number of significant figures* [6]. The number of significant digits; may be different from the number of actual digits, depending on whether the value contains zeros. All non-zero digits are significant. Zero is significant when it appears between two non-zero digits, but it is usually not significant when it appears before all non-zero digits in a number that has a decimal point. For example, one may say that the number 0.0034 has two significant digits; however, the same number may have three significant digits, if it were part of a sequence of numbers with three significant digits, such as 0.0123, 0.0345, When it appears at the end of a number, zero may or may not be significant. For example, the number 53400 may have three, four, or five significant digits, which would be evident if other numbers in the same sequence of results were presented; if indeed one wishes to indicate that this number has three significant digits, it would be preferable to present it in engineering notation, namely 53.4×10^3.

The reported value of a directly measured property must be always rounded to a number of digits that is consistent with the accuracy of measurement. The uncertainty of a measured property should always be presented in one, or at most two, significant digits. A measured value must be reported such that its last significant digit is of the same rank as the last significant digit of its uncertainty (e.g., 256 ± 3, 0.234 ± 0.005). By exception, when a directly measured property is used to estimate indirectly measured properties through mathematical operations, it is recommended to retain one or two more significant digits in order to avoid excessive rounding off of the final estimate.

A number is said to be rounded to N significant digits if all digits to the right of the Nth digit are discarded. The rounding of experimental values should be done according to the following convention. If the discarded $(N + 1)$th digit is between 0 and 4 inclusive, the Nth digit should be left as is; if it is between 6 and 9 inclusive, or if it is 5 followed by at least one non-zero digit, the Nth digit should be incremented by one; if it is 5 followed only by zeros, the Nth digit should be left as is if even, and incremented by one if odd. (For example, 0.2452 should be rounded to 0.245; 3.6751 to 3.68; 0.366 to 0.37; 0.3250 to 0.32; and 0.4350 to 0.44).

When numbers are added or subtracted, it is the number of decimal digits that has to be consistent; the number of decimal digits is not necessarily equal to the number of significant digits. When two or more numbers are added, the more accurate ones should first be rounded to one decimal digit more than the lowest decimal digit contained in the least accurate number, namely the number with the smallest rank of decimal digits. Then the rounded numbers should be added, and the sum should be rounded to the same decimal digit as the less accurate number. For example, among the numbers 33.5, 78974, and 0.335, the least accurate for the purposes of addition is 78974, despite the fact that it has more significant digits than the other two numbers; in order to add these three numbers, one must first round 33.5 to 33.5 (no change) and 0.335 to 0.3; the sum of the three rounded numbers is 79007.8, which must be rounded to 79008 for final presentation. When two numbers are subtracted, the more accurate number should be rounded to the same decimal digit as the less accurate number, and the difference should also be rounded to the same decimal digit as the less accurate number. For example, to subtract 10.5 from 234.456, one should first round the latter to 234.5; the difference, expressed in the correct number of digits, is 224.0.

When two numbers are multiplied or divided, the more accurate number should be rounded to one significant digit more than the number of significant digits of the less accurate number and the product or quotient should be rounded to the same number of significant digits as that of less accurate number. For example, consider the product

0.2567×23; the less accurate number is 23 and has two significant digits; the more accurate number, which is 0.2567, must first be rounded to 0.257; the product $0.257 \times 23 = 5.911$ must be rounded to 5.9, namely to two significant digits.

QUESTIONS AND PROBLEMS

1. Provide an example of a biased, precise measurement process and an example of an unbiased, imprecise one.
2. In a simple experiment, you wish to measure the Joule power consumed by an electrically heated resistor. You have available three instruments, a voltmeter, an ammeter, and an ohmmeter, all of which have the same measuring uncertainty, expressed as a percentage of the measured value. Would the power uncertainty depend on which instruments you use? If so, which instruments do you propose to use?
3. A resistor has a resistance of $3.3 \times 10^3\ \Omega$ and a 5% uncertainty. A voltmeter with an uncertainty of 1 mV is used to measure a voltage of 1.355 V across that resistor. Estimate the current through the resistor, rounded to the proper number of significant figures, and its uncertainty.
4. The discharge coefficient of a concentric orifice-plate meter with corner pressure taps for large Reynolds number flows is given by the empirical relationship

$$C = 0.5959 + 0.0312\beta^{2.1} - 0.1840\beta^{8.0},$$

obtained by curve fitting to a large number of measurement data. The orifice diameter ratio is defined as $\beta = d_t/d$, where d_t is the throat diameter and d is the pipe diameter. It is given that $d_t = 50.00 \pm 0.50$ mm and $d = 65.00 \pm 0.50$ mm (95% confidence uncertainties). Determine the absolute and relative uncertainties of the estimated discharge coefficient for the following two cases: (a) The given diameter uncertainties are independent and include both bias and precision errors, and (b) the given diameter values have been measured with the same instrument, whose total uncertainty is 0.50 mm and whose bias limit is 0.25 mm. For this analysis, it may be assumed that the coefficients in the preceding expression have negligible uncertainty. Discuss the validity of this assumption and suggest a procedure by which you may include the uncertainty of the curve-fitting process into the overall uncertainty of the discharge coefficient.
5. Convective heat transfer in fully developed, turbulent flow through a smooth circular tube can be estimated by the Dittus–Boelter empirical expression,

$$\mathrm{Nu} = 0.023\mathrm{Re}^{0.8}\,\mathrm{Pr}^{0.3},$$

in which the Nusselt, Reynolds, and Prandtl numbers are defined, respectively, as $\mathrm{Nu} = hd/k$, $\mathrm{Re} = Ud/\nu$, and $\mathrm{Pr} = \nu/\gamma$, where h is the heat transfer coefficient, d is the diameter of the tube, k is the thermal conductivity, U is the bulk velocity, ν is the kinematic viscosity, and γ is the thermal diffusivity, with all properties evaluated at the bulk temperature T. Estimate h and its uncertainty for air flow in a tube for which the following results have been obtained by direct measurement: $U = 40.3 \pm 0.1$ m/s, $d = 105 \pm 1$ mm, and $T = 593 \pm 2$ K (95% confidence uncertainties). In this analysis make reasonable estimates of the uncertainties of air properties, explaining your rationale.

REFERENCES

[1] ISO. *Guide to the Expression of Uncertainty in Measurement*. International Organization for Standardization, Geneva, Switzerland, 1993.

[2] Fluid Dynamics Panel Working Group 15, Advisory Group for Aerospace Research and Development. *Quality Assessment for Wind Tunnel Testing*. AGARD Adv. Rep. 304. NATO, Neuilly sur Seine, France, 1994.

[3] H. W. Coleman and W. G. Steele. *Experimentation and Uncertainty Analysis for Engineers* (2nd Ed.). Wiley, New York, 1999.

[4] H. W. Coleman and W. G. Steele. Uncertainty analysis. In R. W. Johnson, editor, *The Handbook of Fluid Dynamics*, pp. 39.1–39.11. CRC Press, Boca Raton, FL, 1998.

[5] A. Papoulis. *Probability, Random Variables and Stochastic Processes* (3rd Ed.). McGraw-Hill, New York, 1991.

[6] E. O. Doebelin. *Measurement Systems Application and Design* (4th Ed.). McGraw-Hill, New York, 1990.

4 Signal conditioning, discretization, and analysis

Electric signals, provided by electric and electronic circuits, constitute the most common form of inputs and outputs of measuring systems and their components. In many cases, these signals have a level, form, or frequency content that is not suitable for observation, recording, or processing and needs to be modified, a process generally referred to as signal conditioning. Although much of signal conditioning as well as observation and recording of signals is still done by means of analogue instrumentation, the overwhelming trend is to transform the signals into a discrete-time series and then further condition and process them with the use of digital computers. This chapter summarizes the definitions and background necessary for understanding the operation of common devices and the procedures used for signal conditioning, discretization, and statistical processing.

4.1 Fundamentals of electric and electronic circuits

Basic properties: Before proceeding with the analysis of electric and electronic circuits, it seems advisable to review their basic properties [1,2]. All properties of electric circuits and components are ultimately related to the *electric charge* of a single proton, considered as positive, or a single electron, equal in magnitude to that of a proton but considered as negative. The value of an elementary electric charge is 1.6×10^{-19} C (coulomb). The movement of an electric charge q along a path is called *electric current* i, defined as

$$i = \frac{\mathrm{d}q}{\mathrm{d}t}. \tag{4.1}$$

In electric circuits involving metallic conductors, the moving charges consist of electrons. The current is considered to be positive if it is in the direction opposite to that of electron movement. Its unit is 1 A (ampere). The work done on an electric charge moving from one position to another, per unit charge, is called *voltage* or *potential difference*, V. The voltage unit is 1 V (volt). An ideal *current source* [Fig. 4.1(a)] is a power source that provides a current that is independent of the voltage across it. An ideal *voltage source* is a power source that provides a voltage that is independent of the current through it.

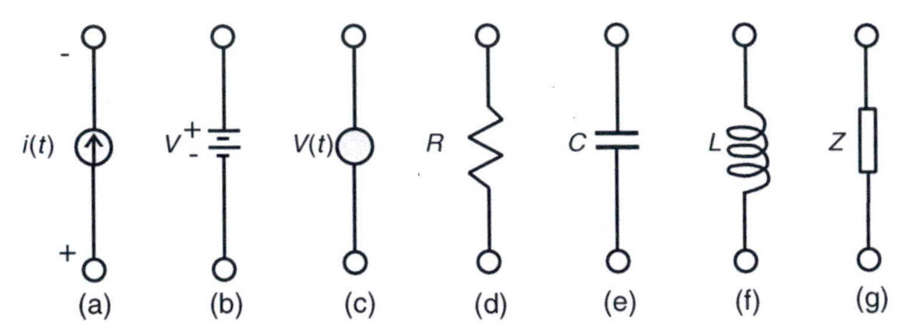

Figure 4.1. Elementary power sources and passive components of linear electric circuits (see text for a description of each part).

Voltage sources could be either *dc* [direct current; Fig. 4.1(b)] or *ac* [alternating current; Fig. 4.1(c)].

The three important, elementary, passive components of electric circuits are the resistance, the capacitance, and the inductance.

A *resistor* with *resistance R* is a device such that the current through it is proportional to the voltage across it (*Ohm's law*), i.e.,

$$R = \frac{V}{i}. \tag{4.2}$$

The unit of resistance is 1 Ω (ohm). The inverse of resistance, $G = 1/R$, is called *conductance*; its unit is 1 S (siemens). A *resistor* [Fig. 4.1(d)] is an electric component assumed to have a pure, lumped, and constant resistance.

A *capacitor* [typically a pair of metallic plates facing each other; Fig. 4.1(e)] is a device that can store an electric charge q when a voltage V is applied across it. Its *capacitance* (assumed to be pure, lumped, and constant) is defined as

$$C = \frac{q}{V}. \tag{4.3}$$

The unit of capacitance is 1 F (farad); its inverse, $S = 1/C$, is called *elastance*. When a time-dependent voltage applies across a capacitor, the current through it can be found as

$$i = C\frac{\mathrm{d}V}{\mathrm{d}t}. \tag{4.4}$$

Finally, an *inductor* [typically a coil of a metallic wire; Fig. 4.1(f)] is a device that produces a *magnetic flux* Ψ when a current passes through it. Its *self-inductance,* or simply inductance (assumed to be pure, lumped, and constant), is defined as

$$L = \frac{\mathrm{d}\Psi}{\mathrm{d}i}. \tag{4.5}$$

The unit of inductance is 1 H (henry). *Faraday's law* connects the voltage to the magnetic flux as

$$V = \frac{\mathrm{d}\Psi}{\mathrm{d}t}, \tag{4.6}$$

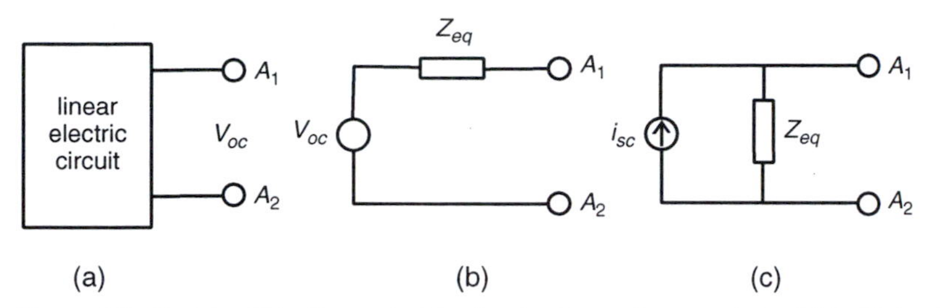

Figure 4.2. Sketch of (a) a linear electric circuit with two free terminals, (b) its Thévenin-equivalent circuit, and (c) its Norton-equivalent circuit.

which leads to the voltage–current relationship for an inductor as

$$V = L\frac{\mathrm{d}i}{\mathrm{d}t}. \tag{4.7}$$

When more than one inductor is connected in a circuit such that their magnetic fields interfere with each other, their voltage–current relationships are coupled through a property M called *mutual inductance*; its unit is also 1 H.

Now let $\mathbf{V}(s)$ and $\mathbf{I}(s)$ be, respectively, the Laplace transforms of the voltage across and the current through a component. Then the *impedance* [Fig. 4.1(g)] of this component is defined as

$$Z(s) = \frac{\mathbf{V}(s)}{\mathbf{I}(s)}. \tag{4.8}$$

The impedance of a component or a circuit section is generally a complex function. The unit of impedance is also 1 Ω. The inverse of impedance, $Y(s) = 1/Z(s)$, is called the *admittance*. The impedances of resistors, capacitors, and inductors are, respectively, given as

$$Z = R, \quad Z = 1/(sC), \quad \text{and} \quad Z = sL. \tag{4.9}$$

The impedances of components in series add up, and it is the admittances of components in parallel that add up. Thus the impedance of a resistor and a capacitor connected in series is $Z = R + 1/(sC)$. In all the preceding expressions, the complex variable s may be replaced with $j\omega$.

Principles of circuit analysis: An analysis of simple electric circuits can be made based on the following two principles. *Kirchhoff's current* (*node*) *law* states that the algebraic sum of all currents entering or leaving a node equals zero. *Kirchhoff's voltage* (*loop*) *law* states that the algebraic sum of all voltage rises and drops around any closed path (loop) equals zero.

The following two theorems permit the transformation of linear electric circuits to equivalent ones that contain a single impedance [2, 3].

Thévenin's theorem states that any electric circuit having two free terminals A_1 and A_2 and comprising current sources, voltage sources, resistors, capacitors, and/or inductors [Fig. 4.2(a)] can be transformed into an equivalent circuit [Fig. 4.2(b)], consisting

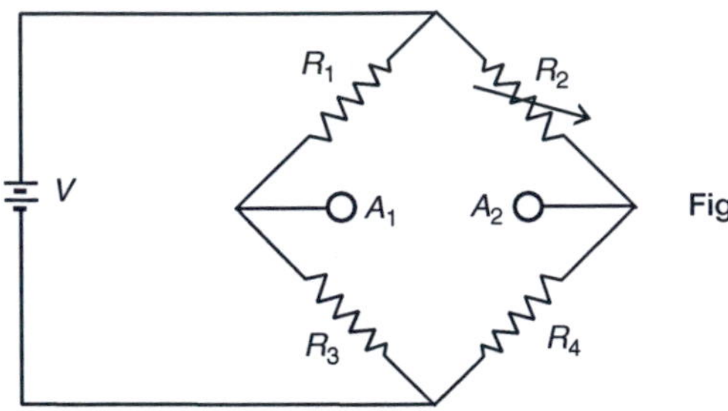

Figure 4.3. Sketch of a Wheatstone bridge.

of an *equivalent voltage source* V_{oc} in series with an *equivalent impedance* Z_{eq}. The value V_{oc} is equal to the voltage across the terminals A_1 and A_2 in the original circuit and is called the *open-circuit voltage*. One can find the equivalent impedance Z_{eq} by short circuiting all voltage sources and leaving open circuited all current sources in the original circuit and then combining all passive components.

Norton's theorem states that any electric circuit having two free terminals A_1 and A_2 and comprising current sources, voltage sources, resistors, capacitors, and/or inductors [Fig. 4.2(a)] can be transformed into an equivalent circuit [Fig. 4.2(c)], consisting of an *equivalent current source* i_{sc} in parallel with an *equivalent impedance* Z_{eq}. The value i_{sc} is equal to the current that would flow between the terminals A_1 and A_2 if they were short circuited in the original circuit and is called the *short-circuit current*. The Norton-equivalent impedance is the same one as the Thévenin-equivalent impedance, and it is also called the *output impedance* of the circuit as seen between the terminals A_1 and A_2. It is related to the Laplace transforms of V_{oc} and i_{sc} as

$$Z_{eq}(s) = \frac{\mathbf{V}_{oc}(s)}{\mathbf{I}_{sc}(s)}. \tag{4.10}$$

As an example of the application of electric circuit principles, let us consider the *Wheatstone bridge*, which is a circuit used quite often in measuring instrumentation. The classical Wheatstone bridge, comprising only resistors, is shown in Fig. 4.3 (a generalized Wheatstone bridge may contain capacitors and inductors in addition to resistors). One of the uses of this bridge is to measure resistance, let's say that of the resistor R_1. Towards this purpose, this resistor is connected in the bridge configuration with three other resistors whose resistances R_2, R_3, and R_4 are known and one of which (let's say R_2) is adjustable. The bridge is connected to a dc voltage source, as shown, and a sensitive voltmeter (e.g., a galvanometer) is connected across the terminals A_1 and A_2. The variable resistance is adjusted such that the voltage shown by the voltmeter vanishes, in which case we say that the bridge is balanced. Then direct application of Ohm's law and Kirchhoff's laws shows that

$$R_1 = \frac{R_2 R_3}{R_4}. \tag{4.11}$$

Let us further determine equivalent circuits for this bridge. If the bridge were balanced, the open-circuit voltage would be zero, so let us assume that the bridge is imbalanced. The Thévenin-equivalent voltage can be easily found as

$$V_{\text{oc}} = V_{A_1} - V_{A_2} = V\left(\frac{R_3}{R_1 + R_3} - \frac{R_4}{R_2 + R_4}\right). \tag{4.12}$$

To find the equivalent impedance, we short the power source and notice that the circuit consists of a series connection of two pairs of resistors connected in parallel. Then,

$$Z_{\text{eq}} = \frac{R_1 R_3}{R_1 + R_3} + \frac{R_2 R_4}{R_2 + R_4}. \tag{4.13}$$

Finally, to determine the short-circuit current, we short A_1 and A_2 and apply Kirchhoff's laws to derive

$$i_{\text{sc}} = V\frac{R_2 R_3 - R_1 R_4}{R_1 R_2 R_3 + R_1 R_2 R_4 + R_1 R_3 R_4 + R_2 R_3 R_4}. \tag{4.14}$$

Alternatively, assuming that V_{oc} and i_{sc} are step functions, as required for their Laplace transforms to exist, we may easily find that

$$\mathbf{V}_{\text{oc}}(s) = \frac{V_{\text{oc}}}{s}, \tag{4.15}$$

$$\mathbf{I}_{\text{sc}}(s) = \frac{i_{\text{sc}}}{s}, \tag{4.16}$$

and verify that, indeed, $Z_{\text{eq}} = \mathbf{V}_{\text{oc}}(s)/\mathbf{I}_{\text{sc}}(s)$.

Operational amplifiers: A serious disadvantage of passive signal conditioning circuits is that, when connected to another circuit, they might alter the voltages and currents through it, a phenomenon identified earlier as loading (Section 2.4). Loading effects may be circumvented by proper selection of the circuit components, a process called *impedance matching*. However, for general-purpose instruments, this is both inconvenient and restrictive. Fortunately, with the advent of vacuum tubes, first, and semiconductors, more recently, it has become possible to essentially decouple electric circuits from each other with the use of active devices, powered by external power supplies. The basic building block for common signal conditioning operations is the *operational amplifier*, or *op-amp* [4–7]. This is an amplifier having a high internal gain, a high input impedance, a low output impedance, and connected in a feedback mode. The equivalent circuit of an op-amp is sketched in Fig. 4.4(a). It can be seen that an op-amp has two inputs, referred to as *inverting input* and *non-inverting input*, and an *output*; the corresponding voltages are denoted as V_-, V_+, and V_0. Power is provided by a *power supply*, usually a dual supply, with voltages $\pm V_s$ and a common ground. Among the most important op-amp specifications are the following:

- *open-loop gain* A, typically of the order of 10^5 or higher;
- *input impedance* Z_i, typically 10^5–10^{11} Ω;
- *output impedance* Z_o, typically 1 to 10 Ω;
- *voltage offset* V_{os}, typically ± 1 mV;
- *bias currents* i_- and i_+, typically 10^{-14}–10^{-6} A;

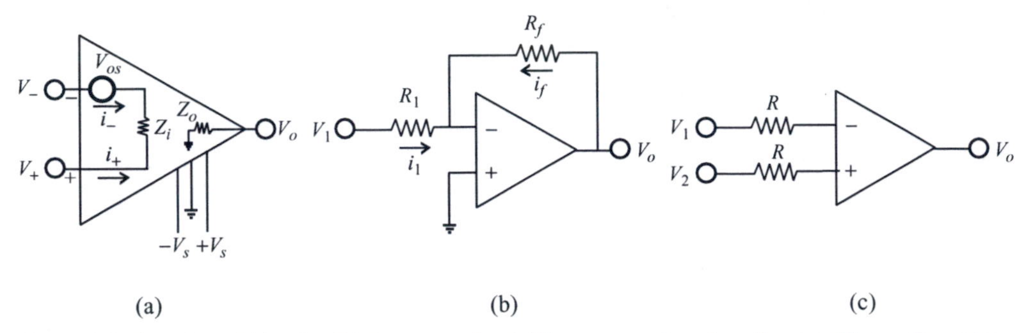

Figure 4.4. Sketches of (a) a simplified op-amp circuit, (b) an op-amp in an inverting amplifier configuration, and (c) an op-amp in a comparator configuration.

- *slew rate*, which indicates how fast the op-amp responds to a change in input voltage; typical values are in the range of volts per microsecond;
- *common-mode rejection ratio* (see Section 4.2), typically in the range 60–120 dB;
- *gain–bandwidth product*, which is the product of the low-frequency gain and the 3-dB high-frequency cutoff (see Section 4.2); the higher the gain is set, the lower the bandwidth of the amplifier would be; a high gain–bandwidth product op-amp would be required for applications involving low-level or high-frequency signals.

Because of their large input impedance and small output impedance, op-amps normally exert negligible loading on circuits connected to their inputs and outputs. A simplified analysis of op-amp operation may be based on the assumption of an *ideal op-amp*, namely a device that has an infinite open-loop gain, an infinite input impedance, and a zero output impedance [6]. Then one may assume that

$$V_- \simeq V_+\,, \quad i_- \simeq 0, \quad i_+ \simeq 0. \tag{4.17}$$

For example, consider a basic op-amp configuration, shown in Fig. 4.4(b), for which the non-inverting input has been grounded and the inverting input has been connected to a voltage source V_1 through a resistor R_1 and to the output through the *feedback resistor* R_f. For an ideal op-amp, the input impedance is infinite, so that there is zero current flowing between the inverting and the non-inverting inputs, while, at the same time, the voltages at these inputs are identical. Kirchhoff's current law at the inverting input gives $i_1 + i_f = 0$, which is equivalent to $V_1/R_1 + V_o/R_f = 0$, or

$$V_o = -\frac{R_f}{R_1} V_1 \tag{4.18}$$

Thus this device is an *inverting amplifier* with a gain R_f/R_1.

The output voltage of an op-amp cannot exceed the range of the power supply voltages. If driven towards higher positive or negative values, V_o will reach a *saturation value* $\pm V_{\text{sat}}$, which cannot be exceeded in magnitude. For the typical power supply voltage of ± 15 V, the saturation voltage would be approximately ± 13 V. One must be careful to avoid saturation, especially when dealing with fluctuating input voltages, in which case the output voltage would be clipped if driven to values exceeding V_{sat} in magnitude. In fact, the op-amp may be configured as a *comparator*, providing an output that is either $+V_{\text{sat}}$, if $V_1 < V_2$, or $-V_{\text{sat}}$, if $V_2 < V_1$ [see Fig. 4.4(c)].

Noise and interference: The electric voltage (*signal*) at the output of a measuring system or component would generally fluctuate in time. Part of the fluctuations may be inherent to the measured property, such as the velocity in an unsteady or turbulent flow, but another part, which is broadly termed as *noise*, is due to various, internal or external, undesirable influences on the measuring device. The noise is generally random and cannot be known precisely, so that its effects are analyzed statistically. The *signal-to-noise ratio* S/N may be defined as the ratio of the average value of the signal and the standard deviation of the noise [7]. It may also be expressed in decibels (see Section 4.2), in which case it is defined as

$$\frac{S}{N}/_{\text{dB}} = 10 \log_{10} \frac{\text{signal power}}{\text{noise power}}. \tag{4.19}$$

A related parameter is the *noise figure*, defined as the signal-to-noise ratio referred to the input normalized by the signal-to-noise ratio referred to the output:

$$F = \frac{S/N_{\text{input}}}{S/N_{\text{output}}}. \tag{4.20}$$

Noise is best considered in the frequency domain. There are many sources of noise, depending on the type of measured property, the measuring system design, and the measuring conditions [7, 8]. A type of noise that is ubiquitous in electric–electronic components is the *thermal* or *Johnson noise*, which is due to fluctuations of the electron gas in a conductor caused by random thermal motions. This type of noise has contributions from fluctuations over a wide range of frequencies, including very high frequencies. A common assumption is to consider that thermal noise is *white*, namely having a uniform power distribution over the entire frequency spectrum (see Section 4.5). An estimate of the rms (*root-mean-square*, i.e., the standard deviation) noise voltage produced by a resistance R at a temperature T and limited to a range of frequencies with a width Δf is

$$\sigma_V = \sqrt{4kTR\Delta f}, \tag{4.21}$$

where $k = 1.38 \times 10^{-23}$ J/K is the *Boltzmann constant*. Clearly the power of thermal noise over the entire frequency spectrum would tend to infinity. Thermal noise cannot be eliminated, and the only way to reduce it is by filtering, usually with the use of a low-pass filter. It is obvious that the narrower the frequency band of the signal, the lower the noise variance. In practice, the optimal filter choice would be such as to produce a signal that contains as much as possible of the fluctuations of the measured property and as little as possible noise. The choice of filter settings may require a detailed analysis of the measuring system followed by trial-and-error-type experimentation.

Another type of noise occurs when the measuring system contains devices that collect electrons at an electrode or divert them over a barrier (examples of such devices are photomultiplier tubes, junction diodes, and junction transistors). This is called *shot noise* and, similarly to thermal noise, it has uniform contributions from all frequencies, so that it may also be termed as white. The collected or diverted electrons form a current with a mean i and a random component that, when limited to the frequency range Δf, has rms value of

$$\sigma_i = \sqrt{2 q_0 i \Delta f}, \tag{4.22}$$

where $q_0 = 1.6 \times 10^{-19}$ C is the elementary electric charge. Low-pass filtering is, once more, the usual approach to remove shot noise.

Measuring systems are sometimes subjected to noise, besides white noise, that is concentrated in particular frequency ranges. A type of noise that is not well understood but is present in semiconductors and seems to decrease with experience in component design and manufacturing is the *$1/f$ noise*. In some cases, the rms voltage of this noise may be estimated with *Hooge's law*:

$$\sigma_V = \sqrt{\frac{\alpha}{N_f} \frac{\Delta f}{f}} \tag{4.23}$$

where $\alpha = 2 \times 10^{-9}$ W/m^3 is an empirical constant, N_f is the density of electrons or holes in the semiconductor, Δf is the frequency range of the signal, and f is the frequency under consideration. It is clear that $1/f$ noise increases with decreasing frequency, and, although it can be reduced by low-pass filtering, its contribution to slowly varying signals cannot be removed.

Another source of distortion of electric signals is various interfering inputs, which could be internal in the circuitry or external [8, 9]. A problem that is especially important in low-voltage applications is *thermoelectric voltages*, created when junctions of different materials (e.g., wire junctions, mating connectors, and even solder used in circuitry) are exposed to different temperatures. This problem can be reduced by a careful selection of materials and by avoiding thermal gradients in the equipment. Another very common problem in electronic circuitry is the generation of *grounding currents* between points that ideally should be at the same electric potential, called the 'ground', but, in practice, are connected to the ground through connectors with a finite resistance. This creates *ground loops*, namely multiple paths towards the ground, with associated undesirable currents. The proper *grounding* of circuits often proves to be a tedious and unpredictable ordeal. Finally, electromagnetic fields, generated by nearby transformers, electric motors, and even radio, television, and other electromagnetic waves, may be coupled to an electric circuit through stray capacitances and inductances (e.g., those that are due to loose connections or looped wires). This would create undesirable fluctuating voltages that interfere with the signals of interest. The most common interference is from electric power lines, at a frequency of 60 Hz in North America or 50 Hz in Europe. When a signal is transmitted in the form of a voltage difference through a pair of conductors, both the positive and the negative sides may be subjected to the same interference, called *common mode*. Thus it may happen that a very weak signal is superimposed upon a very strong common mode. Ideally, a differential amplifier should remove the common mode, but in practice, because input asymmetry and offsets, part of the common mode may

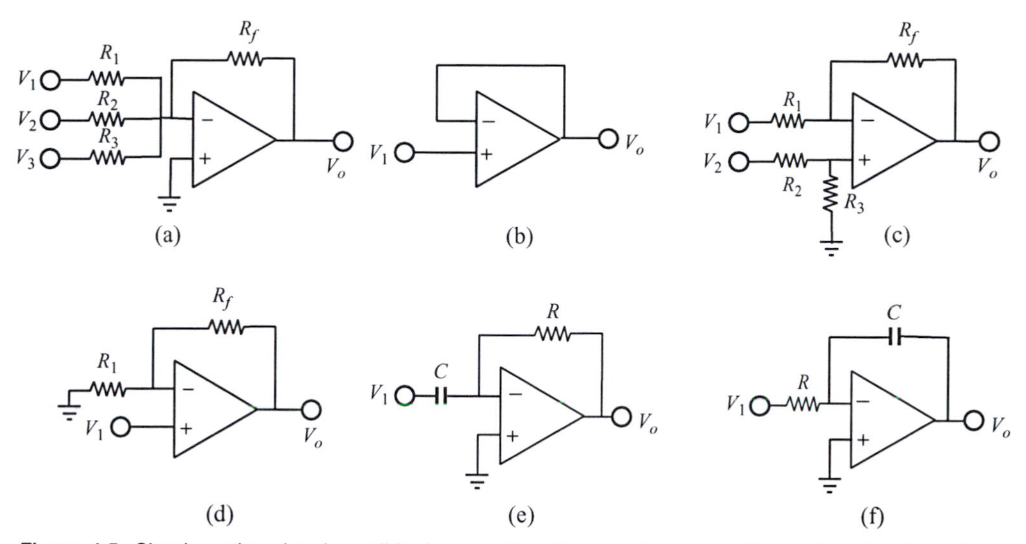

Figure 4.5. Simple active signal conditioning circuits with operational amplifiers: (a) adder, (b) voltage follower, (c) difference amplifier, (d) non-inverting amplifier, (e) differentiator, and (f) integrator.

be amplified and appear at the output. The term *common-mode error* [9] signifies the residual common-mode voltage at the output of the amplifier, divided by the amplifier's gain, so that it is presented in the form of noise superimposed to the input voltage.

The elimination of interference could be a very difficult problem; although suggestions for proper circuit design and other rules of thumb (e.g., shielding, guarding, and grounding [8–10]) are available, this process usually requires trial-and-error-type troubleshooting and adjustment. Even the experienced experimenter must be prepared to spend many hours, to say the least, in tracing and minimizing interference effects.

4.2 Analogue signal conditioning

Analogue signal operations: The most common analogue conditioning operation is the amplification of weak signals. This can be achieved with the use of the inverting amplifier [Fig. 4.4(b)], by selecting $R_f/R_1 > 1$. If $R_f/R_1 < 1$, the same device can be used as an *inverting attenuator* for the purpose of reducing the amplitude of signals that are too strong to be processed. Operational amplifiers can easily be configured in many other types of simple active circuits to perform a variety of signal conditioning tasks; among the most common ones are these:

- *Adder*, or *summing amplifier* [Fig. 4.5(a)]:

$$V_o = -\left(\frac{R_f}{R_1}V_1 + \frac{R_f}{R_2}V_2 + \frac{R_f}{R_3}V_3\right), \tag{4.24}$$

  when all resistors are equal,

$$V_o = -(V_1 + V_2 + V_3). \tag{4.25}$$

- *Voltage follower*, or *buffer* [Fig. 4.5(b)]:

$$V_o = V_1; \tag{4.26}$$

a buffer is inserted between circuits that would load each other if connected directly.

- *Difference amplifier* [Fig. 4.5(c)]:

$$V_o = \frac{R_f R_3}{R_2 + R_3}\left(\frac{1}{R_f} + \frac{1}{R_1}\right) V_2 - \frac{R_f}{R_1} V_1; \tag{4.27}$$

when all resistors are equal, this circuit becomes a *subtractor*:

$$V_o = V_2 - V_1. \tag{4.28}$$

- *Non-inverting amplifier* [Fig. 4.5(d)]:

$$V_o = \left(1 + \frac{R_f}{R_1}\right) V_1. \tag{4.29}$$

- *Differentiator* [Fig. 4.5(e)]:

$$V_o = -RC\frac{\mathrm{d}V_1}{\mathrm{d}t}. \tag{4.30}$$

- *Integrator* [Fig. 4.5(f)]:

$$V_o = -\frac{1}{RC}\int V_1 \mathrm{d}t. \tag{4.31}$$

Before the advent of digital data acquisition and processing, various analogue signal conditioning operations were routinely performed with the use of dedicated devices, including time delays, linearizers, multipliers and dividers, correlators and spectrum analysers. These operations are more conveniently performed by software on discretized signals (see Section 4.3). Even so, specialized integrated circuits, called *function modules*, are available to perform multiplications, divisions, roots, powers, logarithms, trigonometric functions, and other mathematical operations [11], for cases in which it is preferable to condition and process signals by analogue rather than digital means.

Particular care must be taken in the processing of electric signals at very low levels (e.g., of the order of microvolts), such as those often produced by strain gauges, thermocouples, and other transducers. Such signals must be first amplified to an appropriate level before they can be further processed, in which case the amplifier noise as well as interference voltages from various sources would be amplified as well. For an improved signal-to-noise ratio, it is necessary to use dedicated, very low-noise *preamplifiers*. *Instrumentation amplifiers* are particularly suitable as preamplifiers. These are specialized integrated circuits containing op-amps but having a committed response, for example the amplification of a voltage difference with a gain either fixed or selectable by means of a jumper or a single resistor. Instrumentation amplifiers have extremely low internal noise and drift and a moderate gain range. An important quality is their ability to remove a common interference voltage from a differential signal, called the *common-mode rejection* (CMR). A related parameter is the *common-mode rejection ratio* (CMRR), defined as the ratio of the common-mode voltage and the common-mode error voltage, which is referred to the input. CMR is usually expressed in terms of CMRR as [9]

$$\mathrm{CMR_{dB}} = 10 \log_{10} \mathrm{CMRR}. \tag{4.32}$$

Instrumentation amplifiers normally have a large CMR, typically exceeding 60 dB. When very large common-mode voltages are present or for safety purposes (e.g., in medical instrumentation, in which circuits may come in contact with humans), it may be necessary to avoid physical contact between two circuits altogether. This may be achieved by the transmission of fluctuating signals through transformers or optical couplings. Devices that perform such tasks are known as *isolation amplifiers*.

Analogue filters: Filters are used to remove undesirable effects, particularly noise. A convenient presentation of filter performance is in the form of a *Bode diagram*, or *Bode plot*, which is a logarithmic plot of the amplitude and phase response of the filter vs. frequency [1]. At this point, it seems appropriate to differentiate between the *angular frequency* ω, expressed in radians per second (s^{-1}), and the *cyclic frequency* f, expressed in cycles per second [hertz (Hz)]. Their relationship is

$$\omega = 2\pi f = \frac{2\pi}{T}, \tag{4.33}$$

where T is the period of the signal. A filter is designed to have one or more frequency range over which the output is not attenuated significantly, called the *passband*, and one or more frequency range over which the output is attenuated significantly, called the *stopband*. The amplitude ratio over the passband is not necessarily unit, but may have an approximately constant positive value, called the *maximum passband value*. The width of the passband frequency range is called the *bandwidth* of the filter.

The frequency axis in a Bode diagram is divided into *decades*, namely ranges of frequency between end points having a ratio 1:10 (e.g., between 1 and 10 kHz) or *octaves*, namely ranges of frequency between end points having a ratio 1:2 (e.g., between 1 and 2 kHz). The amplitude function, also called the *gain function*, is often expressed in *decibels* (dB), defined as

$$|\mathbf{H}(j\omega)|_{\mathrm{dB}} = 20 \log_{10} |\mathbf{H}(j\omega)|. \tag{4.34}$$

A gain of 0 dB means that the input and output amplitudes at a particular frequency are equal. A positive gain in decibels means that the output amplitude is larger than the input amplitude, whereas a negative gain means the opposite. A characteristic of a filter is the *3-dB cutoff frequency*, which is the frequency at which the output amplitude is reduced to $1/\sqrt{2} \simeq 0.707$ of its maximum passband value. Considering that the power of a signal is proportional to the square of the voltage, we understand that the 3-dB cutoff frequency represents the frequency at which the output power is reduced to 1/2 of its maximum passband value. According to these definitions, a low-pass filter has a *high-frequency cutoff*, a high-pass filter has a *low-frequency cutoff*, and band-pass and band-reject filters have two cutoff frequencies, one low and one high. The *sharpness* of a filter, namely its ability to remove efficiently any components within the stopband, may

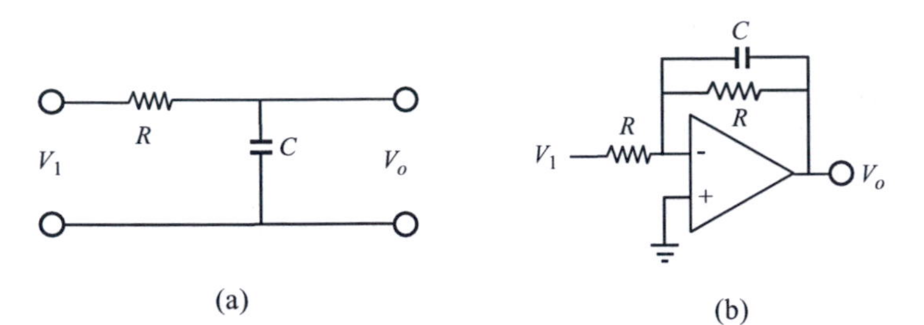

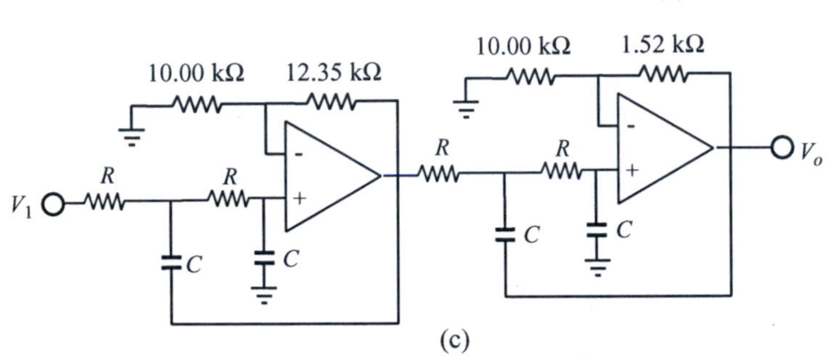

Figure 4.6. Sketches of (a) a passive first-order low-pass filter, (b) an active first-order low-pass filter, and (c) a fourth-order low-pass Butterworth filter.

be quantified by the asymptotic rate of decrease of the amplitude ratio in the stopband (commonly referred to as the *roll-off*), which is usually expressed in *decibels per decade* or *decibels per octave*. The rate of change of the phase can be expressed as degrees per decade or degrees per octave.

Analogue filters are classified as *passive*, which are circuits of exclusively passive components, or *active*, which in addition include op-amps or other active components. An example of a passive filter is the *RC* circuit shown in Fig. 4.6(a). It is easy to show that this circuit satisfies the first-order, time-invariant, ordinary differential equation

$$RC\frac{\mathrm{d}V_o}{\mathrm{d}t} + V_o = V_1, \tag{4.35}$$

and thus has the response of a first-order system with a time constant $\tau = RC$ and a transfer function $\mathbf{H}(j\omega) = 1/(1 + jRC\omega)$. This circuit acts as a low-pass filter with a 3-dB cutoff frequency $\omega_c = 1/(RC)$ [or $f_c = 1/(2\pi RC)$]. The high-frequency ($\omega \gg \omega_c$) roll-off of a first-order filter is -20 dB/decade (equal to approximately -6 dB/octave) and the high-frequency phase shift tends towards $-\pi/2$. The low-frequency ($\omega \to 0$) asymptote is $|\mathbf{H}(j\omega)| = 1$, and the high-frequency ($\omega \to \infty$) asymptote is $|\mathbf{H}(j\omega)| = 1/RC\omega$. The two asymptotes intersect at the *corner frequency* $\omega_{\mathrm{cf}} = 1/(RC)$, which, in this case, is equal to the cutoff frequency ω_c.

An active first-order low-pass filter, using an op-amp, is shown in Fig. 4.6(b). It has a response identical to that of the passive filter, with the exception of a negative sign in its transfer function (inverting filter). Active filters have two general advantages over passive ones: First, they do not load other circuits and, second, they can be *cascaded* (i.e., connected in series) to modify or enhance the filtering action. For example, one may easily construct a first-order band-pass or band-reject filter by cascading a first-order low-pass filter and a first-order high-pass filter; similarly, one may construct a higher-order filter by cascading a series of identical lower-order filters.

A variety of active filter designs are available for different actions. In general, a sharp filter (i.e., one having a steep roll-off) would be desirable; however, this corresponds to a more complicated circuitry as well as an increasing phase shift and thus an increasing *waveform distortion* of the signal. One can construct an nth-order low-pass filter simply by cascading n identical first-order active filters. Its transfer function is

$$\mathbf{H}(j\omega) = \frac{1}{(1 + jRC\omega)^n}, \tag{4.36}$$

which provides an amplitude ratio of

$$|\mathbf{H}(j\omega)| = \frac{1}{\left[\sqrt{1 + (RC\omega)^2}\right]^n}, \tag{4.37}$$

a cutoff frequency of

$$\omega_c = \frac{1}{RC\sqrt{2^{1/n} - 1}}, \tag{4.38}$$

and a high-frequency roll-off of $-20n$ dB/decade. Its corner frequency is always $\omega_{cf} = 1/(RC) \geq \omega_c$. This means, that, if one maintains the same values of R and C, the bandwidth of a cascade filter will increase as the number of stages increases.

Improved designs of filters use different types of polynomials to approximate the corresponding ideal filter, which has a uniform gain within its passband and an infinite roll-off at the cutoff frequency. The most popular designs are based on the Butterworth polynomials. Low-pass Butterworth filters have amplitude ratios given by

$$|\mathbf{H}(j\omega)| = \frac{|K|}{\sqrt{1 + (\omega/\omega_c)^{2n}}}, \tag{4.39}$$

where K is a constant and n is the order of the corresponding polynomial. This expression shows that the bandwidth of these filters is independent of their order. The low-frequency asymptote of the amplitude ratio is $|\mathbf{H}(j\omega)| = |K|$, and the high-frequency asymptote is $|\mathbf{H}(j\omega)| = |K| / (\omega/\omega_c)^n$. Consequently the corner frequency (i.e., the intersection of these asymptotes) is always equal to the cutoff frequency. Like the first-order filter cascades, Butterworth filters have a high-frequency roll-off of $-20n$ dB/decade. However, if one compares two filters of the same order and with the same ω_c, one will find that the Butterworth filter will attenuate its output at a given frequency within the stopband ($\omega > \omega_c$) significantly more than the corresponding cascade filter would. The Butterworth polynomials are selected such as to satisfy the preceding requirements. In general, the Butterworth low-pass transfer function is given as

$$\mathbf{H}(s) = \frac{K}{q_n(s)}, \tag{4.40}$$

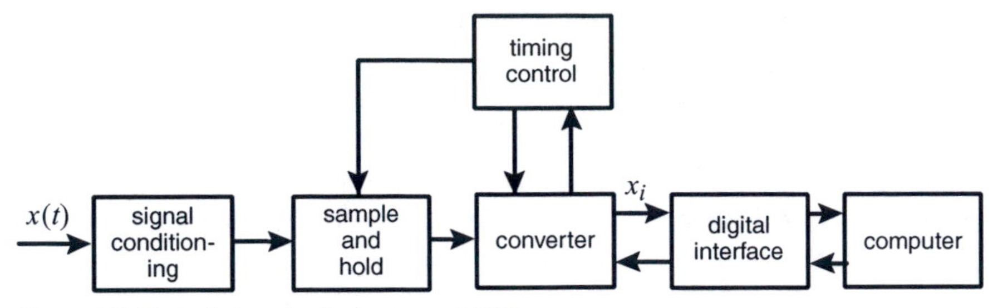

Figure 4.7. Typical steps in single-channel ADC.

where the polynomials $q_n(s)$, up to order four, are given in the following table.

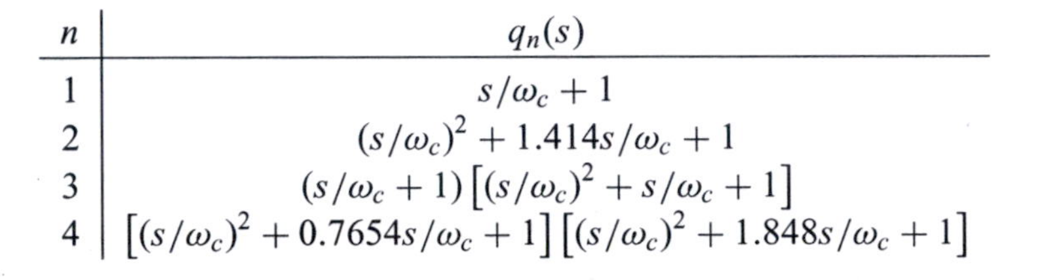

n	$q_n(s)$
1	$s/\omega_c + 1$
2	$(s/\omega_c)^2 + 1.414s/\omega_c + 1$
3	$(s/\omega_c + 1)\left[(s/\omega_c)^2 + s/\omega_c + 1\right]$
4	$\left[(s/\omega_c)^2 + 0.7654s/\omega_c + 1\right]\left[(s/\omega_c)^2 + 1.848s/\omega_c + 1\right]$

As an example, the fourth-order low-pass Butterworth filter is shown in Fig. 4.6(c) [3]. This filter consists of two second-order stages in series, with corresponding low-frequency gains of 2.235 and 1.152, for a total low-frequency gain of 2.575. It has a high-frequency roll-off of -80 dB/decade and a 3-dB cutoff frequency of $\omega_c = 1/(RC)$.

4.3 Discretization of analogue signals

Analogue-to-digital conversion: Most measuring instruments and transducers produce continuous, or *analogue*, electric signals, which are generally time dependent. To some degree, analogue signals can be conditioned, observed, and stored by exclusively analogue instrumentation; however, the use of digital data processing is a far more convenient and accurate approach. The process of converting an analogue electric signal $x(t)$ into a *time series* $x_i, i = 0, \pm 1, \pm 2, \ldots,$ is called *analogue-to-digital conversion* (ADC); the corresponding devices are called *analog-to-digital converters* (also denoted as ADC). The advantage of ADC over analogue processing is that, once the time series is determined, it may be conditioned and processed by use of a digital computer and appropriate software. These are the typical steps in ADC (Fig. 4.7):

1. *Analogue signal conditioning:* This may include amplification, offsetting, or both, in order to optimize the signal level within the range of the ADC. In addition, analogue filtering is best applied just before conversion, in order to remove noise and other undesirable effects most effectively. Low-pass filtering is usually applied at this stage to prevent aliasing (see Section 4.3).
2. *Multiplexing:* For multi-channel conversion, a *multiplexer* is used to select the particular channel that will be discretized at a given time and to establish the sequence of channels to be discretized; a variety of multiplexing methods are available [6].

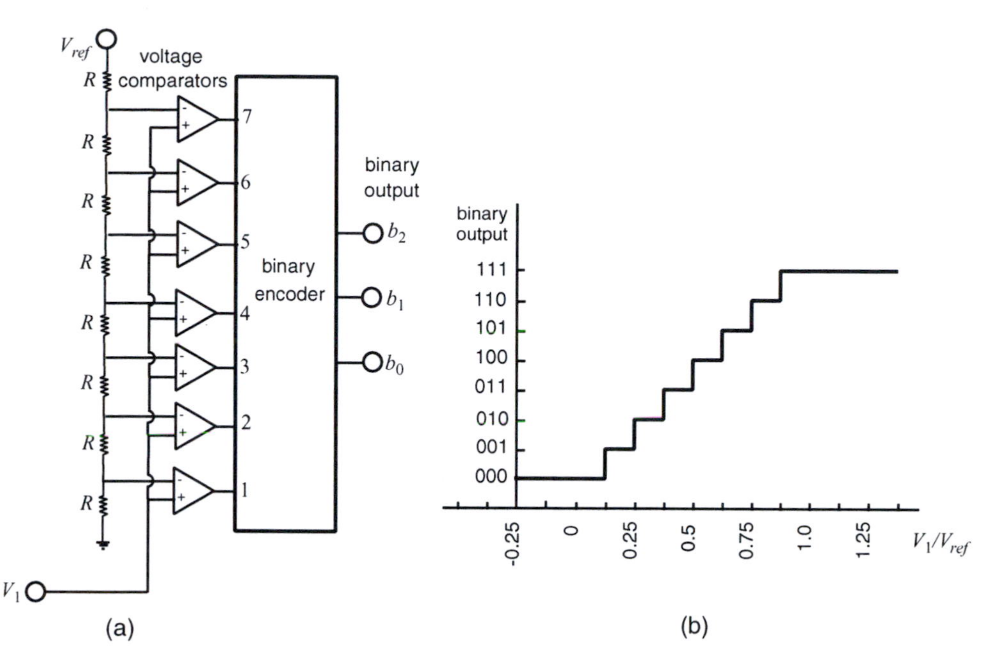

Figure 4.8. (a) Conversion of an analogue voltage V_1 to a binary number $b_2b_1b_0|_{\text{bin}}$ by use of seven voltage comparators and a 3-bit binary encoder, (b) plot of the ideal 3-bit converter output vs. the normalized analogue input.

3. *Sample-and-holding:* The selected signal is fed to a *sample-and-hold circuit*, which continuously tracks ('sample' state) its input voltage, and, when it receives a control signal, it captures the current voltage value by storing it in a capacitor ('hold' state); thus the voltage is maintained essentially constant over a time that is long enough for the conversion to take place.
4. *Conversion:* There are several ADC methods, including the successive approximation method, the countercomparator method, and the dual-ramp method, but, in all cases, the ADC produces a discrete value that approximates the voltage stored in the sample-and-hold circuit. The most common code in which the discrete value is stored is the binary code, in which a number is represented as a series of successive powers of 2, called the *bits*, with each bit having a value of 0 or 1. For example, the digital number 13 can be represented in binary form as $1101|_{\text{bin}} = (1 \times 2^3) + (1 \times 2^2) + (0 \times 2^1) + (1 \times 2^0) = 8 + 4 + 0 + 1 = 13$. An illustration of the ADC process is shown in Fig. 4.8(a) [10], in which a 3-bit ADC is used to discretize the analogue voltage V_1. The input signal is compared, through a series of seven comparators (Section 4.1) connected in parallel, with each of seven voltages equal to, respectively, $V_{\text{ref}}/8$, $V_{\text{ref}}/4$, $3V_{\text{ref}}/8$, $V_{\text{ref}}/2$, $5V_{\text{ref}}/8$, $3V_{\text{ref}}/4$, and $7V_{\text{ref}}/8$. These voltages are obtained from the reference voltage V_{ref} with the use of a voltage divider, consisting of eight identical resistors. Each comparator will give a high value (e.g., $+V_{\text{sat}}$) if V_1, which is connected to the non-inverting input, is greater than the voltage connected to the inverting input and a low value (e.g., $-V_{\text{sat}}$) otherwise. The outputs of all comparators are fed to separate inputs of a *binary encoder*,

which produces a 3-bit binary output $b_2b_1b_0|_{\text{bin}}$, whose value depends on which of the inputs are high or low. For example, if $5V_{\text{ref}}/8 < V_1 < 3V_{\text{ref}}/4$, the encoder will provide the output $101|_{\text{bin}} = 5$. If $7V_{\text{ref}}/8 < V_1$, the output will be $111|_{\text{bin}} = 7$, whereas, if $V_1 < V_{\text{ref}}/8$, the output will be $000|_{\text{bin}} = 0$. Thus, the only possible output values of this ADC are the eight integers between 0 and 7 [Fig. 4.8(a)]; it is obvious that, if V_1 is outside the range 0–V_{ref}, this converter will erroneously produce the constant value 0 or 7, respectively. The *ideal resolution* of an ADC is specified by its range and the number of bits. For example, a 16-bit ADC will produce a digital output equal to an integer between 0 and $2^{16} - 1 = 65{,}535$; if the ADC range is between 0 and 10 V, the ideal ADC resolution would be equal to $10/(2^{16} - 1)\,\text{V} \simeq 0.15\,\text{mV}$. Thus, compared with the analogue input, the discretized signal contains an additional error that is due to the limited resolution of ADC; this error is known as *quantization noise*. Because of this noise, the ADC output will generally fluctuate even if the input is nominally constant. This fluctuation adds the *quantization uncertainty*, approximately equal to one half the ADC resolution (95% confidence level) [10]; in practice this uncertainty may be considerably higher because of various other errors in the ADC process. Another important specification for ADC systems is the *maximum conversion rate*, usually provided in samples per second, or hertz. It is understood that, in multi-channel conversions, this rate must be divided by the number of discretized channels.

5. *Storage:* A *digital interface* transfers the discrete values to the computer memory or a storage device; when high-speed ADC is required, the discretized signals are stored in intermediate *buffers*, which are periodically emptied to their permanent destination.

In multi-channel applications, conversion of different channels can be made in either of two ways.

- *Sequentially:* All channels are connected to a multiplexer, which selects one at a time and feeds it to a single sample-and-hold; after the signal is held, it is fed to the converter, which finishes the conversion before the next channel is fed to the same sample-and-hold; this method requires a single sample-and-hold and is simpler than the next one but suffers from the disadvantage of providing samples of different channels at different times.
- *Simultaneously:* Each channel is fed to a separate sample-and-hold, and a timing circuit triggers all holdings at the same time; when sampling is completed, all sample-and-hold outputs are fed to a multiplexer, which selects one at a time and sends them to the converter. Ideally, this method would provide simultaneous samples. In practice, however, the voltages stored in the sample-and-hold capacitors would leak, causing errors. For this reason, this method requires higher-quality, more expensive components than the previous one. Dedicated converters for each channel may be used to improve the performance of ADC.

Programming of the computer for data acquisition and processing can be done in text-based programming languages, such as assembly language, MATLAB, C^{++}, or

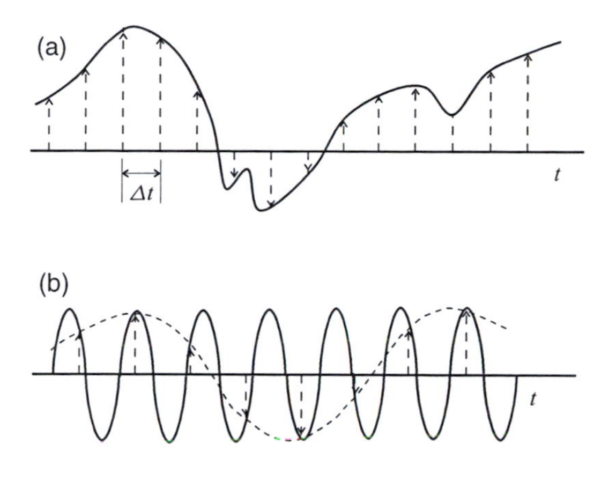

Figure 4.9. (a) Sketch showing an analogue voltage (solid curve) and corresponding equally spaced discrete samples (dashed lines); (b) sketch of a sinusoidal analogue signal with a frequency ω_0 (solid curve), corresponding discrete samples (dashed lines) obtained at a rate $\omega_S < 2\omega_0$, and the perceived analogue signal (dotted curve), which, because of aliasing, has a frequency $\omega_S - \omega_0$.

FORTRAN. A convenient approach would be to use one of several commercial software packages, which employ a graphical representation of the intermediate steps. A widely used package is *LabVIEW*, which uses the graphical programming language G to create programs in block diagram form. LabVIEW is a general-purpose programming system but also includes libraries of functions and development tools designed specifically for data acquisition and control. LabVIEW programs are called *virtual instruments* (VIs), because their appearance and operation imitate those of actual instruments. VIs have an interactive user interface, called the front panel, because it simulates the front panel of an actual instrument (e.g., a voltmeter), displaying images of knobs, pushbuttons, switches, graphs, etc. The VI receives instructions from a block diagram, created by the programmer in G language.

As an illustration of the ADC process, let us consider the simple configuration of a 3-bit, successive-approximation-type converter. The output of the converter will be one of $2^3 = 8$ numbers, namely $0, 1, 2, 3, 4, 5, 6$, or 7. Let us assume that the range of the converter is between 0 and 10 V. Its ideal resolution, equal to the ideal maximum quantization error, would be $10/7$ V. We can further assume that the gain of the entire ADC process is equal to 1, i.e., that there is no amplification or attenuation; we also assume that there is no offsetting of signals during any stage of the process. Then we could multiply the digital output by the conversion factor 10/7 V to recover the converted output in the form of a voltage. The only possible discrete voltage values that can be provided by the converter are 0.0, 1.4, 2.9, 4.3, 5.7, 7.1, 8.6, and 10.0 V; considering that the resolution is 1.4 V, it is clear that there is no point in including more digits in these numbers.

Sampling: Consider an analogue (i.e., continuous) time-dependent electric signal $x(t)$ and a series of equally spaced discrete samples (*time series*) $x_i, i = 0, \pm 1, \pm 2, \ldots,$ such that $x_i = x(i\Delta t)$ [Fig. 4.9(a)]. Δt is called the *sampling time*, corresponding to the *sampling rate* $\omega_s = 2\pi/\Delta t$ (or $f_s = 1/\Delta t$). In general, there are an infinite number of analogue signals that correspond to the same time series. However, the following theorem permits a reversible, at least ideally, representation of analogue signals by time series [12].

Sampling theorem: If the continuous signal $x(t)$ is *band-limited*, namely if $\mathbf{X}(j\omega) = 0$ for $|\omega| > \omega_h$, then it can be uniquely determined by the time series $x_i = x(i\Delta t)$, $i = 0, \pm 1, \pm 2, \ldots$, if Δt is chosen such that $\omega_s = 2\pi/\Delta t > 2\omega_h$ (*Nyquist criterion*). One can exactly reconstruct the signal $x(t)$ by generating a train of impulses with amplitudes equal to the sample values and then low-pass filtering it with an ideal low-pass filter having a gain Δt and a cutoff frequency that is greater than ω_h and smaller than $\omega_s - \omega_h$.

The frequency $2\omega_h$ is called the *Nyquist rate*, and the frequency ω_h itself is called the *Nyquist frequency*. The Nyquist criterion states that equality of the sampling and the Nyquist rates is insufficient for the sampling theorem to hold. In practice, to accurately discretize an analogue signal at a sampling rate of ω_s, one should first low-pass filter it with an analogue filter having a cutoff frequency $\omega_c \leq 0.4\omega_s$. The choice of filter type and ω_h depends on the frequency content of the particular signal, the noise level and frequency content, and the specifications of the available instrumentation.

When the Nyquist criterion is not satisfied, the analogue signal may not be reconstructed by manipulations of the time series. In particular, high-frequency components of the analogue signal appear as lower-frequency components, thus permanently changing the signal composition in the frequency domain. This phenomenon is called *aliasing*; it is illustrated in Fig. 4.9(b), in which the original signal $x(t) = \sin(\omega_0 t)$, sampled at the rate $\omega_s < 2\omega_0$, appears to be $x'(t) = \sin[(\omega_s - \omega_0)t] \neq x(t)$. It must be emphasized that, once aliasing occurs, it cannot be removed by digital filtering or other digital data processing.

The sampling theorem also applies to discrete-time signals [12]. In such cases, a closely spaced discrete-time signal is represented by another that retains only part of the original values, spaced at equal time intervals. This process, which is the inverse of *interpolation*, is known as *decimation*.

Digital filtering [13–15]: Digital filters of various types can be applied to time series in a manner similar to the application of analogue filters to analogue signals, but with some important advantages, to be discussed in what follows. The output of analogue filters has a frequency-dependent amplitude change, compared with the input, but also has a phase shift, which depends on frequency in a non-linear manner. Thus analogue filters would distort the waveform shape of a signal containing various frequency components. For example, if an analogue signal consisting of the sum of two sinusoidal terms with different frequencies were band-pass filtered, not only would the amplitudes of the two terms change by different proportions, but also the appearance of the signal would change as a result of phase shifts of the two terms not in proportion to their frequencies. In some applications, it is essential to preserve the waveform shape, and so a different type of filter would be required.

Digital filters are more flexible than analogue ones because they can utilize parts of or the entire time series at the time of application of the filter. They can also utilize past values of the output and may also be applied in multiple passes in forward or reverse direction. Digital filters applied in the time domain are generally distinguished into two categories:

- *non-recursive* or *finite impulse response* (*FIR*) filters, whose output is not an explicit function of past output values,
- *recursive* or *infinite impulse response* (*IIR*) filters, whose output is a function of past output values, in addition to input values.

There is a great variety of digital filters for general use as well as optimized for specific purposes. In what follows, we discuss only an example of a non-recursive digital filter that introduces zero phase shift at all frequencies and thus causes no waveform distortion of the time series [16]. For the phase shift to vanish, the transfer function $H(\omega)$ of the filter must be real. Let x_n, $n = 1, 2, \ldots,$ be a time series, used as input to the filter, sampled at a time step Δt. Then the output of the filter can be written as

$$y_m = b_0 x_m + \sum_{n=1}^{\infty} b_n (x_{m-n} + x_{m+n}), \tag{4.41}$$

where b_0 and b_n are the cosine Fourier coefficients of the filter transfer function (see Section 4.5), namely

$$b_0 = \frac{\Delta t}{\pi} \int_0^{\frac{\Delta t}{\pi}} H(\omega) \mathrm{d}\omega, \tag{4.42}$$

$$b_n = \frac{\Delta t}{\pi} \int_0^{\frac{\Delta t}{\pi}} H(\omega) \cos(n\omega\Delta t)\, \mathrm{d}\omega, \tag{4.43}$$

such that

$$H(\omega) = b_0 + 2 \sum_{n=1}^{\infty} b_n \cos(n\omega\Delta t). \tag{4.44}$$

Let us now assume that the desired filter has an ideal band-pass response with zero phase shift, with the desired transfer function given by

$$H(\omega) = \begin{cases} 1, & \omega_1 \leq \omega \leq \omega_2 \\ 0, & \text{otherwise} \end{cases}, \tag{4.45}$$

for which the Fourier coefficients become

$$b_0 = \frac{(\omega_2 - \omega_1)\, \Delta t}{\pi}, \tag{4.46}$$

$$b_n = \frac{\sin(n\omega_2 \Delta t) - \sin(n\omega_1 \Delta t)}{n\pi}. \tag{4.47}$$

A complication arises when the Fourier series is truncated to a finite number of terms N, because this process generates oscillations in the transfer function estimate, known as the *Gibbs phenomenon*. To reduce the amplitude of such oscillations, one may employ a 'window' function w_n, such that

$$y_m = b_0 x_m + \sum_{n=1}^{N} b_n w_n \left(x_{m-n} + x_{m+n}\right), \tag{4.48}$$

$$H(\omega) \approx b_0 + 2\sum_{n=1}^{N} b_n w_n \cos\left(n\omega\Delta t\right). \tag{4.49}$$

An example of a suitable window function is the *Hanning window*

$$w_n = \frac{1}{2}\left(1 + \cos\frac{n\pi}{N}\right). \tag{4.50}$$

The sharpness of the filter increases with the number of terms N, but so does the computational time required for the application of the filter.

It is important to emphasize that digital filters cannot be used to remove aliasing caused by insufficient sampling frequency. To prevent aliasing, low-pass analogue filtering must be applied before discretization with a cutoff of $\omega_2/2\pi$ and the sampling time Δt must be less than π/ω_2.

An alternative approach that also results in zero phase shift is to apply a digital filter to a time series twice. First, apply the filter in a forward direction (namely successively applied to input values acquired in increasing time sequence) and afterwards apply the same filter to the output of the first pass in reverse direction. The amplitude ratio of the transfer function of the dual filter would be equal to the square of the single-pass transfer function amplitude ratio.

Digital filters can be also applied in, besides the time domain, the frequency domain [15]. This would require first transforming the time series to the frequency domain through a fast Fourier transform (FFT; see Section 4.5), then multiplying this by the transfer function of the desired filter, and, finally, performing an inverse FFT to recover the filtered time series. This procedure introduces no phase shift but is computationally more intensive than time-domain filtering.

4.4 Statistical analysis of signals

In the following subsections, some fundamental concepts and basic definitions required for the statistical analysis of signals and other forms of data are summarized. Additional examples and details can be found in numerous available references [4, 17–19].

Probability: The term *probability* is used colloquially to express one's belief that something may or may not be true. An example of such use is the assessment 'the probability that it will rain this afternoon is high', based on observation of the morning sky, combined with one's experience and intuition. Clearly, this notion is subjective and non-quantifiable. The classical scientific definition of probability is as the *a priori* ratio of favourable to total number of alternative outcomes of a process (called *events*), assuming that all alternatives are equally likely to occur. For example, one may infer that the probability of obtaining 'heads' in tossing a coin is 50%, because there are only two possible types of events, heads or tails, and each event should be equally likely (beware of fixed coins). The gross inaccuracy of this approach can be illustrated if we consider as events the readings of a thermometer rounded to the closest marking on its scale. If the thermometer scale has 40 markings, the classical definition would lead to all probabilities of all events being equal to 1/41, which is obviously wrong for usual room-temperature monitoring.

To overcome this limitation, probability may be defined as the relative frequency of an event. In this approach, the probability of future events can be predicted from past records of the same process, under the implicit assumption that all past and future events would be subjected to the same external conditions and that there are no particular circumstances that may significantly affect a specific event but not the others. In the coin-tossing experiment, this definition allows the assignment of different probabilities to the 'heads' and 'tails' events, depending on their relative frequencies in past repeat trials. An example showing that this approach may also lead to errors is the prediction of the probability that an expectant mother will bear twins as the relative frequency of twin births with respect to total births in a certain country over a certain period of years. This estimate disregards the effect of genetic background of the parents, the possible use of fertility medication, and other factors, which may substantially affect the conception of twins.

The preferable definition of probability is axiomatic, namely it avoids assumptions or reference to past events. Consider the set of all possible events $\mathcal{A}_i, i = 1, 2, \ldots,$ in a repeatable experiment. Two events are called *mutually exclusive* if it is impossible for both to occur at the same trial. The *compound event* $\mathcal{A}_i + \mathcal{A}_j, i \neq j$, is the event that occurs when either $\mathcal{A}_i$ or $\mathcal{A}_j$ or both occur at a given trial. Finally, let $\mathcal{S}$ denote the *certain event*, namely the event that occurs in every trial. Then the probability of an event $\mathcal{A}_i$ is defined as a number $P(\mathcal{A}_i)$ that obeys the following three postulates:

- $P(\mathcal{A}_i) \geq 0$,
- $P(\mathcal{S}) = 1$,
- if $\mathcal{A}_i$ and $\mathcal{A}_j, i \neq j$, are mutually exclusive, then $P(\mathcal{A}_i + \mathcal{A}_j) = P(\mathcal{A}_i) + P(\mathcal{A}_j)$.

Thus the probability is a non-negative real number, less than or equal to one. As an example, consider the coin-tossing experiment. Two mutually exclusive events are 'heads', $\mathcal{A}_1$, and 'tails', $\mathcal{A}_2$. Let us neglect the possibility that the coin will stand up on its side, be snatched by a bird before it reaches the ground, etc. Then the event $\mathcal{S} = \mathcal{A}_1 + \mathcal{A}_2$ is a certain event. If the probabilities of obtaining heads or tails are denoted as $P(\mathcal{A}_1) = p$ and $P(\mathcal{A}_2) = q$, respectively, the only requirements of the axiomatic definition are that $0 \leq p, q \leq 1$, and $p + q = 1$. Moreover, this definition does not provide a means of estimating p and q. For comparison, the classical definition would predict that $p = q = 1/2$, whereas the relative frequency definition would require the experiment to be repeated many times and actually estimate p and q in future tosses from past records, which may produce unequal values of p and q, as a result of uneven weight distribution, technique of the experimenter, or other reasons.

Random variables, distribution functions, and probability density functions: A *real random variable* $\mathbf{x}$ is a real function of the events ζ, such that the set $\{\mathbf{x}(\zeta) \leq x\}$ is an event for any real number x and $P\{\mathbf{x} = \infty\} = P\{\mathbf{x} = -\infty\} = 0$. In this notation, boldfaced characters, such as $\mathbf{x}$, denote the random variable, whereas italic characters, such as x, denote the real values that this random variable takes. Random variables could be discrete, continuous, or mixed.

The *distribution function* (or *cumulative distribution function*) of a random variable $\mathbf{x}$ is defined as

$$F_x(x) = P\{\mathbf{x} \leq x\}. \tag{4.51}$$

It has the following obvious properties:

$$F_x(-\infty) = 0, \tag{4.52}$$

$$F_x(+\infty) = 1, \tag{4.53}$$

$$F_x(x_1) \leq F_x(x_2) \text{ if } x_1 < x_2. \tag{4.54}$$

The *pdf* $f_x(x)$ of a random variable $\mathbf{x}$ is the derivative of its distribution function, i.e.,

$$f_x(x) = \frac{\mathrm{d}F_x(x)}{\mathrm{d}x}, \tag{4.55}$$

which is equivalent to

$$F_x(x) = \int_{-\infty}^{x} f_x(x)\mathrm{d}x. \tag{4.56}$$

The *mean* (or *average*, or *expectation*, or *expected value*) of a random variable $\mathbf{x}$ is defined as

$$\langle \mathbf{x} \rangle = \int_{-\infty}^{\infty} x f_x(x)\mathrm{d}x. \tag{4.57}$$

The *variance* of a random variable $\mathbf{x}$ is defined as

$$\sigma_x^2 = \langle \mathbf{x} - \langle \mathbf{x} \rangle \rangle = \int_{-\infty}^{\infty} (x - \langle \mathbf{x} \rangle)^2 f_x(x)\mathrm{d}x. \tag{4.58}$$

The positive square root σ_x of the variance is called the *standard deviation*. The terms 'variance' and 'standard deviation' are often used alternately with the terms *mean-squared value* and *rms value*, respectively. However, the definition of the latter two terms does not involve removal of the mean, which may cause ambiguity in certain contexts.

In general, a *moment of* nth *order* of a random variable is defined as

$$m_n = \langle \mathbf{x}^n \rangle = \int_{-\infty}^{\infty} x^n f_x(x)\mathrm{d}x, \tag{4.59}$$

and a *central moment of* nth *order* is defined as

$$\mu_n = \langle (\mathbf{x} - \langle \mathbf{x} \rangle)^n \rangle = \int_{-\infty}^{\infty} (x - \langle \mathbf{x} \rangle)^n f_x(x)\mathrm{d}x. \tag{4.60}$$

It is obvious that $\mu_0 = 1, \mu_1 = 0$, and $\mu_2 = \sigma_x^2$. Two other commonly used statistical properties are the *skewness factor* and the *flatness factor*, respectively defined as

$$S = \frac{\mu_3}{\sigma_x^3}, \tag{4.61}$$

$$F = \frac{\mu_4}{\sigma_x^4}, \tag{4.62}$$

with the latter parameter sometimes presented as the *kurtosis*:

$$K = \frac{\mu_4}{\sigma_x^4} - 3. \tag{4.63}$$

Given the distribution function or the pdf of a random variable, one may compute all its moments and central moments. In reverse, the distribution function or the pdf of a random variable may be computed only if all its moments are known. Therefore, in general, a random variable cannot be described statistically by a finite number of moments.

Now consider two random variables $\mathbf{x}$ and $\mathbf{y}$. Their *joint distribution function* is defined as

$$F_{xy}(x, y) = P\{\mathbf{x} \leq x \text{ and } \mathbf{y} \leq y\}. \tag{4.64}$$

It has the following obvious properties:

$$F_{xy}(-\infty, y) = F_{xy}(x, -\infty) = 0, \tag{4.65}$$

$$F_{xy}(+\infty, +\infty) = 1. \tag{4.66}$$

The distribution functions F_x and F_y of the individual random variables are called *marginal* and satisfy the relationships

$$F_{xy}(x, \infty) = F_x(x), \tag{4.67}$$

$$F_{xy}(\infty, y) = F_y(y). \tag{4.68}$$

The *joint pdf* of $\mathbf{x}$ and $\mathbf{y}$ is defined as

$$f_{xy}(x, y) = \frac{\partial^2 F_{xy}(x, y)}{\partial x \partial y}. \tag{4.69}$$

Two random variables are *statistically independent* if

$$f_{xy}(x, y) = f_x(x) f_y(y). \tag{4.70}$$

An indicator of possible interdependence between two random variables is their *correlation coefficient*,

$$\rho = \frac{\langle (x - \langle \mathbf{x} \rangle)(y - \langle \mathbf{y} \rangle) \rangle}{\sigma_x \sigma_y}, \tag{4.71}$$

which spans the range $[-1, 1]$. Two random variables are called *uncorrelated* if $\rho = 0$; otherwise they are called *correlated*, in which case they are also statistically dependent on each other. On the other hand, two uncorrelated random variables do not necessarily have to be statistically independent. A value $|\rho| = 1$ is proof that the two variables are proportional to each other and not simply statistically dependent.

The Gaussian random variable: A random variable **x** is called *normal* or *Gaussian* if its pdf is given by

$$f_x(x) = \frac{1}{\sqrt{2\pi}\sigma_x} e^{-\frac{(x-\langle \mathbf{x}\rangle)^2}{2\sigma_x^2}}. \tag{4.72}$$

The values of this function, together with values of the normal distribution function, are given in many handbooks and textbooks of probability [20]. Because the pdf of a normal random variable depends on two parameters alone (namely the mean and the standard deviation), all its moments can be expressed as functions of these two parameters. It is easy to show that all odd central moments of a Gaussian random variable vanish ($\mu_{2i+1} = 0$, i integer) and that its even central moments can be expressed as

$$\mu_{2i} = 1 \times 3 \times 5 \times \cdots \times (2i - 1)\sigma_x^{2i}, \quad i \text{ integer} \tag{4.73}$$

Both the flatness factor and the kurtosis of Gaussian random variables vanish, and deviations from these reference values are often used to detect systematic statistical features of random variables.

Two random variables are called *jointly normal* or *jointly Gaussian* if their joint pdf is

$$f_{xy}(x, y) = \frac{1}{2\pi\sigma_x\sigma_y\sqrt{1-\rho^2}} e^{-\frac{1}{2(1-\rho^2)}\left[\frac{(x-\langle \mathbf{x}\rangle)^2}{\sigma_x^2} - \frac{2\rho(x-\langle \mathbf{x}\rangle)(y-\langle \mathbf{y}\rangle)}{\sigma_x\sigma_y} + \frac{(y-\langle \mathbf{y}\rangle)^2}{\sigma_y^2}\right]}. \tag{4.74}$$

Uncorrelated jointly Gaussian variables are also statistically independent.

Random processes: Consider a time-dependent experiment, and let ζ represent each of its realizations, assumed to be repeated independently from each other. A *random process* or *stochastic process* is the process of assigning, according to a specified rule, a time-dependent function $\mathbf{x}(\zeta, t)$ to each realization ζ. Therefore a random process is a function of two variables, the particular realization ζ and the time t. Associated with a particular realization ζ_i is the time function $\mathbf{x}(\zeta_i, t)$, which is called a *time series*. The values $\mathbf{x}(\zeta, t_0)$ of all realizations at a fixed time instance t_0 constitute a random variable. The set of all time series is called an *ensemble*, and each time series is a *member* of the ensemble. A few realizations of a general random process are illustrated in Fig. 4.10(a).

As an example, consider a large number of vertical tubes supplied with water by the same large head-tank and such that the flow in each tube does not affect the flow in any other. Let the velocity at the centre of the exit plane of each tube be continuously measured with an instrument that can follow accurately velocity variation in time. In this example, the random process is the magnitude of velocity in the tube system previously specified; the velocity measurement in each tube represents one realization of

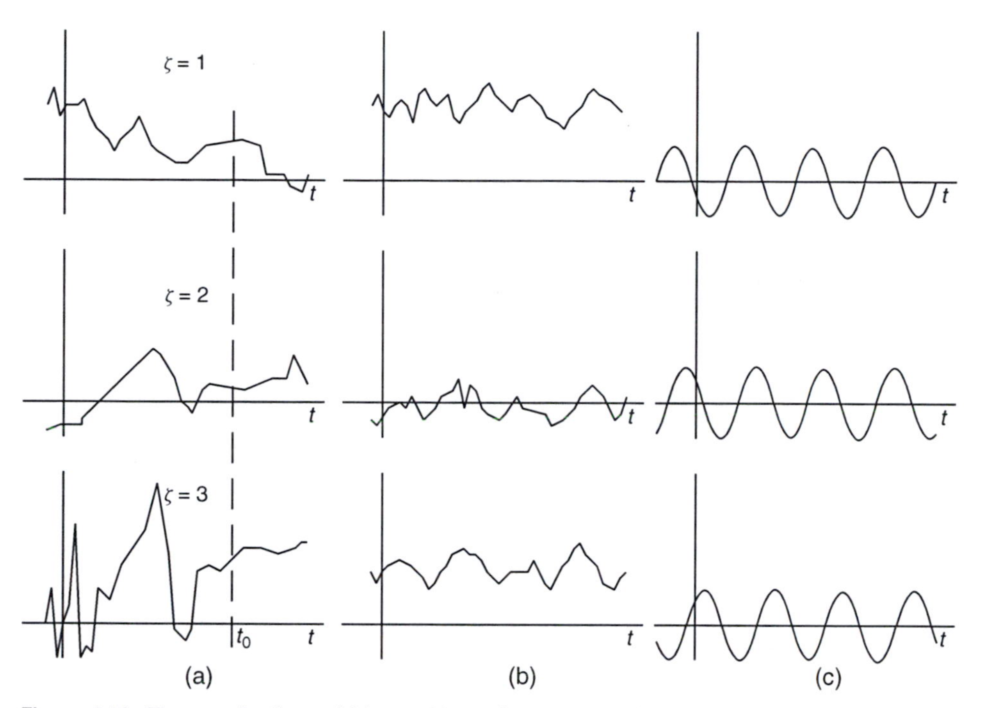

Figure 4.10. Three realizations of (a) an arbitrary (non-stationary) random process, (b) a stationary but non-ergodic random process, and (c) the stationary and ergodic random process defined by Eq. (4.92).

the experiment; the temporal record of the velocity variation in a particular tube is a time series; and the set of velocity values in all tubes at a given time instant constitutes a random variable. So far, no assumption has been made concerning the diameters and lengths of the tubes. One can construct a related but distinct random process by considering a single tube connected to a large head-tank. One realization of the experiment is to open the valve connecting the tube to the tank and to measure velocity as a function of time up to a certain time instance when the valve is closed again. One can obtain additional independent realizations by repeating this process after adding or removing fluid in the tank and allowing the fluid in the tank and tube to become still. Compared with the duration of the multiple-tube experiment, which, in principle may extend to infinite times, the duration of the single-tube experiment would be finite, a limitation that is common to all physical experiments. In addition, all realizations in the single-tube experiment would be constrained by the use of the same tube, whereas all realizations of the multiple-tube experiment would be constrained by the application of the same tank head at a given instant. Thus some caution is required when a single physical setup is used as representative of a general random process.

Consider a random process $\mathbf{x}(\zeta, t)$ and the corresponding random variable at a fixed time instant t. Then the *first-order distribution function* of this random process at time instant t is defined as

$$F_x(x, t) = P\{\mathbf{x}(\zeta, t) \leq x\}. \tag{4.75}$$

The *first-order pdf* of the same random process is defined as

$$f_x(x,t) = \frac{\partial F_x(x,t)}{\partial x}. \tag{4.76}$$

Note that both functions just defined depend on two variables, the real number x and time t. Now consider the two random variables $\mathbf{x}(\zeta, t_1)$ and $\mathbf{x}(\zeta, t_2)$, corresponding to times t_1 and t_2. Then one can define the *second-order distribution function* of the random process $\mathbf{x}(\zeta, t)$ as

$$F_{xx}(x_1, x_2, t_1, t_2) = P\{\mathbf{x}(\zeta, t_1) \leq x_1 \text{ and } \mathbf{x}(\zeta, t_2) \leq x_2\} \tag{4.77}$$

and the *second-order pdf* as

$$f_{xx}(x_1, x_2, t_1, t_2) = \frac{\partial^2 F_{xx}(x_1, x_2, t_1, t_2)}{\partial x_1 \partial x_2}. \tag{4.78}$$

In a similar manner, one can define the nth-*order distribution function* and the nth-*order pdf*. A random process is said to be *statistically determined* if all its distribution functions are known.

The *mean* of a random process is a time function defined as

$$\langle \mathbf{x}(t) \rangle = \int_{-\infty}^{\infty} x f_x(x,t) \mathrm{d}x. \tag{4.79}$$

The *autocorrelation function* and the *autocovariance function* of a random process are, respectively, defined as

$$R_x(t_1, t_2) = \int_{-\infty}^{\infty} \int_{-\infty}^{\infty} x_1 x_2 f_x(x_1, x_2, t_1, t_2) \mathrm{d}x_1 \mathrm{d}x_2, \tag{4.80}$$

$$C_x(t_1, t_2) = \int_{-\infty}^{\infty} \int_{-\infty}^{\infty} [x_1 - \langle \mathbf{x}(t_1) \rangle] [x_2 - \langle \mathbf{x}(t_2) \rangle] f_x(x_1, x_2, t_1, t_2) \mathrm{d}x_1 \mathrm{d}x_2. \tag{4.81}$$

A *two-dimensional random process* consists of two random processes $\mathbf{x}(\zeta, t)$ and $\mathbf{y}(\zeta, t)$. For this process, one may easily extend the preceding definitions to define the *second-order joint distribution function* $F_{xy}(x_1, y_2, t_1, t_2)$, the *second-order joint pdf* $f_{xy}(x_1, y_2, t_1, t_2)$, and joint functions of higher orders, for example,

$$F_{xy}(x_1, y_2, t_1, t_2) = P\{\mathbf{x}(\zeta, t_1) \leq x_1 \text{ and } \mathbf{y}(\zeta, t_2) \leq y_2\}. \tag{4.82}$$

Furthermore, one can also define the *cross-correlation function* and the *cross-covariance function*, respectively, of the two random processes as

$$R_{xy}(t_1, t_2) = \int_{-\infty}^{\infty} \int_{-\infty}^{\infty} x_1 y_2 f_x(x_1, y_2, t_1, t_2) \mathrm{d}x_1 \mathrm{d}y_2, \tag{4.83}$$

$$C_{xy}(t_1, t_2) = \int_{-\infty}^{\infty} \int_{-\infty}^{\infty} [x_1 - \langle \mathbf{x}(t_1) \rangle] [y_2 - \langle \mathbf{y}(t_2) \rangle] f_{xy}(x_1, y_2, t_1, t_2) \mathrm{d}x_1 \mathrm{d}y_2. \tag{4.84}$$

Two random processes are called *uncorrelated* if $C_{xy}(t_1, t_2) = 0$ for any pair of times (t_1, t_2). They are called *independent* if, for any two sets of times $\{t_1, t_2, \ldots\}$ and $\{t_1', t_2', \ldots\}$, the set of random variables $\{\mathbf{x}(t_1), \mathbf{x}(t_2) \ldots\}$ is independent of the set $\{\mathbf{y}(t_1'), \mathbf{y}(t_2') \ldots\}$.

Stationary and ergodic random processes: A random process $\mathbf{x}(\zeta, t)$ is called *stationary* (in the strict sense) if its statistical properties are not affected by a shift in the time origin, i.e., if the statistical properties of $\mathbf{x}(\zeta, t)$ are the same as the corresponding properties of $\mathbf{x}(\zeta, t+\tau)$ for any time increment τ. An example of a stationary random process is shown in Fig. 4.10(b). A physical example of a stationary random process is the velocity in the multiple-tube system described in the preceding subsection, if the tank head is maintained constant and after sufficient time is allowed for the effects of valve opening to vanish (the latter requirement obviously poses some limitations to the ranges of t and τ). A related non-stationary process would be the case of the same multiple tubes when supplied with fluid from a tank with a time-varying head.

The following properties of stationary random processes are direct consequences of the preceding definition:

- The first-order pdf of a stationary random process is independent of time.
- Its mean is a constant.
- Its second-order pdf is a function of the time difference $\tau = t_2 - t_1$ only and not of the specific times t_1 and t_2.
- Its autocorrelation function is a function of only the time difference τ; thus it may be denoted as $R_x(\tau)$.

The requirement of stationarity for a physical random process is quite severe and cannot be precisely verified experimentally. A more flexible requirement is that of *weak stationarity*. A random process is called weakly stationary if its mean is a constant and its autocorrelation function is a function of the time difference τ only.

Two random processes, $\mathbf{x}(\zeta, t)$ and $\mathbf{y}(\zeta, t)$, are called *jointly stationary* if their joint statistical properties at any time t are the same as the corresponding properties at $t + \tau$ for any time increment τ.

The autocorrelation and cross-correlation functions of stationary random processes have the following properties (see Ref. [17] for a proof):

$$R_x(\tau) = R_x(-\tau), \text{ i.e., } R_x(\tau) \text{ is an even function,} \tag{4.85}$$

$$-R_x(0) \leq R_x(\tau) \leq R_x(0), \tag{4.86}$$

$$R_{xy}^2(\tau) \leq R_x(0)R_y(0), \tag{4.87}$$

$$2\left|R_{xy}(\tau)\right| \leq R_x(0) + R_y(0). \tag{4.88}$$

As mentioned earlier, the set of realizations $\{\zeta_1, \zeta_2, \ldots,\}$ of a repeated time-dependent experiment is represented by a set of functions $\{\mathbf{x}(\zeta_i, t), i = 1, 2, \ldots,\}$ that constitutes a random process, whereas the function $\mathbf{x}(\zeta_i, t)$ for each realization constitutes a time series. Assume that the random process is stationary. Then one may define a *time average* for each time series $\mathbf{x}(\zeta_i, t)$ as

$$\overline{x}_i = \lim_{T\to\infty} \frac{1}{2T} \int_{-T}^{T} \mathbf{x}(\zeta_i, t)\mathrm{d}t. \tag{4.89}$$

Similarly, one may define time-averaged moments and central moments of different orders for each time series. In general, the time-averaged properties of each member of the ensemble would be different from those of any other member. To obtain the statistical properties of the random process, one must perform *ensemble averaging*, namely averaging across all (or, at least, a sufficiently large number of) members of the ensemble, which may prove to be a tedious procedure. In many experiments, however, there is sufficient control of the process such that time averages do not vary from one ensemble member to any other. Such random processes are called *ergodic*. Therefore a stationary random process would be ergodic if all its statistical properties could be determined from a single member of the ensemble. In other words, for stationary and ergodic random processes, ensemble averages would be equal to corresponding time averages. Notice that a stationary random process may or may not be ergodic, but stationarity is a prerequisite for ergodicity, otherwise, one would not be able to define time averages. The velocity variation in the previously described tube–head-tank system would be a stationary and ergodic random process only if the tank head were maintained constant and if all tubes were identical. If the tubes were different from each other, then the process would be non-ergodic and statistical information from one tube would not represent statistics of the ensemble. One must be warned against taking ergodicity for granted, because many processes in the environment and in technology are non-ergodic, even if they can be considered as approximately stationary.

From the preceding definitions, the mean and the autocorrelation function of a stationary and ergodic random process can be computed from any single member ζ of the ensemble as, respectively,

$$\overline{x} = \lim_{T\to\infty} \frac{1}{2T} \int_{-T}^{T} \mathbf{x}(\zeta, t)\mathrm{d}t, \tag{4.90}$$

$$R_x(\tau) = \lim_{T\to\infty} \frac{1}{2T} \int_{-T}^{T} \mathbf{x}(\zeta, t)\mathbf{x}(\zeta, t+\tau)\mathrm{d}t.$$

In a similar fashion, given two stationary random processes $\mathbf{x}(\zeta, t)$ and $\mathbf{y}(\zeta, t)$, one calls them *jointly stationary and ergodic* if all their joint statistical properties can be defined from any single pair of members of the ensemble. Then their cross-correlation function can be computed as

$$R_{xy}(\tau) = \lim_{T\to\infty} \frac{1}{2T} \int_{-T}^{T} \mathbf{x}(\zeta, t)\mathbf{y}(\zeta, t+\tau)\mathrm{d}t. \tag{4.91}$$

Many measurement processes are non-stationary, even when considered within a relatively short time interval. A simple qualitative test that can be used to detect non-stationarity in a random signal or time series is to inspect its variation over the time of interest, trying to identify possible systematic upward or downward trends. For long signals and time series, it would be more convenient to divide them into a number of sequential or overlapping blocks, compute the means for each block, plot these means vs. time, and try to identify trends in the block means, rather than in the entire signal. The absence of a trend in the mean does not necessarily preclude non-stationarity, as it is possible that, even if the mean remains constant within a time interval, other moments may be time dependent. To increase the confidence in a qualitative assessment of stationarity, one may inspect the variation of additional block statistics, such as standard deviations and skewness and flatness factors, noting that the test sensitivity generally increases with the order of the moment. A more reliable approach would be to apply one of the available quantitative statistical non-stationarity tests to various block statistics. A relatively simple test is the *reverse arrangement test* [4], introduced in Section 2.2, which is capable of detecting monotonic trends. Assume that the signal or time series has been separated into a number of blocks and that statistical properties have been calculated and may be assumed to be independent of those in any other block. Then one may apply this test to any such statistical block property, which itself is a random variable. In cases in which a monotonic trend has been detected for the mean but not for higher statistical moments, and that trend can be plausibly attributed to spurious effects (e.g., instrumentation drift), one may replace the original time series with a corresponding *quasi-stationary* time series, which one determines by fitting a straight line to a plot of the block means and subtracting the values provided by this line at all times from all corresponding measured values. More sophisticated approaches have also been suggested. For example, a time series may be restored to a quasi-stationary form by the application of a *moving-average filter*, namely by subtracting from each value the average of a number of its neighbouring points, either equally weighted or weighted by a factor that decreases inversely to proximity. In such cases, one must be aware that, in addition to spurious effects that may introduce non-stationarity, it is possible to remove signal variations that are essential features of the physical process under study.

Example: Sinusoidal random process with random phase. To assist with the understanding of the preceding definitions and relationships among various statistical properties, consider the random process defined by

$$\mathbf{x}(\boldsymbol{\theta}, t) = a \cos(2\pi f_0 t + \boldsymbol{\theta}), \tag{4.92}$$

where a is a positive constant number, f_0 is a constant frequency, and the phase $\boldsymbol{\theta}$ is a random variable, which is uniformly distributed in the interval $[-\pi, \pi]$, i.e.,

$$f_\theta(\theta) = \begin{cases} \frac{1}{2\pi}, & -\pi \le \theta \le \pi \\ 0, & \pi < |\theta| \end{cases} \tag{4.93}$$

Each realization of this random process is a sinusoidal function, corresponding to a fixed value of the phase $\boldsymbol{\theta}$ [see Fig. 4.10(c)]. We obtain ensemble averaging by fixing the value of time t and averaging over all realizations. It is evident that $f_x(x, t)$ should be independent of time, as, at any given time, the random variable $\mathbf{x}$ would take values corresponding to all possible values of the phase $\boldsymbol{\theta}$. Therefore the process is stationary, and, as a result, its statistics could be computed for an arbitrary choice of time origin. For convenience, select the time origin such that $t = 0$. Then, we are concerned about the properties of the random variable

$$\mathbf{x}(\boldsymbol{\theta}) = a \cos \boldsymbol{\theta}. \tag{4.94}$$

For any value $\boldsymbol{\theta} = \theta$, we get the value $\mathbf{x}(\boldsymbol{\theta}) = x = a\cos\theta$, which is equivalent to $\theta = \cos^{-1}(x/a)$. Let us first compute $F_x(x)$ and $f_x(x)$. Starting with the definition of $F_x(x)$, we get

$$F_x(x) = P\{\mathbf{x} \leq x\} = \begin{cases} 0, & x \leq -a \\ P\{-\pi < \theta < -\theta\} + P\{\theta < \theta < \pi\}, & -a < x < a \\ 1, & a \leq x \end{cases}$$

$$= \begin{cases} 0, & x \leq -a \\ \frac{-\theta-(-\pi)}{2\pi} + \frac{\pi-\theta}{2\pi}, & -a < x < a \\ 1, & a \leq x \end{cases} = \begin{cases} 0, & x \leq -a \\ 1 - \frac{1}{\pi}\cos^{-1}(x/a), & -a < x < a. \\ 1, & a \leq x \end{cases}$$

Then,

$$f_x(x) = \frac{\mathrm{d}F_x(x)}{\mathrm{d}x} = \begin{cases} 0, & x \leq -a \\ \frac{1}{\pi}\frac{1}{\sqrt{a^2-x^2}}, & -a < x < a. \\ 0, & a \leq x \end{cases} \tag{4.95}$$

Notice that, although $f_x(x) \to \infty$, as $|x| \to a$, this is not a concern, because $f_x(a) = f_x(-a) = 0$. Because $f_x(a)$ is an even function, the mean of $\mathbf{x}$, as well as all its odd-order moments, would vanish. One may proceed in this manner to compute additional statistical properties through application of the corresponding definitions. One can significantly simplify the analysis, however, by noticing that this random process is also ergodic. This is evident because each realization is identical to any other with the exception of a phase shift, which does not affect time integration. Then one may compute the autocorrelation function as

$$\begin{aligned} R_x(\tau) &= \lim_{T\to\infty} \frac{1}{2T}\int_{-T}^{T} \mathbf{x}(,t)\mathbf{x}(\boldsymbol{\theta}, t+\tau)\mathrm{d}t \\ &= a^2 \lim_{T\to\infty} \frac{1}{2T}\int_{-T}^{T} \cos(2\pi f_0 t + \boldsymbol{\theta})\cos[2\pi f_0(t+\tau) + \boldsymbol{\theta}]\,\mathrm{d}t \\ &= \frac{a^2}{2}\lim_{T\to\infty} \frac{1}{2T}\int_{-T}^{T} [\cos(4\pi f_0 t + 2\pi f_0\tau + 2\boldsymbol{\theta}) + \cos(2\pi f_0\tau)]\,\mathrm{d}t \\ &= \frac{a^2}{2}\cos(2\pi f_0\tau). \end{aligned} \tag{4.96}$$

4.5 Frequency analysis of signals

Physical properties are commonly expressed by time series or continuous functions as functions of time. Their analysis, with time used as the independent variable, is said to be conducted in the *time domain*. An alternative approach is to transform these properties, through a series expansion or an integral transform, into counterparts that are functions of frequency. Then we may analyse these properties by using frequency as the independent variable. Such analyses are said to be conducted in the *frequency domain*. In the following subsections, we shall adopt the usual convention that the term frequency denotes both the parameter f, measured in hertz, and the parameter $\omega = 2\pi f$, also called *angular frequency*, measured in inverse seconds.

Fourier analysis: *Fourier analysis*, also called *harmonic analysis*, is the representation of a function or time series in terms of sinusoidal ('harmonic') functions [21–23]. Consider a *periodic function* $s(t)$, with a *period* T, such that $s(t) = s(t + T)$. Then this function can be represented by a *Fourier series*, as

$$s(t) = \sum_{i=0}^{\infty} \left[a_i \cos\left(2\pi i \frac{t}{T}\right) + b_i \sin\left(2\pi i \frac{t}{T}\right)\right], \tag{4.97}$$

where the *Fourier coefficients* can be determined as

$$a_i = \frac{2}{T} \int_t^{t+T} s(t) \cos\left(2\pi i \frac{t}{T}\right) \mathrm{d}t, \tag{4.98}$$

$$b_i = \frac{2}{T} \int_t^{t+T} s(t) \sin\left(2\pi i \frac{t}{T}\right) \mathrm{d}t. \tag{4.99}$$

The only additional requirement for this representation is that $s(t)$ satisfy the *Dirichlet conditions*, which is always the case for measurement data. The first term in the series is a constant equal to a_0. The second term oscillates with the *fundamental frequency* $f_1 = 1/T$, whereas the following terms, called *harmonics*, oscillate with frequencies $f_i = i/T, i = 2, 3, \ldots$.

Given any real or complex function $s(t)$ of a real variable t that satisfies the condition $\int_{-\infty}^{\infty} |s(t)|\, \mathrm{d}t < \infty$, one may define its *Fourier transform* as

$$F(f) = \int_{-\infty}^{\infty} s(t) e^{-j2\pi f t} \mathrm{d}t, \tag{4.100}$$

which, in general, is a complex function of the real variable ('frequency') f. The function $s(t)$ can be computed through the *inverse Fourier transform*

$$s(t) = \int_{-\infty}^{\infty} F(f) e^{j2\pi f t} \mathrm{d}f. \tag{4.101}$$

The Fourier transform can be defined only if $s(t) \to 0$ as $|t| \to \infty$. This condition is not satisfied by stationary functions; however, in practice, values of physical processes are available within only a finite time interval, say $[-T, T]$, and may be considered as having zero values outside this interval. Then one may define the *finite-interval Fourier transform* as

$$F(f, T) = \int_{-T}^{T} s(t) e^{-j2\pi f t} \mathrm{d}t. \tag{4.102}$$

Examples: Consider the simple sinusoidal function

$$s(t) = \cos(2\pi f_0 t), \tag{4.103}$$

where f_0 is a constant frequency. Its Fourier transform is

$$F(f) = \int_{-\infty}^{\infty} s(t) e^{-j2\pi f t} \mathrm{d}t$$
$$= \int_{-\infty}^{\infty} \cos(2\pi f_0 t) \cos(2\pi f t) \, \mathrm{d}t - j \int_{-\infty}^{\infty} \cos(2\pi f_0 t) \sin(2\pi f t) \, \mathrm{d}t. \quad (4.104)$$

Notice that the last integral vanishes because the integrand is an odd function of f. Furthermore, the cosine product in the preceding integral can be transformed into the sum of two harmonic functions by use of a trigonometric identity. Then,

$$F(f) = \frac{1}{2} \int_{-\infty}^{\infty} \{\cos[2\pi(f - f_0)t] + \cos[2\pi(f + f_0)t]\} \, \mathrm{d}t$$
$$= \frac{1}{2} [\delta(f - f_0) + \delta(f + f_0)] \, . \quad (4.105)$$

The easiest way to verify the preceding expression is to use the inverse transformation, Eq. (4.101), combined with the property of Dirac's delta function, Eq. (2.25).

The finite Fourier transform of the function described by Eq. (4.103) is

$$F(f, T) = \frac{1}{2} \int_{-T}^{T} \{\cos[2\pi(f - f_0)t] + \cos[2\pi(f + f_0)t]\} \, \mathrm{d}t$$
$$= \frac{1}{2\pi} \left\{ \frac{\sin[2\pi(f - f_0)T]}{f - f_0} + \frac{\sin[2\pi(f + f_0)T]}{f + f_0} \right\}. \quad (4.106)$$

As can be seen by a comparison of expressions (4.105) and (4.106), the infinite-interval Fourier transform of a sinusoidal function has non-zero values at only the two frequencies f_0 and $-f_0$, whereas the finite-interval Fourier transform has values extending into the entire frequency domain. The latter is of oscillatory nature with positive and negative peaks, called *sidelobes*, whose amplitudes are diminishing as $|f|$ increases.

Finally, consider the *rectangular* or *boxcar function*

$$w(t) = \begin{cases} 1, & -T \le t \le T \\ 0, & T < |t| \end{cases} . \quad (4.107)$$

Its Fourier transform can be easily found as

$$F_w(f) = \frac{\sin(2\pi f T)}{\pi f}. \quad (4.108)$$

Convolution theorem: Consider two functions $s_1(t)$ and $s_2(t)$ of the real variable t. Their *convolution* (indicated by the $*$) is defined as

$$s_1(t) * s_2(t) = \int_{-\infty}^{\infty} s_1(t') s_2(t - t') \mathrm{d}t' = \int_{-\infty}^{\infty} s_1(t - t') s_2(t') \mathrm{d}t'. \quad (4.109)$$

The *convolution theorem* states that the Fourier transform of the convolution of two functions is equal to the product of the Fourier transforms of these functions. An alternative form of the same theorem states that the Fourier transform of the product of two functions is equal to the convolution of the Fourier transforms of these functions. One can apply this theorem to the computation of finite-interval Fourier transforms from corresponding infinite-interval ones by noticing that the finite-interval Fourier transform of a function $s(t)$ can be considered as the finite-interval Fourier transform of the product of the function $s(t)$ and the rectangular function $w(t)$ as

$$\begin{aligned} F(f, T) &= \int_{-T}^{T} s(t) e^{-j2\pi f t} \mathrm{d}t = \int_{-\infty}^{\infty} s(t) w(t) e^{-j2\pi f t} \mathrm{d}t \\ &= \left[\int_{-\infty}^{\infty} s(t) e^{-j2\pi f t} \mathrm{d}t\right] * \left[\int_{-\infty}^{\infty} w(t) e^{-j2\pi f t} \mathrm{d}t\right] \\ &= F(f) * \left[\frac{\sin(2\pi f T)}{\pi f}\right]. \end{aligned} \tag{4.110}$$

As an example, consider the sinusoidal function described by Eq. (4.103). Its finite Fourier transform can be computed as

$$\begin{aligned} F(f, T) &= F(f) * \left[\frac{\sin(2\pi f T)}{\pi f}\right] \\ &= \int_{-\infty}^{\infty} \frac{1}{2} \left[\delta(f - f_0 - f') + \delta(f + f_0 - f')\right] \frac{\sin(2\pi f' T)}{\pi f} \mathrm{d}f' \\ &= \frac{1}{2\pi} \left[\frac{\sin[2\pi(f - f_0)T]}{f - f_0} + \frac{\sin[2\pi(f + f_0)T]}{f + f_0}\right]. \end{aligned} \tag{4.111}$$

Fast Fourier transform: A form of the finite-interval Fourier transform that is suitable for a real or a complex time series $s_i, i = 0, 1, 2, \ldots, N - 1$, sampled at a rate of $f_s = 1/\Delta t$ over a time interval $T = N\Delta t$, is the *discrete Fourier transform (DFT)*, which results in N discrete complex values at frequencies $f_k = k/T$, computed as

$$F_k = \frac{T}{N} \sum_{i=0}^{N-1} s_i e^{-2\pi j \frac{ik}{N}}, \quad k = 0, 1, 2 \ldots, N - 1. \tag{4.112}$$

Only the first $N/2$ values (i.e., for $k = 0, 1, 2, \ldots, N/2 - 1$) are independent; the others may be found by symmetry conditions. The frequency increment among consecutive terms of the DFT is $\Delta f = 1/T$. To compute all terms in a DFT according to Eq. (4.112), one would have to perform N^2 complex multiplications and additions, which would require excessive computer time for large values of N. Fortunately, efficient algorithms have been developed for the computation of a DFT, requiring $2N \log_2 N$ complex operations, which are significantly less than N^2. Such algorithms are known as *FFT*s. Since the original invention of the first FFT by Cooley and Tukey in 1965 [24], a variety of FFTs have been devised and have become a standard component of signal analysis software. Their common requirement is that the number of discrete values in the time series must be equal to a power of 2, because they all involve a number of intermediate computation steps that arrange the data in groups of numbers of data that are powers of 2. If the number of values in a time series is not equal to a power of 2, one must simply add zeros to increase the number of values up to the closest power of 2. As an example of the efficiency of the FFT, assume that $N = 2^{13} = 8192$. Then computation of the DFT by application of the definition, Eq. (4.112), would require $N^2 = 67{,}108{,}864$ complex operations, whereas that by FFT would require $2N \log_2 N = 212{,}992$ operations, a reduction by a factor of 315.

Frequency spectra: In this subsection, we consider exclusively stationary random processes. We further assume that the (constant) mean of the process has been computed and subtracted from all values, or, equivalently, that the random process has *zero mean*. In such case, the autocorrelation function would coincide with the autocovariance function and the cross-correlation function would coincide with the cross-covariance function. The *frequency spectrum*, also called *power-spectrum* or *power spectral density function* of a mean-free stationary random process is defined as the Fourier transform of its autocorrelation function, i.e., as

$$S_x(f) = \int_{-\infty}^{\infty} R_x(\tau) e^{-j2\pi f \tau} \mathrm{d}\tau = 2 \int_0^{\infty} R_x(\tau) \cos(2\pi f \tau) \mathrm{d}\tau, \tag{4.113}$$

where $j = \sqrt{-1}$ and the variable f has dimensions of frequency (inverse time). The *cross-spectrum* or *cross-spectral density function* of two mean-free jointly stationary random processes is defined as

$$S_{xy}(f) = \int_{-\infty}^{\infty} R_{xy}(\tau) e^{-j2\pi f \tau} \mathrm{d}\tau. \tag{4.114}$$

Inversely, the correlation functions can be computed from the corresponding spectra by use of the inverse Fourier transform, as

$$R_x(\tau) = \int_{-\infty}^{\infty} S_x(f) e^{j2\pi f \tau} \mathrm{d}f = 2 \int_0^{\infty} S_x(f) \cos(2\pi f \tau) \mathrm{d}f, \tag{4.115}$$

$$R_{xy}(\tau) = \int_{-\infty}^{\infty} S_{xy}(f) e^{j2\pi f} \mathrm{d}f. \tag{4.116}$$

Letting $\tau = 0$, one gets

$$R_x(0) = \int_{-\infty}^{\infty} S_x(f) \mathrm{d}f, \tag{4.117}$$

$$R_{xy}(0) = \int_{-\infty}^{\infty} S_{xy}(f) \mathrm{d}f. \tag{4.118}$$

Because $R_x(\tau)$ is a real, even function, the power spectrum is a real, non-negative, even function, i.e.,

$$S_x(f) = S_x(-f) \geq 0, \tag{4.119}$$

whereas the cross spectrum is generally a complex function, consisting of a real part, called the *coincident* or *co-spectral density function*, and an imaginary part, called the *quadrature* or *quad-spectral density function*. A real-valued quantity, which is related to the cross spectrum, is the *coherence function* $\gamma_{xy}(f)$, defined as

$$\gamma_{xy}^2(f) = \frac{|S_{xy}(f)|^2}{S_x(f)S_y(f)}. \tag{4.120}$$

Notice that the preceding definitions apply equally to ergodic and non-ergodic random processes. When a process is both stationary and ergodic, the power spectrum can also be computed from the Fourier transform of a single realization (time series) $s(t)$, as

$$S_x(f) = \lim_{T\to\infty} \frac{1}{2T} \left| \int_{-T}^{T} s(t) e^{-j2\pi f t} \mathrm{d}t \right|^2, \tag{4.121}$$

which is, in practice, approximated by its finite-interval estimate as

$$S_x(f, T) = \frac{1}{2T} \left| \int_{-T}^{T} s(t) e^{-j2\pi f t} \mathrm{d}t \right|^2 = \frac{1}{2T} |F(f, T)|^2. \tag{4.122}$$

This approach is useful in random signal analysis because it permits the use of FFT algorithms, which greatly reduces the required computational time. The discrete power spectrum of a time series s_i, $i = 0, 1, 2 \ldots, N-1$, is

$$S_k = \frac{1}{N} \sum_{i=0}^{N-1} \left| s_i e^{-2\pi j \frac{ik}{N}} \right|^2, \quad k = 0, 1, 2 \ldots, N/2 - 1. \tag{4.123}$$

Note that only values at the lower $N/2 - 1$ frequencies need be considered, as the remainder are related to them through symmetry.

Besides its use for spectral estimates, FFTs can be used for the determination of autocorrelation and cross-correlation functions. For this, one needs to apply the FFT once to compute the spectrum, from which one can find the corresponding correlation functions by applying an inverse FFT.

Examples: As an example of power-spectral calculations, one may compute the power spectrum of the sinusoidal random process [Eq. (4.92)] from its autocorrelation function as

$$S_x(f) = \frac{a^2}{4} \left[\delta(f - f_0) + \delta(f + f_0)\right]. \tag{4.124}$$

Another example that was introduced in Section 4.1 is the *white-noise random process*, assumed to be stationary and ergodic. Let this random process be denoted as $\mathbf{n}(\zeta, t)$ and defined such that it has a power spectrum that is uniform over the entire frequency range:

$$S_n(f) = n, \quad -\infty < f < \infty, n = \text{const.} \tag{4.125}$$

Then its autocorrelation function can be computed as

$$R_n(\tau) = a\delta(\tau), \tag{4.126}$$

which indicates that white noise at any instant is uncorrelated with itself at any other instant. The preceding notion of white noise implies that there is activity in the process even at infinite frequency, which is a non-physical requirement. A more realistic process is that of *band-limited white noise*, whose power spectrum is, by definition,

$$S_n(f) = \begin{cases} a, & -f_n < f < f_n \\ 0, & |f_n| < f \end{cases}. \tag{4.127}$$

The autocorrelation function of this process can be easily computed as

$$R_n(\tau) = \frac{a \sin(2\pi f_n \tau)}{\pi \tau}, \tag{4.128}$$

which is an oscillatory function with diminishing amplitude. Its maximum value is $2af_n$ and occurs at $\tau = 0$.

Finite-interval effect: Spectral estimates from finite-interval Fourier transforms, Eq. (4.122), are distorted compared with those computed with the general definition, Eq. (4.121). As mentioned earlier, the finite-interval Fourier transform is equal to the convolution of the infinite-interval Fourier transform and the Fourier transform of the rectangular function. Thus, for example, the finite-interval power spectrum of the sinusoidal function, Eq. (4.103), is

$$\begin{aligned} S(f, T) &= \frac{1}{2T} |F(f, T)|^2 \\ &= \frac{1}{2T} \left\{ \frac{\sin[2\pi(f - f_0)T]}{2\pi(f - f_0)} + \frac{\sin[2\pi(f + f_0)T]}{2\pi\,(f + f_0)} \right\}^2, \end{aligned} \tag{4.129}$$

rather than the delta-function-like $S(f)$. Although both $S(f)$ and $S(f, T)$ have peaks at $f = f_0$ and $f = -f_0$, the former has no energy except at these two peaks, whereas the latter has energy distributed over the entire frequency range, and, in fact, it has a sequence of secondary peaks (sidelobes) of slowly diminishing amplitudes. Thus, it appears that energy of the peaks at $\pm f_0$ 'leaks' to other frequencies, distorting the spectrum. This distortion is a consequence of the abrupt termination of the signal at the boundaries of the interval $[-T, T]$ and can be reduced by the introduction of a smooth transition to zero near these boundaries. This is effected by multiplication of the signal by a smooth *window function*, instead of the rectangular function $w(t)$, discussed previously. The most commonly used window function is the *Hanning window*, defined as [see also Eq. (4.50)]

$$h(t) = \begin{bmatrix} \frac{1}{2}\left[1 - \cos\left(\frac{2\pi t}{T}\right)\right] & -T \le t \le T \\ 0 & |t| > T \end{bmatrix}. \tag{4.130}$$

This window results in much lower sidelobe amplitudes compared with those introduced by the rectangular window. Because it reduces the total energy of the spectrum, a correction factor of $8/3$ must be applied to the spectral values.

Spectral analysis of randomly spaced time series [25–27]: Direct spectral analysis of discrete-time series through a FFT requires that the samples be evenly spaced. Certain measuring systems, however, with the notable example of laser Doppler velocimeters, provide discrete data at randomly spaced intervals. Two general approaches have been applied successfully to the spectral analysis of such signals. The first one, known as the *slot correlation method*, separates all possible pairs of sample points into bins according to the difference in their times of arrival. If the width of each bin is relatively narrow, the average product of all sample pairs within the bin would be a good estimate of the covariance of the signal for a time lag equal to the midtime of the bin. In this way, one could estimate the autocorrelation of the time series, from which one could estimate its spectrum. This method does not introduce additional distortion to the spectrum but requires a large number of samples, which may necessitate impractically long measurement times. A more practical approach is *resampling at a constant rate*. This means that the randomly spaced time history is replaced with a continuous function, which is subsequently resampled at a fixed rate and to which a FFT can be applied. Among the various types of interpolation between samples that can be used to construct the continuous function, the most widely used is the simple method of holding the value of each sample until the next sample becomes available. This procedure distorts the spectral estimates in two ways. First, it acts as a first-order low-pass filter with a cutoff frequency equal to $N/(2\pi T)$, where N is the number of original samples over the sampling time T. Second, it introduces a white noise that is due to the step-like character of the resampling process; the variance of this noise decreases as $(N/T)^{-3}$. Thus, for data rates that are significantly higher than the frequencies of interest, the resampling method would produce acceptable spectral estimates.

Wavelet analysis: *Wavelets* have been used widely in recent years in a variety of applications, including data compression and detection of periodicity. Their mathematical basis and practical aspects have been documented in several sources [28–34]. In the present discussion, we focus on those aspects that are used in *time-frequency analysis* of signals, namely in the identification of frequencies with significant energy present in non-stationary signals at a given time. To illustrate this concept, consider a sinusoidal signal that changes frequency at some time t_0, as [Fig. 4.11(c)]

$$s(t) = \sin(2\pi f_1 t), \quad t < t_0;$$
$$s(t) = \sin(2\pi f_2 t), \quad t \geq t_0. \tag{4.131}$$

Spectral analysis of this signal over a time interval containing t_0 will give an energy spectrum with two distinct peaks at f_1 and f_2, which cannot be distinguished from peaks in the spectrum of the stationary signal $s'(t) = \sin(2\pi f_1 t) + \sin(2\pi f_2 t), -\infty < t < \infty$. Wavelets are capable of extracting the peak signal frequency as a function of time. To do so, one has to apply an appropriate *wavelet transform*. Both discrete and continuous wavelet transforms of different mathematical forms are available. The most commonly used wavelet for time-frequency analysis of signals is the *Morlet wavelet*, which is a complex harmonic function modulated by a Gaussian envelope. The Morlet wavelet [35]

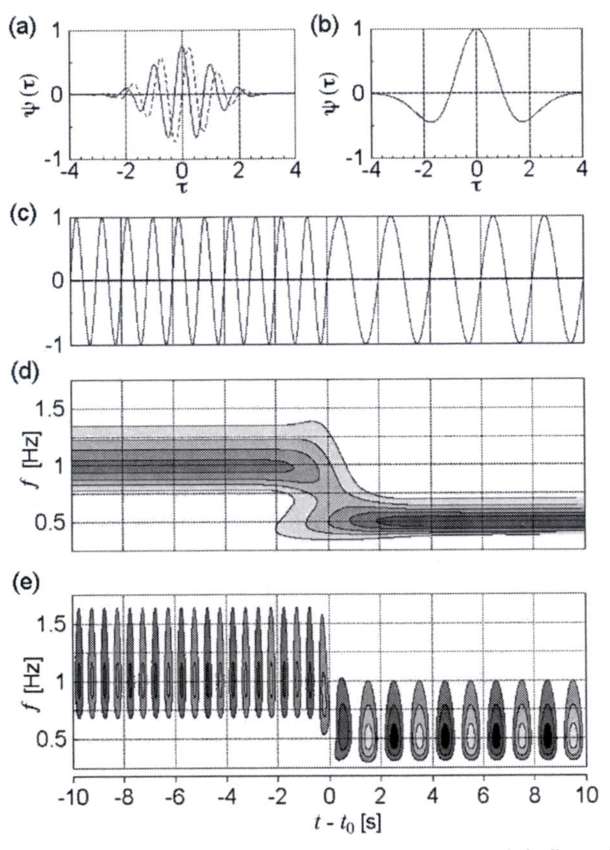

Figure 4.11. (a) Morlet wavelet function (the real part is indicated by a solid curve and the imaginary part by a dashed curve); (b) Mexican hat wavelet function; (c) sinusoidal function with a step change in frequency; (d) Morlet wavelet map of this function; and (e) Mexican hat wavelet map of the same function [in (d) and (e), the value of the wavelet spectrum amplitude is indicated by the grey tone, so maxima and minima can be identified by grey-tone extremes].

can be defined as

$$\psi(\tau) = \pi^{-1/4} e^{j2\pi\gamma\tau} e^{-\tau^2/2}, \tag{4.132}$$

where $j = \sqrt{-1}$, τ is a dimensionless variable (e.g., a normalized time) and γ is the *dimensionless centre frequency* of the wavelet. Rigorous definition of the Morlet wavelet requires the addition of a correction term to the right-hand side of Eq. (4.132). However, in practice, the value of γ is taken to be sufficiently large for this term to be negligible. The value of γ also determines the amplitudes of the successive peaks of the harmonic function relative to that of the central peak at $\tau = 0$. In his original work, Morlet selected γ such that the second peak of the real part of Eq. (4.132) would be equal to half the value of central peak. For simplicity, we choose the value $\gamma = 1$, which gives second, third, and fourth peak amplitudes equal to 0.606, 0.135, and 0.011 times the central peak amplitude, respectively. This wavelet is illustrated in Fig. 4.11(a), which shows that it essentially vanishes after three oscillations. The Morlet wavelet is most appropriate for identifying the presence of a wavetrain (i.e., a set of periodic cycles) in a random signal. Another popular wavelet is the *Mexican hat wavelet*, which is the second derivative of a Gaussian function, defined as

$$\psi(\tau) = 0.867\left(1 - \tau^2\right) e^{-\tau^2/2}. \tag{4.133}$$

This wavelet, illustrated in Fig. 4.11(b), is most appropriate for identifying isolated maxima and minima or discontinuities in the signal. The shape of Eq. (4.133) defines a dimensionless centre frequency as $\gamma \approx 0.25$. Because of the decaying exponential, both wavelets just discussed have a 'localized' character, because their magnitudes decrease rapidly with increasing $|\tau|$. Both wavelet functions have been scaled by appropriate coefficients that correct the distortion of the original signal by the wavelet transform (see subsequent discussion).

The *continuous wavelet transform* of a function $s(\tau)$ of the dimensionless variable τ consists of its convolution by a wavelet function as

$$W(a, \tau) = \int \frac{1}{\sqrt{a}} s(\tau') \psi^* \left(\frac{\tau' - \tau}{a}\right) \mathrm{d}\tau', \tag{4.134}$$

where the asterisk indicates a complex conjugate and a is the *scale* or *width* of the wavelet. The function $W(a, \tau)$ is commonly called the *wavelet coefficient* and is generally complex and expressed by its amplitude $|W(a, \tau)|$ and phase $\varphi(a, \tau)$. The variable $|W(a, \tau)|^2$ is referred to as the *wavelet power spectrum*. Variation of the scale adjusts the width of the wavelet so that its frequency matches as best as possible the local frequency of the fluctuating signal. For a given τ and wavelet function, the function $W(a, \tau)$ would depend on the scale a alone. The common way of presenting wavelet transform results is the *wavelet map*, in which contours of constant amplitude $|W(a, \tau)|$ are plotted vs. time and frequency axes. If a nearly periodic wavetrain is present in the signal, then $|W(a, \tau)|$ would present a maximum at a certain value of a from which the signal peak frequency can be calculated. It is customary to normalize the value of $|W(a, \tau)|$ by its maximum within the plot and then to discard contours with values lower than a certain percentage (e.g., 40%) of the maximum in order to reduce clutter.

In signal analysis, the wavelet transform would be applied to a discrete-time series $s_i, i = 0, 1, 2 \ldots, N - 1$, sampled at a rate $f_s = 1/\Delta t$, over an interval $T = N\Delta t$. Then the wavelet transform would be defined as

$$W_i(a) = \sum_{i'=0}^{N-1} \frac{1}{\sqrt{a}} s_{i'} \psi^* \left[\frac{(i' - i)}{a}\right]. \tag{4.135}$$

The signal frequency corresponding to a particular wavelet scale a can be computed as

$$f = \frac{\gamma}{a\Delta t}. \tag{4.136}$$

As an illustration of the wavelet transform analysis, we apply the Morlet and Mexican hat wavelet transforms to a sinusoidal signal that changes frequency, according to Eqs. (4.131), in which $f_1 = 1$ Hz and $f_2 = 0.5$ Hz. The corresponding wavelet maps are shown in Figs. 4.11(d) and 4.11(e). It is clear from these figures that the Morlet wavelet is most appropriate for identifying the frequency of oscillation at a given time, whereas the Mexican hat wavelet also identifies positive and negative peaks in the signal.

Spatial vs. temporal analysis: All definitions and properties concerning time-dependent signals and time series can be extended to the one-dimensional space domain if time t, as the independent variable, is replaced with the length r along a straight axis. Then one would consider the random process $\mathbf{x}(\zeta, r)$ and should appropriately modify all definitions to reflect this change. The equivalent of a stationary process in the space domain would be a *homogeneous random process*, namely a process whose statistical properties are independent of coordinate origin translations, such that the properties of $\mathbf{x}(\zeta, r)$ are the same as those of $\mathbf{x}(\zeta, r + \Delta r)$, where Δr is the *separation distance*. In the absence of a different term, one may also use the term *ergodic* to specify a homogeneous random process whose statistical properties can be computed by appropriate space averaging in one member of the ensemble. The autocorrelation function $R_x(\Delta r)$ of a homogeneous random process would be a function of the separation distance and its power spectrum $S_x(\kappa)$ would be defined in terms of the spatial Fourier transform as

$$S_x(\kappa) = \int_{-\infty}^{\infty} R_x(\Delta r) e^{-j\kappa \Delta r} \mathrm{d}r. \tag{4.137}$$

The parameter κ, which has a dimension of inverse length, is called the *wavenumber* and is analogous to the frequency. Thus spectral and wavelet analyses can be performed in the one-dimensional space domain by direct analogy to the time domain. When, however, one considers two- or three-dimensional space, the direct analogy ceases to exist, because the position vector $\vec{r}$ is of higher dimension than time t. Even so, one-dimensional space analysis can be extended to higher dimensions by the modification of definitions and relationships. In this case, properties such as the autocorrelation function and power spectrum would have to be replaced with tensors, whose components would in general depend on orientation. The properties of two- and three-dimensional Fourier and wavelet transforms are beyond the scope of the present text.

QUESTIONS AND PROBLEMS

1. Applying Kirchhoff's laws and assuming ideal op-amp operation, derive algebraic or differential equations for the circuits shown in Fig. 4.5. Then determine the transfer function for each circuit; in cases of multiple inputs, consider one input at a time, connecting all other inputs to the ground.
2. Analyse the fourth-order, low-pass Butterworth filter shown in Fig. 4.6(c). Determine its transfer function. Construct a Bode plot of its amplitude and phase response.
3. Consider three ADCs with the same input range between −5 and 5 V but with 8-, 12- and 16-bit conversions, respectively. Determine their ideal resolutions and quantization uncertainties.
4. Consider the repeated casting of a pair of dice, which may be assumed to be fair. Define a random variable as the sum of the values indicated by the two dice. Determine

the probability that this random variable will be equal to 7. Also determine the value of the distribution function of this random variable at 7.

5. Plot isocontours of the joint pdf of two jointly normal random variables with correlation coefficients equal to 0, 0.5, −0.5, and 1. Discuss the shapes of these contours.
6. Consider a square-wave function with amplitude A and period T, which may be represented as

$$s(t) = A(-1)^{2t/T}.$$

Determine its Fourier series expansion, taking advantage of the fact that it is an odd function.

7. Verify that the power spectra listed below are related to the corresponding autocorrelation functions and plot both sets of functions.

a.

$$R(\tau) = e^{-\lambda|\tau|}, \quad S(f) = \frac{2\lambda}{\lambda^2 + (2\pi f)^2}.$$

b.

$$R(\tau) = e^{-\lambda\tau^2}, \quad S(f) = \sqrt{\frac{\pi}{\lambda}}e^{-\frac{(2\pi f)^2}{4\lambda}}.$$

c.

$$R(\tau) = \begin{cases} 1 - \frac{|\tau|}{T}, & |\tau| \leq T \\ 0, & T < |\tau| \end{cases}; \quad S(f) = \frac{4\sin^2\frac{2\pi f T}{2}}{(2\pi f)^2 T}.$$

8. Consider a random process that is the sum of a sinusoidal function with uniformly distributed random phase and band-limited white noise:

$$\mathbf{x}(\theta, t) = c\cos(2\pi f_0 t + \boldsymbol{\theta}) + \mathbf{n}(\zeta, t).$$

Assume that the noise is statistically independent of the sinusoidal function. Compute and plot the autocorrelation function and the power spectrum of this random process for different combinations of the ratios f_n/f_0 and a/c.

9. Plot the finite-interval power spectrum of a sinusoidal function having a frequency $f_0 = 10/T$ in the interval $[-T, T]$. Compute and plot the spectrum of this function after the Hanning window has been applied to it. Compare the two results.

REFERENCES

[1] R. E. Thomas and A. J. Rosa. *The Analysis and Design of Linear Circuits* (2nd Ed.). Prentice-Hall, Upper Saddle River, NJ, 1998.

[2] A. R. Cohen. *Linear Circuits and Systems*. Regents, New York, 1965.

[3] J. Millman and C. H. Halkias. *Integrated Electronics: Analog and Digital Circuits and Systems*. McGraw-Hill, New York, 1972.

[4] R. A. Gayakwad. *Op-Amps and Linear Integrated Circuits* (3rd Ed.). Prentice-Hall, Englewood Cliffs, NJ, 1993.

[5] W. D. Stanley. *Operational Amplifiers with Linear Integrated Circuits* (3rd Ed.). Merrill, New York, 1994.

[6] D. H. Sheingold (Editor). *Analog–Digital Conversion Handbook*. Analog Devices, Inc., Norwood, MA, 1972.
[7] M. Sayer and A. Mansingh. *Measurement, Instrumentation and Experiment Design in Physics and Engineering*. Prentice-Hall of India, New Delhi, 2000.
[8] H. W. Ott. *Noise Reduction Techniques in Electronic Systems*. Wiley, New York, 1976.
[9] D. H. Sheingold (Editor). *Transducer Interfacing Handbook*. Analog Devices, Inc., Norwood, MA, 1980.
[10] T. G. Beckwith, R. D. Marangoni, and J. H. Lienhard V. *Mechanical Measurement* (5th Ed.). Addison-Wesley, Reading, MA, 1993.
[11] D. H. Sheingold (Editor). *Nonlinear Circuits Handbook*. Analog Devices, Inc., Norwood, MA, 1976.
[12] A. V. Oppenheim and A. S. Willsky. *Signals and Systems* (2nd Ed.). Prentice-Hall, Upper Saddle River, NJ, 1996.
[13] S. D. Stearns. *Digital Signal Analysis*. Hayden, Rochelle Park, NJ, 1975.
[14] D. J. Stearns. *Digital Signal Processing with Examples in MATLAB*. CRC Press, Boca Raton, FL, 2004.
[15] J. S. Bendat and A. G. Piersol. *Random Data Analysis and Measurement Procedures* (3rd Ed.). Wiley, New York, 2000.
[16] S. Tavoularis and S. Corrsin. Experiments in nearly homogeneous turbulent shear flow with a uniform mean temperature gradient. Part 2. The fine structure. *J. Fluid Mech.*, 104:349–367, 1981.
[17] A. Papoulis. *Probability, Random Variables and Stochastic Processes* (3rd Ed.). McGraw-Hill, New York, 1991.
[18] W. Feller. *An Introduction to Probability Theory and Its Application* (3rd Ed.). Wiley, New York, 1968.
[19] K. A. Brownlee. *Statistical Theory and Methodology* (2nd Ed.). Wiley, New York, 1965.
[20] W. H. Beyer (Editor). *CRC Handbook of Mathematical Sciences* (6th Ed.). CRC Press, Boca Raton, FL, 1978.
[21] P. Bloomfield. *Fourier Analysis of Time Series: An Introduction*. Wiley, New York, 1976.
[22] H. S. Carlslaw. *An Introduction to the Theory of Fourier's Series and Integrals* (3rd Ed.). Dover, New York, 1950.
[23] L. A. Pipes and L. R. Harvill. *Applied Mathematics for Engineers and Physicists*. McGraw-Hill, New York, 1970.
[24] J. W. Cooley and J. W. Tukey. An algorithm for the machine calculation of complex fourier series. *Math. Comput.*, 19:297–301, 1965.
[25] R. J. Adrian and C. S. Yao. Power spectra of fluid velocities measured by laser doppler velocimetry. *Exp. Fluids*, 5:17–28, 1987.
[26] A. Host-Madsen and C. Caspersen. Spectral estimation for random sampling using interpolation. *Signal Process.*, 46:297–313, 1995.
[27] L. H. Benedict, H. Nobach, and C. Tropea. Estimation of turbulent velocity spectra from laser doppler data. *Meas. Sci. Technol.*, 11:1089–1104, 2000.
[28] I. Daubechies. *Ten Lectures on Wavelets*. Society for Industrial and Applied Mathematics, Philadelphia, 1992.
[29] A. Teolis. *Computational Signal Processing with Wavelets*. Birkhaeuser, Boston, 1998.
[30] J. Lewalle. Three lectures on the application of wavelets to experimental data analysis. In F. A. E. Breugelmans, editor, Von Karman Institute Lecture Series on Advanced Measurement Techniques, April 6–9, 1998, VKI LS 1998-06.
[31] M. Farge. Wavelet transforms and their application to turbulence. *Annu. Rev. Fluid Mech.*, 24:395–457, 1992.

[32] C. Torrence and G. P. Compo. A practical guide to wavelet analysis. *Bull. Am. Meteorol. Soc.*, 79:61–78, 1998.

[33] D. Jordan and R. W. Miksad. Implementation of the continuous wavelet transform for digital time series analysis. *Rev. Sci. Instrum.*, 68:1484–1494, 1997.

[34] S. V. Gordeyev and F. O. Thomas. Temporal subharmonic amplitude and phase behaviour in a jet shear layer: Wavelet analysis and hamiltonian formulation. *J. Fluid Mech.*, 394:205–240, 1999.

[35] J. Morlet, G. Arens, E. Fourgeau and D. Giand. Wave propagation and sampling theory. Part I: Complex signal and scattering in multilayered media. *Geophys.*, 47:203–221; Part II: Sampling theory and complex waves. *Geophys.*, 47:222–236, 1982.

5 Background for optical experimentation

Visual and optical techniques occupy a prominent role in experimental fluid mechanics. Understanding them, even at an elementary level, requires some familiarity with concepts of light propagation that, in a typical undergraduate engineering curriculum, are taught as part of general physics without focus on issues of present concern. This chapter briefly reviews some definitions and other background material that is necessary for explaining the visual and optical methods introduced in later chapters. Although the various phenomena associated with light propagation have, over the centuries, been the subject of several theories of increasing complexity, some important aspects remain to be explained in an entirely satisfactory manner. Consequently the present discussion is of limited scope and focussed on the needs of our subject. First, the fundamental properties of light and the principles of light emission and propagation through media are reviewed; next, some common instrumentation used for the generation, conditioning, detection, and recording of light in presented; and, finally, the optical and dynamic characteristics of materials that serve as flow markers are discussed.

5.1 The nature of light

Light as waves: According to *classical electromagnetic theory*, *light* is considered to be radiation that propagates through vacuum in free space in the form of electromagnetic waves, both oscillating transversely to the direction of wave propagation and normal to each other, as illustrated in Fig. 5.1 [1,2]. The intensities of the electric and magnetic fields, E_y and B_z, respectively, oscillate harmonically both in time t and along their direction of propagation x, as, respectively,

$$E_y(x,t) = E_{y0} \sin 2\pi \left(\frac{x}{\lambda} - \frac{t}{T} \right), \tag{5.1}$$

$$B_z(x,t) = B_{z0} \sin 2\pi \left(\frac{x}{\lambda} - \frac{t}{T} \right), \tag{5.2}$$

in which λ is called the *wavelength* and T is called the *period* of oscillation. The reciprocal of the period, $\nu = 1/T$, is called the *frequency* of oscillation, and the reciprocal

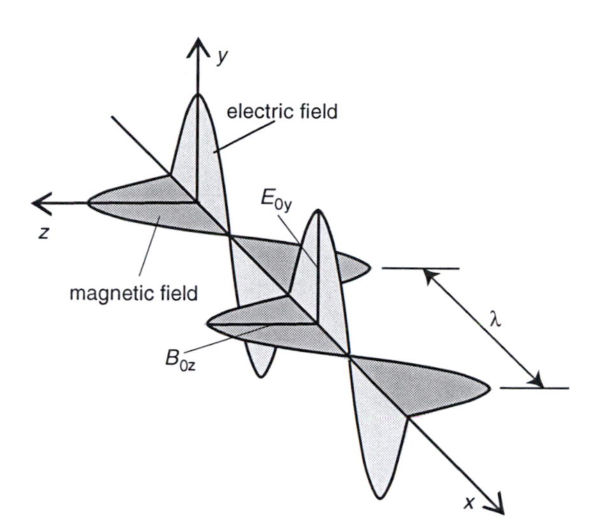

Figure 5.1. Sketch of light viewed as travelling electromagnetic waves.

of the wavelength, $\kappa = 1/\lambda$, is called the *wavenumber*. Under certain conditions, the same model may be used to describe light propagation through various media. The speed of propagation v of any point in the wave that maintains a constant phase difference from a reference point in the cycle (e.g., a crest or a zero) is called the *phase speed* and can be found as

$$v = \frac{\lambda}{T}. \tag{5.3}$$

Thus frequency and wavelength are related through the wave speed as

$$v = \nu\lambda. \tag{5.4}$$

The *speed of light propagation in vacuum* is equal to $c = 2.998 \times 10^8$ m/s, or roughly 300,000 km/s, which is the maximum possible speed.

Now, consider light propagating along different paths in three-dimensional space. The locus of all points along the different paths that have the same phase is a surface called a *wave front*. If all wave fronts are plane, then the light is considered to be a *plane wave*. Light may propagate in the form of not just plane waves, but also are *spherical* or *cylindrical waves*.

As mentioned earlier, light propagation is associated with electric and magnetic fields. These are in phase and their amplitudes are related as

$$E_{y0} = cB_{z0}. \tag{5.5}$$

It is usually sufficient to analyse electromagnetic waves by considering only the electric field. The term *polarization* is associated with the orientation of the plane of oscillation of the electric field. If the oscillating electric field lies on a single plane at all times, then the light is called *plane* or *linearly polarized*. When two light waves with the same frequency but out-of-phase travel along the same path and are both linearly polarized but on two mutually perpendicular planes, the resulting light wave is called *elliptically polarized*. If the two waves have the same amplitudes but a phase difference of $\pi/2$, the resulting wave is called *circularly polarized*. If the amplitudes are equal but the phase

Table 5.1. Summary of wavelength ranges of light; the boundaries of the different ranges reported by different sources vary somewhat

Colour	Wavelength range
Ultraviolet	0.85 nm < λ < 380 nm
Violet	380 nm < λ < 424 nm
Blue	424 nm < λ < 491 nm
Green	491 nm < λ < 575 nm
Yellow	575 nm < λ < 585 nm
Orange	585 nm < λ < 647 nm
Red	647 nm < λ < 750 nm
Infrared	750 nm < λ < 1 μm

changes randomly with time, the wave is called *unpolarized* or, more appropriately, *randomly polarized*. Natural light is essentially randomly polarized.

The colours: *Visible light* consists of radiation with wavelengths in the range 380–750 nm (1 nm = 10^{-9} m), which corresponds to the frequency range between 4.0×10^{15} and 7.9×10^{15} Hz. The colours, as perceived by a 'standard' human eye, are customarily defined as radiation with wavelengths in the ranges specified in Table 5.1 [3].

For comparison, the typical ranges of other types of electromagnetic radiation, in order of magnitude, are summarized in Table 5.2.

Light as photons: Now let us go one step back and discuss the mechanism of light emission from different materials [1,4]. First consider an *isolated atom*, which is unaffected by other particles and influences, as in the case of gases at very low pressure. This atom is said to be at its *ground state* when its nucleus is surrounded by its electrons at their lowest energy levels. The ground state is the normal, stable state of an atom and will last indefinitely, unless the atom is disturbed by the 'pumping' of energy to it, in the form

Table 5.2. Wavelength ranges of different types of radiation

Radiation type	Wavelength range
Cosmic rays	$\lambda < 10^{-4}$ nm
Gamma rays	10^{-4} nm < λ < 10^{-1} nm
X rays	10^{-2} nm < λ < 10^{2} nm
Disinfecting radiation	10 nm < λ < 380 nm
Visible light	380 nm < λ < 750 nm
Space heating	750 nm < λ < 10^{7} nm
Microwaves	10^{6} nm < λ < 10^{9} nm
Radar	10^{7} nm < λ < 10^{9} nm
Radio and television	10^{8} nm < λ < 10^{13} nm
Electrical power waves	10^{14} nm < λ < 10^{17} nm

of a collision with another atom, approach of an electron, or absorption of radiation energy. In general, the electrons of atoms may exist only at distinct and well-defined energy levels. When an atom absorbs energy, one or more electrons may move to energy states above the ground state, called the *excited states*. An electron may undertake a *quantum jump* from the ground state to an excited state, following absorption of a fixed amount of energy, equal to the difference in energy levels of the two states. The excited states are unstable and tend to undergo *transition* to a lower energy state extremely rapidly, typically within 10^{-9} to 10^{-8} s. One type of transition is the return to the ground state or any other lower energy state by emission of a *quantum* of radiant energy, called a *photon*. For atoms having many electrons, it is only the outermost electrons that emit photons. The motion of a photon may be reconciled with the propagation of an electromagnetic pulse if each photon is viewed as a *wavetrain* of extremely short duration, whose energy is proportional to the frequency of oscillation, as

$$\mathcal{E} = h\nu, \tag{5.6}$$

where $h = 6.624 \times 10^{-34}$ Js is the *Planck constant*. Thus an isolated atom would absorb and emit radiation at distinct frequencies only, called *resonant frequencies* and comprising the *atomic spectrum*. Each resonant frequency actually consists of a narrow spectral band, rather than a discrete value. This may be due to various causes, for example *Doppler broadening* that is due to thermal motion of the atom. Isolated atoms may be encountered only in rarefied gases. In liquids and solids, atoms are relatively close to each other, affecting each other's energy states and further broadening the frequency bands. Thus emitted and absorbed radiation would extend over essentially continuous, relatively broad, frequency ranges.

Geometrical optics: Electromagnetic wave theory and quantum electrodynamic theory provide explanations for different phenomena associated with light propagation through media [1]. For the sake of simplicity, however, one can also analyse a number of light-related phenomena by assuming that light propagates through a medium in the form of *rays*, which are lines normal at all their points to the wave fronts. This approach is known as *geometrical optics* and describes the phenomena of light transmission, refraction, reflection, and dispersion at a macroscopic level.

5.2 Light propagation through media

This section applies mainly simple geometrical optics concepts to describe light propagation through media and at interfaces between different media. A deeper understanding of these phenomena may be based on light-scattering considerations, briefly discussed in Section 5.4.

Refractive index: The *index of refraction* or *refractive index* n of a medium is defined as

$$n = \frac{c}{v}, \tag{5.7}$$

Table 5.3. Refractive indices of some common materials for light at a wavelength of 589 nm; gases are at 273 K and standard atmospheric pressure; liquids and solids are at 293 K

Gases	n	Liquids	n	Solids	n
Air	1.00029	Water	1.333	Fused quartz	1.46
He	1.00036	Ethyl alcohol	1.361	Pyrex glass	1.47
CO_2	1.00045	Turpentine	1.472	Crown glass	1.52
H_2	1.00013	Benzene	1.501	Flint glass	1.57–1.89
				Plexiglas	1.51
				Lexan	1.58
				Polystyrene	1.59
				Sapphire	1.77
				Zircon	1.92
				Diamond	2.42

where c is the speed of light in free space (vacuum) and v is the speed of light in the medium. Because c is the highest possible speed, it follows that $n \geq 1$. Considering that $v = \nu\lambda$ and that the frequency ν of light is not affected by the propagation medium, one may conclude that

$$n = \frac{\lambda_0}{\lambda}, \tag{5.8}$$

where λ_0 is the wavelength of monochromatic light in free space and λ is the wavelength of the same light propagating through a medium.

For air and other gases, the refractive index is only slightly greater than unity, whereas, for liquids and solids, it usually varies in the range 1.3–3.5. Typical values of n for some common materials are listed in Table 5.3 [1,2].

The index of refraction of a material generally increases slightly with decreasing wavelength of light. This phenomenon, known as *dispersion*, is further discussed later in this section. The composition of a substance and its density are also factors that affect the refractive index. A general relationship between the refractive index of transparent materials and their density is the Lorentz–Lorenz (or Clausius–Mosotti) expression

$$\frac{1}{\rho}\frac{n^2 - 1}{n^2 + 2} = K_L : \text{const}, \tag{5.9}$$

in which the coefficient K_L depends on material properties and the wavelength of light. For a given material and wavelength, this expression gives

$$n = \sqrt{\frac{1 + 2K_L\rho}{1 - K_L\rho}} \tag{5.10}$$

For gases, the *refractivity* $n - 1$ is extremely small and the expression (5.10) can be simplified to the *Gladstone–Dale formula*

$$n - 1 \approx K\rho, \tag{5.11}$$

Table 5.4. Values of the Gladstone–Dale constant for common gases at 273 K and a light wavelength of 589 nm

Gas	K (m^3/kg)
O_2	0.190×10^{-3}
N_2	0.238×10^{-3}
He	0.196×10^{-3}
CO_2	0.229×10^{-3}
Air	0.226×10^{-3}

in which K is the *Gladstone–Dale constant*. Typical values of K for gases consisting of neutral (i.e., non-ionized) molecules are given in Table 5.4 [5].

For gas mixtures, the constant K may be found as the sum of the constants of each component weighted by the corresponding mass fraction. Note that ionized gases have different values of K compared with those of neutral gases. An empirical expression for the dependence of the refractive index of water on temperature T_c (given in degrees Celsius and valid in the range 20–34 °C) for $\lambda = 632.8$ nm is [6, 7]

$$(n - 1.332156) \times 10^5 = -8.376(T_c - 20) - 0.2644(T_c - 20)^2 + 0.00479(T_c - 20)^3. \tag{5.12}$$

In view of the preceding discussion, variations in the refractive index inside a fluid could be caused if the fluid is heated or if two streams of the same fluid but different temperatures are mixed. For common gases under constant pressure, density variations are related to temperature variations ΔT through the perfect-gas law as

$$\frac{\Delta \rho}{\rho} \sim -\frac{\Delta T}{T}. \tag{5.13}$$

The mixing of two different fluids would also result in a fluctuating refractive-index field. For compressible flows, density variations occur any time other properties change. For example, for an isentropic flow of a perfect gas, it is easy to show that

$$\frac{\Delta \rho}{\rho} \sim \frac{1}{\gamma}\frac{\Delta p}{p}, \tag{5.14}$$

where p is the pressure and γ is the ratio of specific speeds.

Refraction: When light propagates through a homogeneous medium, its path would be straight, whereas, if the medium is non-homogeneous or if the light crosses from one medium to another, the path may change direction gradually or abruptly. This change of direction of propagation of light is called *refraction*. The refraction of light crossing the interface of two media with refractive indices n_1 and n_2 obeys the *law of refraction*, or *Snell's law* [Fig. 5.2(a)]:

$$\frac{\sin \varphi_1}{\sin \varphi_2} = \frac{n_2}{n_1} \tag{5.15}$$

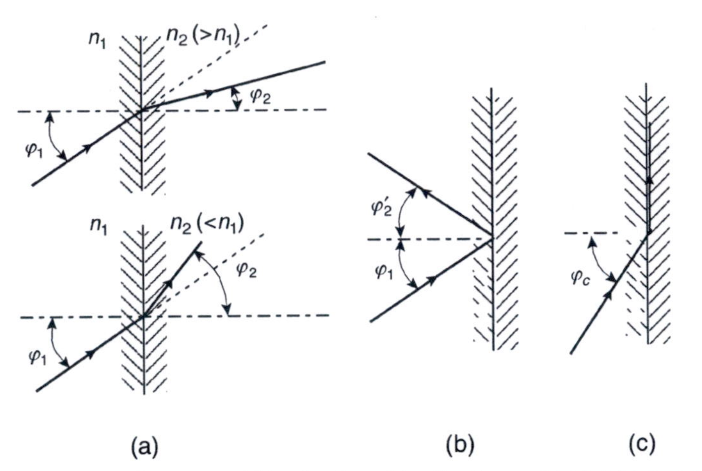

Figure 5.2. Sketches of (a) refraction, (b) reflection and (c) total internal reflection of a light ray crossing the interface between two media.

where φ_1 and φ_2 are, respectively, the angles between the directions of the ray in the two media and the normal to the tangent plane on the interface. Thus a ray will be refracted towards the normal ($\varphi_2 < \varphi_1$) if it enters an *optically denser* medium ($n_1 < n_2$) and away from the normal otherwise. The refractive index of fluids depends on their density, which in turn depends on their composition and temperature. Therefore a ray of light propagating in a medium with a non-uniform density will follow a curved path. The actual path of a ray between two points is determined by *Fermat's principle*, which states that it will be such that the time it takes for the light to traverse this path will be minimum. Refraction may distort the appearance of objects immersed partly in each of two fluids in contact or in non-homogeneous fluids; examples include the perceived 'bending' of a spoon in a glass of water or of an oar in the sea, the 'dancing' of shadows over a radiation heater or over a metallic surface exposed to the sun, and 'mirages' appearing in deserts or over a hot asphalt road.

As a simple application of refraction, consider the deflection of rays of light by convergent and divergent glass lenses. Consider a thin ray of light in air, entering one side of the lens while parallel to the lens's axis and then exiting to air from the other side. Assume that both lens surfaces have spherical shapes. Then the convergent lens

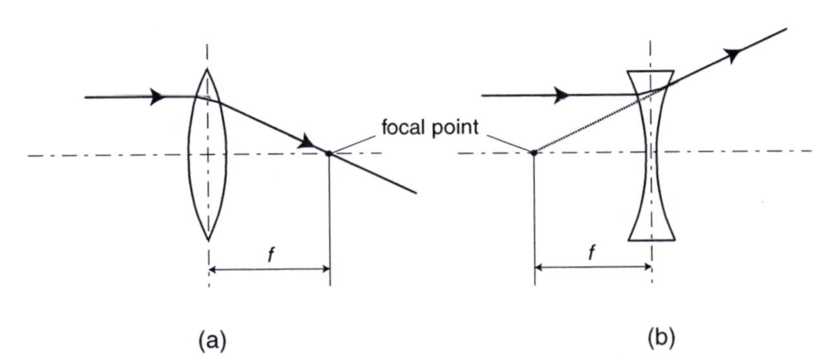

Figure 5.3. Illustration of light refraction through (a) a convergent lens and (b) a divergent lens.

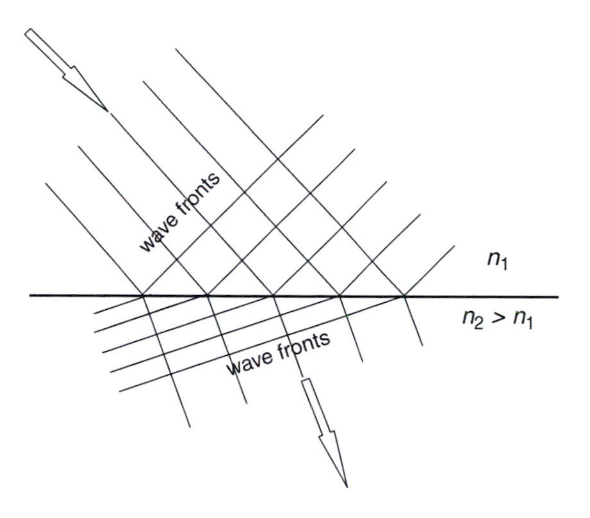

Figure 5.4. Interpretation of light refraction at the interface between two media as the rotation of wave fronts.

[Fig. 5.3(a)] will direct the ray towards its focal point, whereas the divergent lens [Fig. 5.3(b)] will direct the ray away from its focal point. The distance f of the focal point from the centre of the lens is called the *focal distance* of the lens.

A physical interpretation of refraction can be based on the wave-like nature of light. A light ray represents waves propagating in a direction normal to plane wave fronts, the distance between which is equal to the wavelength of the wave. When the ray enters an optically denser medium, i.e., a medium with a higher refractive index, its speed decreases and the distance between wave fronts decreases as well. Thus, as shown in Fig. 5.4, the ray appears to get deflected towards the normal.

Very recently, artificial materials displaying a *negative refractive index* have been constructed [8]. This means that the refracted beam would be on the same side of the normal to the interface as the incident beam [see Fig. 5.2(a)]. Both the theory and the application of this phenomenon are still at an early development stage.

Reflection: In general, when a beam of light propagating through a medium reaches a smooth interface with a second medium, part of its energy will be transmitted through the second medium according to the law of refraction. Another part, however, will be turned back into the first medium. This phenomenon is called *reflection* and follows the *law of reflection* [Fig. 5.2(b)]:

$$\varphi_2' = -\varphi_1. \tag{5.16}$$

When a ray travelling in a medium having a refractive index n_1 reaches an interface with a second medium having a refractive index $n_2 < n_1$ at the *critical angle,*

$$\varphi_1 = \varphi_c = \arcsin(n_2/n_1), \tag{5.17}$$

the law of refraction gives $\varphi_2 = \pi/2$, which means that the refracted ray will be parallel to the interface, so that it will not enter the second medium [Fig. 5.2(c)]. At incidence angles $\varphi_1 > \varphi_c$, the ray will be reflected back into the first medium. This phenomenon is called *total internal reflection.* When viewed from such angles, the second medium will be invisible. Typical values of critical angles are 42° for glass–air interfaces and 62° for glass–water interfaces.

Refractive-index matching: In view of the preceding discussion, serious image distortion may occur when light propagates in non-homogeneous materials or encounters interfaces between different materials. A common situation in experimental fluid mechanics is the use of transparent walls or windows, made of glass or plastics, for viewing a contained gas or liquid flow. In some cases, models and internal sections of the apparatus are also made of transparent solids to allow optical access. When fluids are viewed through curved walls, their images get particularly distorted. In some cases, this may result in entire flow regions becoming invisible or multiple images of the same region appearing simultaneously. These distortions can be reduced by use of walls as thin as possible and avoidance of curved sections. Corrections for the refraction of laser beams and other collimated light beams through curved walls are also available [9, 10]; such corrections are difficult to apply to broad images.

An effective method to reduce or eliminate optical distortion is to *match the refractive indices* of the contained fluid and the transparent wall. As shown in Table 5.3, the refractive indices of air and water are appreciably smaller than those of glass and other transparent wall materials, so there is no possibility of matching the refractive index of these fluids. The same applies to all gas flows. On the other hand, a number of liquids and solutions are available with refractive indices in the same range as those of common wall materials. Thus it is possible to precisely match the refractive indices of glass and acrylic materials by either mixing different liquids, for example glycerol and water or various natural and mineral oils (e.g., silicon oil, xylene, naphtha, turpentine) or by dissolving various salts (e.g., ammonium thiocyanate at near-saturation concentrations) in water or other solvents [11]. This procedure is by no means routine, as most liquids with a relatively large refractive index happen to be flammable, volatile, toxic, corrosive, foul smelling, unsafe, or a combination of any of these. Besides a careful consideration of potentially hazardous effects of the selected liquid on the experimenter and destructive effects on the apparatus and the laboratory, consideration must be given to their clarity, density, viscosity, sensitivity to temperature, environmental impact, and price. Because the refractive index of materials is sensitive to many factors, it may also be necessary to conduct on-site measurements of the refractive index of both the wall material and the liquid. This can be achieved with the use of a *refractometer*.

Even when a contained liquid is matched optically with the surrounding walls, optical distortion will be inevitable if light propagating through air encounters a curved wall (e.g., a circular tube) before it enters the liquid. In such cases, it is advisable to create a plane wall–air interface, either by machining the transparent wall, if this is possible, or by immersing the viewing section into a rectangular *viewing tank*, filled with the same liquid, but still and not communicating with the liquid inside the test section. In the latter case, one must take care to eliminate possible temperature differences between the still and flowing liquids. If such differences persist, they may introduce optical distortion that is due to refractive-index variations; in addition, temperature-related density variations may generate convection currents, which could cause flow distortions, especially in very low-Reynolds-number flows.

When dealing with two-phase or variable-density flows, one encounters differences in refractive index not only between fluids and surrounding walls, but also within the fluid itself. In most cases, these differences are impossible to eliminate, and one has to work within the limitations of the adopted visual or optical technique. Under certain conditions, however, one can match refractive indices by selecting specific materials. For example, dense solid suspensions in liquids have been simulated by use of liquid–particle combinations consisting of chemicals carefully mixed such as to match their refractive indices as well as having a desired density ratio, including unity [12].

Absorption: When light or other radiation is transmitted through a material along a path with length l, it will be absorbed by the molecules of this material according to *Beer's law*:

$$I = I_0 e^{-\alpha l} \tag{5.18}$$

where I is the radiant intensity (see Section 5.3) of the passing light, I_0 is the radiant intensity of the incident light, and α is the *absorption* or *attenuation coefficient*, which depends on the material and the wavelength of radiation [1]. The length $1/\alpha$, called the *skin* or *penetration depth*, represents the thickness of this material that will absorb 63% of the incident-light energy. Opaque materials have an extremely large value of α, whereas transparent materials have a relatively small value. Metals, in general, have a very small penetration depth and reflect most of the incident light. For example, copper has a penetration depth that varies between 0.6 nm for $\lambda = 100$ nm (ultraviolet) and 6 nm for $\lambda = 10000$ nm (infrared); thus a sheet of copper with a thickness of 2 nm will act as a high-pass light filter.

Birefringence [1]: *Birefringence*, also known as *double refraction*, is the separation of light into two linearly polarized components, an 'ordinary' ray and an 'extraordinary' one, which have polarization planes normal to each other and travel through the medium at different speeds, thus having different indices of refraction. It is exhibited by some crystalline solids as well as by some liquid polymers and colloidal solutions. Thus, when illuminated with linearly polarized light of the appropriate frequency, such materials could be opaque in one polarization direction while being transparent in another. Birefringent solids have uniaxial crystalline structure, with hexagonal, tetragonal, and trigonal crystals; the most commonly used such material is calcite. Materials with biaxial crystals, including orthorhombic, monoclinic, and triclinic ones, exhibit three indices of refraction, thus being *trirefringent*. Combinations of calcite prisms joined at different angles have been used as *polarizers*, removing one ray by total reflection (*Nicol prism*, *Glan–Foucault prism*), and as *beam splitters*, splitting non-polarized light into two diverging, linearly polarized rays (*Wollaston prism*).

5.3 Illumination

Visual and optical experiments usually require specialized light sources and illumination techniques. The characteristics of a light source that are important are its brightness, the duration of light it produces, and the distribution of power of light over its wavelength

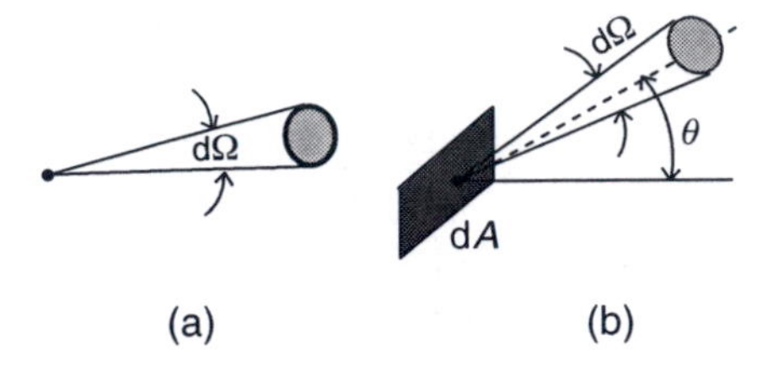

Figure 5.5. Sketches of (a) a point source of radiation and (b) a plane source of radiation.

bandwidth (*light spectrum*). Most light sources produce light in the visible range, which may be directly observed and recorded by conventional and electronic cameras. There are occasions, however, on which infrared or ultraviolet radiation may be preferable; such radiation, although invisible, can be detected and recorded by certain types of cameras and photodetectors. The measurable properties of electromagnetic radiation, in general, are the subject of *radiometry*, whereas the term *photometry* refers exclusively to visible radiation [2, 5].

Light sources include *thermal sources* and *lasers*. They are classified into *continuous-wave* (*CW*) sources, which produce radiation continuously, and *pulsed* sources, which produce single or repetitive radiation pulses of short duration.

Visual and optical experimentation requires, in addition to the choice of a suitable light source, an optimized illumination arrangement, which is appropriate for the observation and/or recording methods to be employed. Although this procedure can be guided by theory and previous experience, it is always worthwhile to make final adjustments by trial-and-error optimization that is specific to each experimental setup.

Radiometric and photometric definitions: A *point source* of light is an idealized source of electromagnetic radiation, which is concentrated at a point in space and radiates uniformly in all directions. Its *radiation power* Φ_e is defined as the total emitted radiation energy per unit time. The *radiant intensity* I_e (elsewhere, this parameter is denoted as I, for simplicity) of this source is defined as the radiation power per unit solid angle Ω [Fig. 5.5(a)], i.e.,

$$I_e = \frac{\mathrm{d}\Phi_e}{\mathrm{d}\Omega}. \tag{5.19}$$

A *plane source* of light emits energy uniformly from all points on a plane surface. A small plane source with area $\mathrm{d}A$ is characterized by its *radiance* L_e, defined as

$$L_e = \frac{\mathrm{d}^2\Phi_e}{\mathrm{d}\Omega \mathrm{d}A \cos\theta}, \tag{5.20}$$

where $\mathrm{d}\Omega$ is a solid angle centered at the center of the source and θ is the angle between the axis of this solid angle and the direction normal to the source plane [Fig. 5.5(b)]. The *spectral radiance* $L_{e\lambda}$ of this source is defined as the radiance per unit wavelength of the emitted radiation, i.e.,

$$L_{e\lambda} = \frac{\mathrm{d}L_e}{\mathrm{d}\lambda}. \tag{5.21}$$

Now consider a plane surface element with area $\mathrm{d}A$, receiving an amount of radiation power $\mathrm{d}\Phi_e$. The *irradiance* E_e of this element is defined as the irradiation power per unit area, i.e.,

$$E_e = \frac{\mathrm{d}\Phi_e}{\mathrm{d}A}. \tag{5.22}$$

Notice that the irradiance is independent of the orientation of the surface with respect to the direction of oncoming radiation.

All the preceding properties have conventional dimensions and units, called *radiometric units*; for example, the radiation power is measured in watts, and the radiant intensity is measured in watts per steradian. When visible radiation is dealt with, however, it is common practice to use *photometric units*, defined in terms of the response of a 'standard' human eye. Although both radiometry and photometry deal with the same physical properties, they use different names and symbols to avoid possible confusion by the mixing of different units.

The power of visible radiation sensed by the standard human eye is called *luminous power* or *luminous flux* Φ_v, and it is measured in *lumens* (lm). The following luminous properties are defined in a manner that is analogous to the definition of the corresponding radiation properties.

Luminous intensity:

$$I_v = \frac{\mathrm{d}\Phi_v}{\mathrm{d}\Omega}; \tag{5.23}$$

its unit is the candela (cd; 1 cd = 1 lm/sr).

Luminance (commonly referred to as brightness):

$$L_v = \frac{\mathrm{d}^2\Phi_v}{\mathrm{d}\Omega \mathrm{d}A \cos\theta}. \tag{5.24}$$

Spectral luminance:

$$L_{v\lambda} = \frac{\mathrm{d}L_v}{\mathrm{d}\lambda}. \tag{5.25}$$

Illuminance:

$$E_v = \frac{\mathrm{d}\Phi_v}{\mathrm{d}A}; \tag{5.26}$$

the unit of illuminance is the lux (1 lux = 1 lm/m^2); if the power of the oncoming radiation is contained over more than one wavelength, the total illuminance may be found by integration of the narrow-band illuminance over the entire visible spectrum.

The human eye: To present the correspondence between radiometric and photometric units, it is necessary to summarize a few facts concerning the human eye. The eye is enclosed within three membranes. Starting from the outer one, these are the *cornea–sclera*,

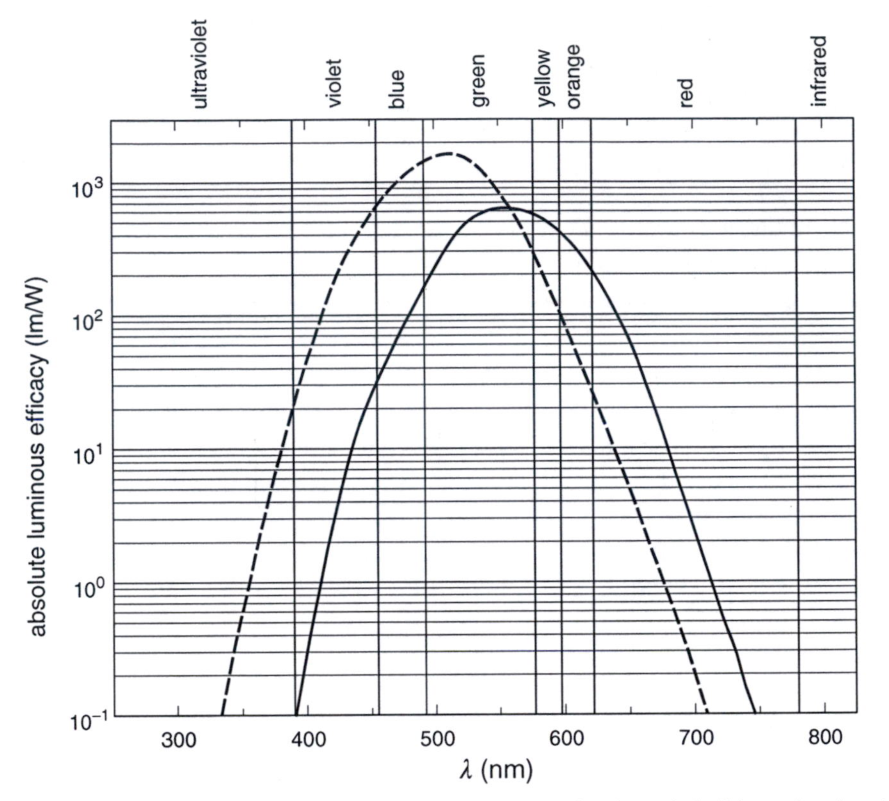

Figure 5.6. Luminous efficacies of the standard human eye for photopic (solid curve) and scotopic (dashed curve) visions [3].

the *choroid*, and the *retina*. A *lens*, which is flexible and adjusts its curvature for focussing purposes, images the received radiation onto the retina, which is lined with a large number of receptors sensitive to light. There are two types of such receptors: the *cones* (roughly 7 million), which respond only to bright light and are sensitive to colour, and the *rods* (of the order of 100 million), which are sensitive to dim light but cannot separate the different colours. A part of the retina, called the *fovea*, contains a particularly large concentration of cones and therefore has the highest resolution; it is on this part that images are focussed when highest clarity of vision is achieved. The sensitivity of the eye to different colours depends on the brightness of light. When the light is bright, the cones are activated, resulting in *photopic* or *bright-adapted* vision, whereas, when the light is dim, such as during twilight, vision is provided mainly by the rods, and it is called *scotopic* or *dark-adapted* vision. Thus, the ratio of luminous power to radiant power, called *luminous efficacy* and measured in lumens per watt, is different for photopic and scotopic visions. When normalized by the corresponding maximum value, the luminous efficacy is called *spectral sensitivity* or *luminous efficiency*. The luminous efficacies of the two types of vision are plotted in Fig. 5.6 [3]. It may be seen that the maximum eye sensitivity to bright light is 673 lm/W at about $\lambda = 555$ nm (green colour), whereas the maximum sensitivity to dim light is 1725 lm/W at about $\lambda = 510$ nm (blue-green colour). Notice also that scotopic vision is insensitive to yellow, orange, and red colours, thus explaining why colourful objects appear to be bluish under dim illumination. To convert radiant power of light with $\lambda = 555$ nm to luminous power for photopic vision in lumens, one has to multiply the former, measured in watts, by 673. At other wavelengths, 1 W of radiation power will produce less than 673 lm of luminous power; one can find the exact amount of the latter by multiplying the radiant power by the value indicated by the solid curve in Fig. 5.6 at the corresponding wavelength. For example, 1 W of radiation power at $\lambda = 650$ nm (red) will produce roughly 65 lm of luminous power, which is less than 10% of that at $\lambda = 555$ nm (green).

The human eye can distinguish variations in brightness over an enormous range (10 orders of magnitude) of luminous powers, bounded upwards by the *glare limit* and downwards by the *scotopic threshold* [13]. The *subjective brightness* of the collected light is a logarithmic function of the luminous power, following two different curves, one for photopic and one for scotopic vision. The range of brightness differences that can be recognized simultaneously is much narrower, as it takes time for the eye to adapt to each average brightness level. The range of recognizable colours also depends on the brightness and the adaptation time. Another important parameter that concerns the recognition of visual patterns is the sensitivity of the eye to changes in contrast. Typically, the minimum detectable difference between the brightness of a spot and a uniform background brightness is about 2% for a wide range of brightness levels. For varying background brightness, this limit could be significantly higher [13].

Colour-related terminology [14]: Terms such as colour, hue, and brightness are used routinely in everyday life and are commonly utilized by personal-computer-based image processing software. Such terms are understood intuitively, although rather vaguely, as one might be at a loss to define them in a general and unambiguous fashion. Because the acquisition, processing, display, and interpretation of images are common activities in experimental fluid mechanics, it seems advisable to clarify some relevant concepts. In fact, colour-related metrics and terminology can be quite complex and have been the object of standardization by the International Commission on Standardization (CIE – Commission Internationale de l'Eclairage) [15]. The CIE defines *colour* as an attribute of visual perception consisting of any combination of *chromatic* and *achromatic* content. The former includes common colour names, such as yellow, red, etc., whereas the latter includes names, such as white, black, and grey, and qualifications, such as bright or dim and light or dark. Thus colour is subjective and does not exist independently of the observer. *Related colours* are those perceived to belong to an area of an object seen in relation to other colours, whereas *unrelated colours* are those perceived to belong to an area of an object seen in isolation from other colours. Brown and grey can be perceived only as related colours and do not exist in isolation. *Hue* indicates whether an area appears to be similar to one of the perceived colours: red, yellow, green, and blue, or to a combination of two of them. *Brightness* signifies whether an area appears to emit more or less light. *Lightness* is the brightness of an area judged relative to the brightness of a similarly illuminated area that appears to be white or highly transmitting. *Colourfulness* indicates whether the perceived colour of an area appears to be more or less chromatic, whereas *saturation* is the ratio of the colourfulness and the brightness of an area. Thus the vividness of a colour is indicated by its saturation, whereas the intensity of a colour is indicated by its lightness. Hue, saturation, and lightness (HSL) have been identified as the parameters that determine colour perception, according to the popular *HSL model.* When displaying or reproducing colour images, one needs to simulate human perception of colour. Computer and television monitors display colours as mixtures, in different proportions, of the three primary colours red, green and blue (*RGB model*). Mixing equal amounts of all three produces grey, mixing maximum amounts of all produces white, and mixing zero amounts of all produces black. More complex models are also available, such as the *CMYK* (cyan, magenta, yellow, and black) *model*, utilized by high-quality printers.

Thermal radiation: An idealized source of thermal radiation is the *blackbody*, which radiates at all wavelengths. The spectral radiance of the blackbody is given by *Planck's radiation law*

$$L_{e\lambda} = \frac{2\pi hc^2}{\lambda^5} \frac{1}{e^{hc/(\lambda k_B T)} - 1}, \tag{5.27}$$

where h is *Planck's constant* and $k_B = 1.38042 \times 10^{-23}$ J/K is *Boltzmann's constant.* For the infrared–visible–ultraviolet wavelength range and temperatures in the range below 10^4 K, unity may be neglected compared with the exponential term in the denominator of Eq. (5.27), and Planck's radiation law may be simplified to *Wien's radiation law,*

$$L_{e\lambda} = \frac{2\pi hc^2}{\lambda^5} e^{-hc/(\lambda k_B T)}. \tag{5.28}$$

Differentiation of this equation with respect to λ provides a maximum spectral radiance at

$$\lambda_{\text{max}} = \frac{0.002898}{T}, \tag{5.29}$$

where λ_{max} is given in metres and T in degrees Kelvin. This relationship, called *Wien's displacement law*, shows that the peak of the radiation spectrum shifts towards lower wavenumbers as the temperature increases. Approximate integration of Planck's radiation law over the entire wavelength range gives the total radiation power emitted by a blackbody as

$$\Phi_e = \sigma A T^4, \tag{5.30}$$

where $\sigma = 5.67033 \times 10^{-8}$ W/m^{2}K is the *Stefan–Boltzmann constant, A* is the radiating area, and T is the absolute temperature. This expression is called the *Stefan–Boltzmann law.*

Although some opaque objects and dense gases may be treated approximately as blackbodies, most objects may not. However, the radiation emitted by many objects may be referred to the blackbody radiation by introduction of a correction factor ε, called the *total emissivity*, as

$$\Phi_e = \varepsilon\sigma A T^4. \tag{5.31}$$

The emissivity of shiny metallic surfaces could be as low as 0.02 to 0.03, whereas that of black, flat surfaces may exceed 0.95, approaching the blackbody emissivity of 1. In addition to the total emissivity, the *monochromatic emissivity* ε_λ may also be defined as the ratio of the spectral radiance of the body and the spectral radiance of a blackbody at the same wavelength and temperature. In general, ε_λ is a function of λ, T, and surface conditions. Objects whose monochromatic emissivity is independent of wavelength, $\varepsilon_\lambda = \varepsilon$, are called *grey bodies.*

Thermal light sources: Thermal light sources emit electromagnetic radiation as a result of being heated to a high temperature [5, 16]. The important characteristics of thermal light sources are their radiation power and spectral radiance. A variety of thermal light sources with spectra in the visible, ultraviolet, and infrared ranges are available. They are distinguished as *line sources*, which produce radiation at one or more narrow spectral bands, and *continuum* sources, which produce wideband radiation.

Among the CW thermal light sources, the most commonly used are the following ones:

Incandescent lamps: These contain an electrically heated tungsten filament in an evacuated glass container. They have a smooth continuous spectrum across the visible range, following Planck's law, Eq. (5.27), with a peak at 900 nm at a temperature of 2854 K. *Halogen lamps* are filled with a halogen (iodine or bromine) compound, which prolongs the life of the tungsten filament and allows it to operate at higher temperatures.

Electric discharge lamps: Common-type *fluorescent lamps* are filled with mercury vapour at low pressure and utilize an electric discharge through it to produce light in the ultraviolet range, which, through fluorescence, is converted to the visible range. They have a higher efficiency compared with that of incandescent lamps and have a spectrum that is comparable with that of natural light. *High-pressure arc lamps* consist of a quartz container filled with a gas, usually mercury, xenon, or a mixture of the two, under high pressure, and produce light at the arc between two tungsten electrodes. This light has a wideband continuous spectrum having superimposed spectral lines. Mercury lamps can be excited to emit at one or only a few wavelengths. A commonly used, nearly monochromatic, electric discharge lamp is the *low-pressure sodium lamp*, which emits in the yellow ($\lambda = 589$ and 589.6 nm) and also has the highest efficiency of operation.

The following are thermal light sources that provide radiation pulses of short duration:

Flash lamps: These are tubes containing a noble gas, usually xenon, krypton, or argon. For their operation, high voltage stored in a capacitor is discharged through the gas, producing a highly luminous *corona discharge*. Single-flash and stroboscopic devices are available. The duration of each light pulse varies typically between 1 μs and 1 ms.

Sparks: These are produced by the electric breakdown of a gas (helium, neon, argon, or air) during an electric discharge between two electrodes. The choice of different electrodes produces sparks of different shapes.

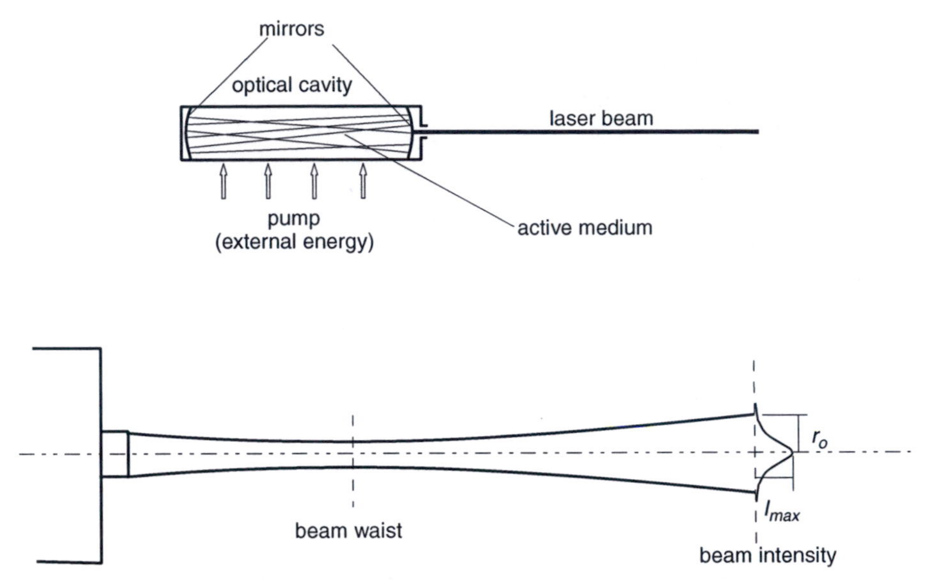

Figure 5.7. Schematics of a basic laser and a typical laser beam.

Exploding wires: These are metallic (e.g., Cu–Mg) wires, which evaporate explosively when an extremely high current (of the order of 10^5–10^6 A/mm^2) causes them to heat rapidly to temperatures of tens of thousands degrees Kelvin [17].

Explosive flashes: These consist of small amounts of explosive materials in a noble-gas environment, detonated to produce shock waves, which are accompanied by short-duration (10–100 μs) high luminosity.

Plasma-focus discharges: These techniques utilize high-temperature plasmas to produce short pulses in the ultraviolet range.

Lasers: The word laser is an acronym for 'light amplification by stimulated emission of radiation'. Since the invention of the ruby laser in 1958, followed by the He–Ne laser in 1960, continuous development has produced a great variety of lasers and rendered them indispensable to many different applications. Compared with thermal light sources, lasers have several important advantages: Laser beams are highly *coherent* (with all light wave fronts in phase), *collimated* and *concentrated* (essentially parallel, with a small cross-sectional area), and *monochromatic* (with spectral energy concentrated in one or more extremely narrow band). In experimental fluid mechanics, lasers are used routinely as light sources for flow visualization, but also in various techniques for measuring velocity, pressure, temperature, and composition, in both liquids and gases.

In simple terms, the principle of operation of a laser is as follows (Fig. 5.7) [2,18,19]. The radiation energy is produced by an *active medium*, which could be a gas, crystal, semiconductor, or liquid solution. The medium consists of particles (atoms, ions, or molecules), containing electrons that may exist at only specific, quantized energy levels. Now consider a photon, having energy $h\nu$, approaching the particle. If the photon energy matches a quantum jump of the particle, the photon may be *absorbed*, causing an electron to be raised temporarily to a higher energy level; after some time, not of any particular length, the particle would return to its original state by *spontaneous emission* of a photon of the same energy as that of the absorbed one, but in a random direction. However, a situation quite distinct from absorption and spontaneous emission is also possible: An atom that is already at a higher energy level may become excited by an incident photon (which may have, for instance, been generated by spontaneous emission), and, without absorbing it, it may be stimulated to return to a lower energy level by *coherent emission* of another photon, namely one that is identical in energy (frequency), phase, and direction with the incident photon. This process, called *stimulated emission*, results in amplifying the original photon energy while maintaining its frequency and phase. For the process to be effective, the population of atoms at the higher energy state must be larger than that at the lower energy state, a situation referred to as *population inversion*, because it contradicts the normal state of equilibrium of matter, in which lower energy states occur with a higher probability than higher ones. Obviously external power is required for maintaining a sufficient population of atoms at the higher energy state. This is supplied by the *energy pump source*, in the form of electromagnetic or chemical energy provided by an electric discharge, a flash lamp, or another laser. The preceding light amplifier is turned into an oscillator by being placed in an *optical cavity*, namely a tube that has two plane or concave mirrors at its two ends. The distance between the mirrors is adjusted precisely to an integral multiple of the light wavelength, which produces standing light waves in the cavity. One of the mirrors is slightly transparent, so that some of the light escapes in the form of a laser beam.

A performance characteristic of the optical cavity is the *quality factor*, or Q, which characterizes energy decay within the cavity (high Q means better quality). A common method to produce extremely large light pulses is *Q-switching*. This method consists of a temporary disruption of stimulated emission within the cavity by various means (e.g., by rotating or blocking one mirror), which results in a large increase of inverted population of atoms; then the cavity is suddenly restored and a massive emission of coherent photons is produced.

The intensity of light in the laser beam is not uniform, and it depends on the number and types of *modes* of standing waves present in the optical cavity. There are two types of modes: *longitudinal modes*, forming along the length of the cavity, and *transverse modes*, normal to it. For a longitudinal mode with wavelength λ to be present, the distance between mirrors must be an integral multiple of $\lambda/2$. Thus mode selection can be made by adjustment of the mirrors. The most commonly used mode is the *fundamental mode*, in which the light intensity amplitude I_a in the beam decreases monotonically with radial distance r from the axis, following approximately a Gaussian-type shape, as

$$I_a(r) = I_{a\,\max} e^{-2(2r/d_e)^2}, \tag{5.32}$$

where the laser beam diameter d_e is defined as the diameter of the circle on whose perimeter the intensity amplitude is $e^{-2} I_{a\,\max}$, namely 13.5% of the maximum intensity on the axis. Other modes have intensity profiles with peaks off the axis; for example, the first-order transverse mode can be made to have an approximately annular intensity distribution (*doughnut mode*). One should further note that the laser beam is not cylindrical but it first converges to a section of minimum diameter (*beam waist*), where the average intensity is maximum, and then it diverges (Fig. 5.7), with the exact shape depending on the arrangement of the mirrors in the optical cavity. Although multi-mode operation of the laser would provide higher total power than the Gaussian mode alone would, the latter is generally preferable, as it is spatially coherent, it results in the smallest possible beam diameter, it can be focussed to the smallest possible spot, and it has the lowest beam divergence; a typical divergence angle for the fundamental mode is of the order of 1 mrad or less. Special lens systems are available to modify the Gaussian intensity profile to an approximately uniform one or other shape, as required in specific applications.

One disadvantage of laser light is its *diffraction* at dust particles inside the optical cavity, resulting in a beam distorted by fringes and patterns, called *speckles*. Such diffraction patterns may be partly removed by the passing of the beam through a *pinhole*, which acts as a spatial filter.

There is a great variety of lasers that use many different media. The ones most commonly used in experimental fluid mechanics and combustion research are as follows:

Helium–neon (He–Ne) lasers: These are of the CW type, producing powers between 0.3 and 15 mW, at $\lambda = 633$ nm (red). Because of their easy operation and relatively low cost, they are frequently used for flow visualization purposes and in a variety of other applications. The active medium is a mixture of helium and neon atoms, and the energy pump source is a high-voltage (~2000-V) electric field. Free electrons detached from a cathode are accelerated by the electric field and occasionally collide with helium atoms, which are excited by the electron's kinetic energy to a higher energy state. This state is referred to as *metastable*, because it is relatively long lasting, although it eventually returns to the ground state by spontaneous emission. Some of these excited helium atoms collide with neon atoms at their ground state and excite them to their higher energy state; this process works because the excited energy states of the helium and neon atoms are essentially identical. Thus a population inversion of more neon atoms at the excited state than at lower states is generated. When a photon of a particular energy (corresponding to $\lambda = 633$ nm) strikes the excited neon atom, it triggers the stimulated emission of an identical photon, resulting in reducing the energy state of the atom to a lower level, which very quickly returns to the ground state. Doppler spreading, which is due to the random motion of neon atoms, causes some widening of the emitted spectrum around the $\lambda = 633$ nm line.

Argon-ion (Ar-ion) lasers: These are also CW, producing powers between 100 mW and 10 W or more, at seven wavelengths, with the strongest peaks at $\lambda = 488.0$ nm (green) and 514.5 nm (blue). They have an extremely low power efficiency, and, while some air cooled laser at 100 mW power is available, the higher power ones require continuous water cooling. These are the most commonly used lasers in laser Doppler velocimetry (LDV) and are also used for flow visualization and phase Doppler particle analysis (PDPA, or PDA for short), as well as for particle image velocimetry (PIV) in low-speed flows. The active medium is argon atoms maintained at the ion state through collisions with electrons accelerated by an electric field, all contained within a *plasma tube*. Some of these ions are further excited to a higher energy state by collisions with electrons; stimulated emission is triggered by interactions of the excited ions with photons at the appropriate frequency.

Nd:YAG lasers: These are solid-state lasers, containing the rare earth neodymium (Nd^{+3}) as the active medium, incorporated, as an impurity, into a crystal of yttrium aluminum garnet (YAG), which serves as a host. Energy pumping is usually produced optically by a flash lamp. In a single-pulse mode, these lasers produce pulses with energy between 100 and 400 mJ and duration of the order of 100 ps to 10 ns. They generally require a Q-switch to generate a short intense laser pulse. Their main emission is at $\lambda = 1064$ nm (infrared), but a frequency-doubled output at 532 nm or a frequency-tripled output in the ultraviolet range may be produced with the use of a special crystal. Dual configurations are available for PIV, in which two lasers are fired between 0 and 10 ms apart, at repetition rates of up to 30 Hz. With pumping provided by a CW diode laser (see subsequent subsection on laser diodes), Nd:YAG lasers can give a train of pulses at repetition rates exceeding 1 kHz, although the energy of each pulse would diminish with increasing repetition rate.

Copper-vapour (Cu) lasers: They typically produce repetitive pulses of duration 15–60 ns, energy of 10 mJ per pulse, and rate of 5–15 kHz, mainly at $\lambda = 510.6$ nm (green) and 578.2 nm (yellow). Unlike most other lasers, instead of requiring cooling, they are insulated to achieve a high operating temperature. These lasers are particularly suitable for particle tracking, flow visualization, and PIV.

Dye lasers: Their active medium consists of complex, multi-atomic, organic molecules, with essentially continuous emission spectra over relatively wide wavelength bands in the range from 200 to 1500 nm; their name derives from the fact that such substances are used for dying fabrics. They are pumped by flash lamps or other types of lasers. When pumped by a Nd:YAG laser, in combination with frequency filters contained in the optical cavity, they can be 'tuned' to emit radiation at narrow bands matching the resonant frequencies of various molecules contained in the fluid of interest. They are used for the measurement of species concentration in gas mixtures and combustion products.

Excimer lasers [4]: Their active medium is diatomic molecules, whose atoms are bound attractively only if one of them is at an excited electronic state (*exciplexes*), whereas they dissociate if it is at the ground state; examples include KrF and XeCl. Excimer lasers emit in the ultraviolet range and provide high-energy pulses at a relatively high repetition rate. They are used in specific measurement techniques in combustion research.

CO_2 lasers: They oscillate at $\lambda = 10.6$ μm (infrared), either continuously or in pulsed mode. They are used for heating materials, cutting, and welding. The active medium is CO_2 molecules at their ground electronic state. Population inversion of CO_2 molecules to higher vibrational energy states is caused by collisions with nitrogen molecules at a metastable, excited vibrational mode.

Laser diodes: These are semiconductor devices that emit coherent radiation in the visible or infrared ranges when current passes through them. They are used extensively in optical-fibre communication systems, compact disc players, laser printers, remote controls, and intrusion detection systems. They are much smaller than conventional lasers, and they have a much lower power requirement. High-power laser diodes have been used in fluid mechanics research for flow visualization, LDV, and PIV.

Diffraction: *Diffraction* is the transverse spreading of a light beam when it passes through a small opening or near the edge of an object [1, 20, 21]. This phenomenon is against geometrical optics concepts, but can be explained by consideration of the wave-like character of light. Consider a beam of radiation produced by a light source and passing through a small *aperture*, such as a slit or a hole on an opaque surface. If the radiation is then projected on a screen, it will form a *diffraction pattern*. Depending on the arrangement of the basic components, the diffraction process can be classified into one of the following two classes:

Fraunhofer or far-field diffraction, which occurs when both the light source and the screen are at essentially infinite distances from the aperture; in practice, this process can be produced when the light of the source is collimated with a collimator lens and the light passing through the aperture is projected to the screen through a collecting lens focussed on the screen.

Fresnel or near-field diffraction, which occurs when either the source or the screen or both are at finite distances from the aperture.

The description of Fresnel diffraction is more complex than the description of Fraunhofer diffraction. In what follows, we deal exclusively with the latter. First consider a parallel, monochromatic, coherent light beam with an intensity I_0 and a wavelength λ, approaching an opaque plane at an angle θ_i and passing through a single slit with a width w [Fig. 5.8(a)]. The intensity of the light that is diffracted by the slit would vary with the diffraction angle θ_d according to the relationship

$$I = I_0 \left(\frac{\sin\beta}{\beta} \right)^2 , \quad \beta = \frac{\pi w}{\lambda} (\sin\theta_i + \sin\theta_d) . \tag{5.33}$$

Thus the intensity of the diffracted light would alternate between relative maxima, at angles having $\tan\beta = \beta$, and relative minima equal to zero, at angles having $\beta = \pm k\pi$, $k = 1, 2, 3\ldots$. The absolute maximum $I = I_0$ (*principal maximum*) would occur at $\beta = 0$, i.e., when $\theta_d = -\theta_i$. In the case of normal incidence, for which $\theta_i = 0$, the absolute maximum would be at $\theta_d = 0$ and minima would occur at $\theta_d = \pm\sin^{-1}(k\lambda/w)$. The light pattern on the far screen would be a set of parallel bright fringes, corresponding to intensity maxima, and dark fringes corresponding to intensity minima, with the principal maximum being much brighter than any other. A simple explanation of this phenomenon can be based on *Huygens–Fresnel principle* [1], which states that 'every unobstructed point of a propagating wave front at a given instant serves as the source of spherical secondary wavelets with the same frequency as that of the primary wave;

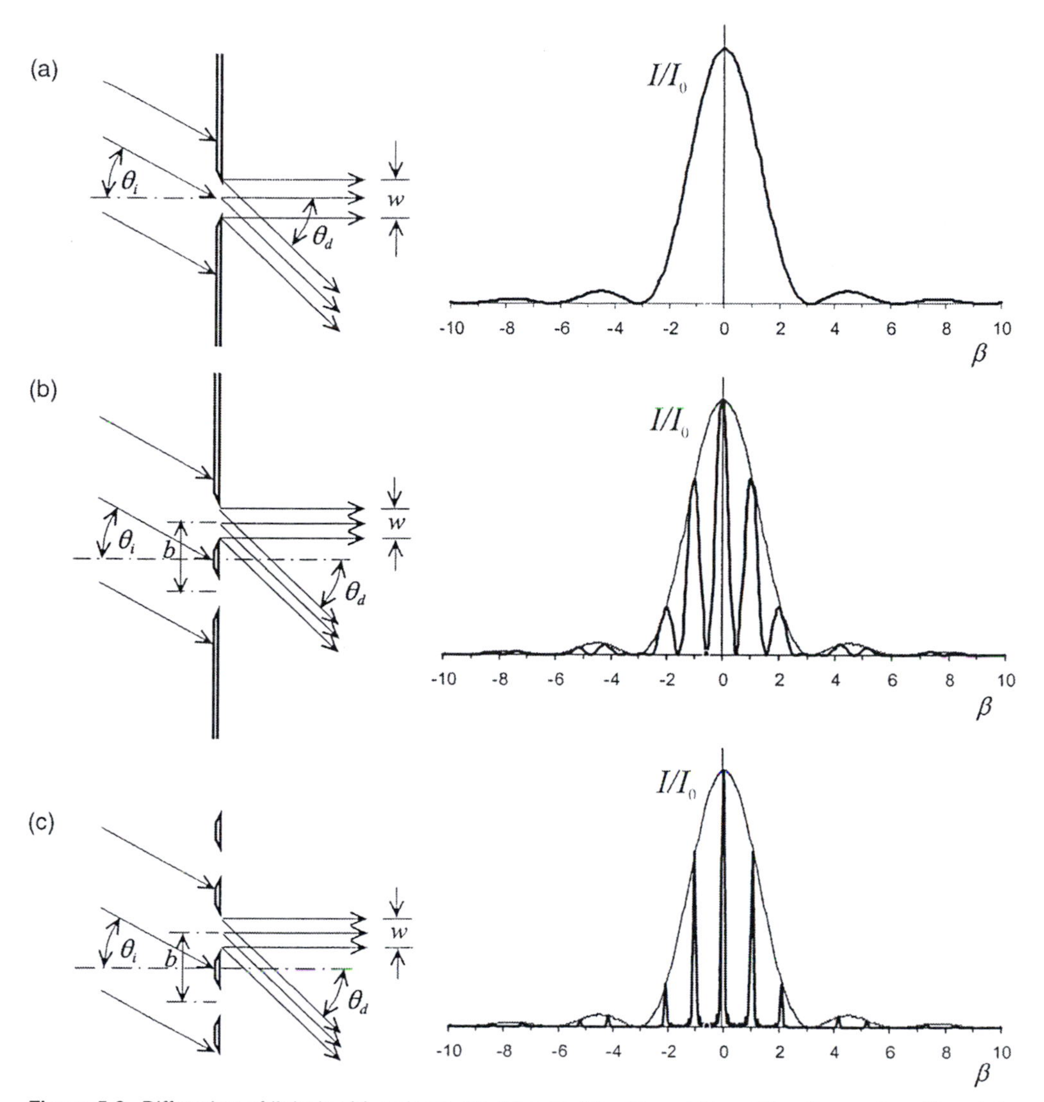

Figure 5.8. Diffraction of light by (a) a single slit, (b) a double slit, and (c) a diffraction grating. The plots show the variation of the relative intensity I/I_0 for normal incidence ($\theta_i = 0$) through (a) a single slit, (b) a double slit with $b/w = 3$, and (c) a diffraction grating with $b/w = 3$ and $N = 8$. Envelopes in plots (b) and (c) indicate the corresponding single-slit relative intensity variation, in case (c) multiplied by N^2.

the amplitude of the optical field at any point beyond is the superposition of all these wavelets considering their amplitudes and relative phases'. At normal incidence, all light would reach the slit plane in-phase. Each point across the slit can be viewed as the source of spherical wavelets, which propagate in all directions and interfere with wavelets produced by other points. The light wave at a location after the slit would be the result of superposition of all wavelets emitted by points between the two edges of the slit, which reach that location with different phases. Integration of these wavelets across the slit leads to Eq. (5.33). One may interpret the parameter β at some location as one-half the phase difference between the two wavelets that reach this location, having been emitted at the same time by the two edges of the slit.

A similar situation arises when monochromatic coherent light passes through a circular aperture. If this happens at normal incidence, and the aperture diameter is d, the intensity of the diffracted light would be

$$I = I_0 \left[\frac{2J_1(\beta)}{\beta} \right]^2, \quad \beta = \frac{\pi d}{\lambda} \sin\theta_d, \tag{5.34}$$

where J_1 is the first-order Bessel function. The pattern on the far screen would be a bright central spot, called *Airy's disk*, surrounded by dark and bright rings. The first dark ring would occur at $\beta = 1.22\pi$.

Next, consider coherent light passing through two slits of equal widths w and separated by a distance b centre-to-centre [Fig. 5.8(b)]. The diffracted light intensity beyond the slit plane would be

$$I = I_0 \left(\frac{\sin\beta}{\beta} \right)^2 \cos^2\gamma, \quad \beta = \frac{\pi w}{\lambda}(\sin\theta_i + \sin\theta_d), \quad \gamma = \frac{\pi b}{\lambda}(\sin\theta_i + \sin\theta_d), \tag{5.35}$$

where the reference for measuring angles would be the axis of symmetry of the double slit. Thus the single-slit angular distribution will be modulated by the $\cos^2\gamma$ factor, so that the principal fringe in the single-slit configuration would now be replaced with a number of fringes, with maxima at $\gamma = k\pi$, $k = 0, 1, 2 \ldots$.

A number of parallel, equidistant slits with equal widths is called a *diffraction grating* [Fig. 5.8(c)]. The intensity of monochromatic coherent light passing through a diffraction grating with N slits with widths w and separation distances b would be

$$I = I_0 \left(\frac{\sin\beta}{\beta} \right)^2 \frac{\sin^2 N\gamma}{\sin^2\gamma}, \quad \beta = \frac{\pi w}{\lambda}(\sin\theta_i + \sin\theta_d), \quad \gamma = \frac{\pi b}{\lambda}(\sin\theta_i + \sin\theta_d). \tag{5.36}$$

The maxima of the modulating function $\sin^2 N\gamma / \sin^2\gamma$ are equal to N^2 and occur when $\gamma = k\pi$, $k = 0, 1, 2 \ldots$. This function represents the interference among wavelets produced by the N slits. Although the locations of these peaks are independent of N and the same as for the double-slit configuration, their strengths increase with increasing number of slits and their widths decrease accordingly. The diffraction angles θ_d at which the peaks occur would be such that

$$\sin\theta_i + \sin\theta_d = k\frac{\lambda}{b}, \quad k = 0, 1, 2, \ldots, \tag{5.37}$$

namely they depend on the wavelength of light. Thus one can determine the wavelength of incident light by measuring the distances between the principal fringes on the screen. For this reason, diffraction gratings are used as *spectrometers*, a term denoting devices that measure radiation wavelength.

The slit array is only one of several kinds of diffraction gratings. Owing to its function, it is known as the *transmission amplitude grating*. A related device is the grating produced when parallel grooves are scratched on a clear-glass plate. Thickness variations would create phase differences across a light beam transmitted through the

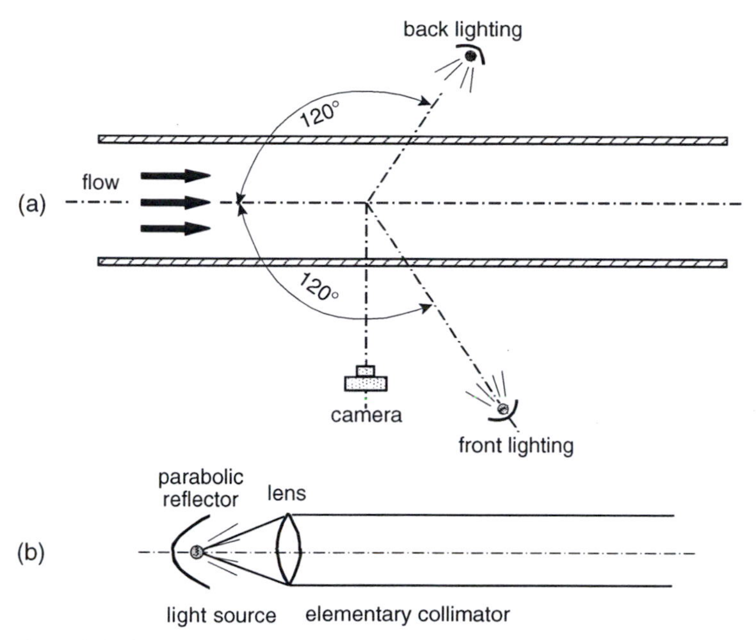

Figure 5.9. Sketches showing (a) optimal illumination arrangements and (b) an elementary light collimator.

plate and so the spherical wavelets emitted from the surface of the plate would interfere, creating bright and dark fringes. This is known as the *transmission phase grating*. The latter device may also serve as a *phase reflection grating*, if light is reflected on it, rather than being transmitted through it. Reflection gratings can be produced on opaque, reflective materials, such as aluminium. Most modern diffraction gratings are of the *blazed reflection phase grating* type, with a sawtooth-like reflecting surface.

Illumination techniques: Although sufficient illumination may often be achieved by simple means, it is worthwhile to spend some time refining the selected technique. Thermal light sources are usually operated in a floodlight arrangement. Even so, the position of the light source with respect to the visualized part of the flow and the observation – recording plane may be optimized by trial and error in order to obtain the highest clarity and contrast possible. For example, when markers are introduced into a flow, it is best to view it along a line perpendicular to its direction, while illuminating it along an axis inclined by 120°, either from the front or the back of the apparatus [Fig. 5.9(a)].

Collimators [Fig. 5.9(b)], which are combinations of lenses, such as the lens of a slide projector, may be used to produce a cylindrical or slightly diverging light beam, with increased intensity and possible to direct towards specific areas to reduce reflections. Combinations of collimators and slits or cylindrical lenses may be used to produce a "sheet of white light" (see subsequent discussion), with a thickness of a few millimetres or centimetres. One may achieve further reduction of light reflections and increased contrast by painting illuminated solid surfaces white or black and draping parts of the apparatus.

A laser beam provides high-intensity illumination, albeit of only a narrow volume of fluid, with a typical cross-sectional diameter of the order of 1 mm. Whether the beam

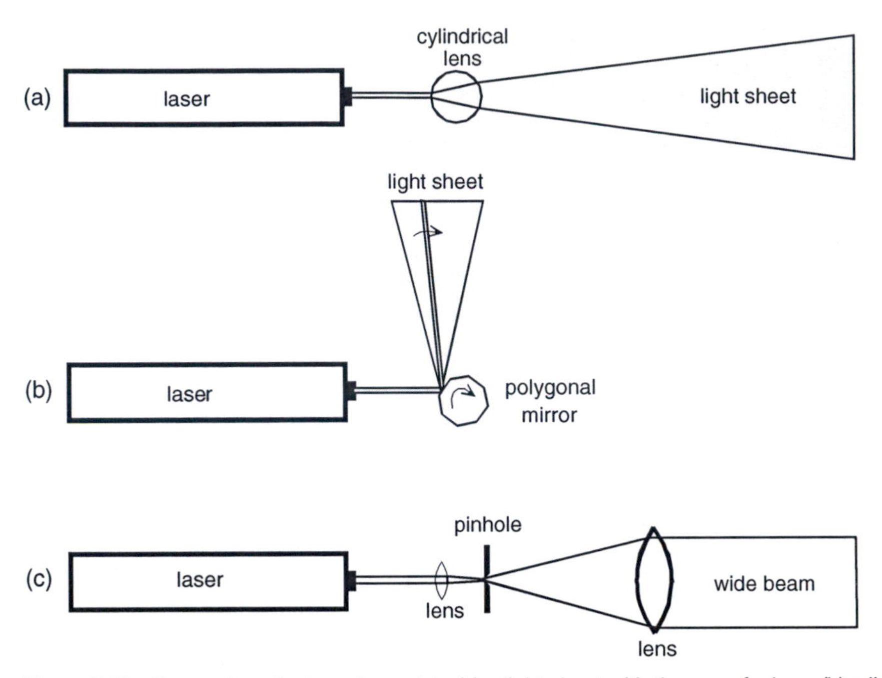

Figure 5.10. Conversion of a laser beam into (a) a light sheet with the use of a lens, (b) a light sheet with the use of a polygonal mirror, and (c) a wide beam.

path in the fluid and intersected solids will be straight or not depends on whether the refractive index of materials along this path is uniform and on the angle of intersection of surfaces. A number of optical arrangements have been utilized to increase the volume of laser-light illuminated fluid. These can be distinguished to *plane-illumination* and *volume-illumination* techniques.

A commonly used and very effective method of illumination is the *laser-sheet* technique [22]. In its simplest form, a laser beam is passed through a cylindrical lens (e.g., a glass rod), which fans the beam into a diverging sheet of light of a thickness of a few millimetres [Fig. 5.10(a)]. Further use of a converging lens may reduce the sheet thickness to typically 1 mm or less, whereas passing the sheet through a plano-convex lens will eliminate its lateral divergence and turn a triangular sheet to a rectangular one. Specially shaped lenses (*Powell lenses*) are also available to compensate for the variation of the ideal light intensity across the beam and produce a light sheet of nominally uniform intensity. Even so, imperfections inside the laser and on the various lenses create non-uniformities in the intensity of the sheet, usually in the form of darker and brighter stripes. The use of an inexpensive, rotating, cylindrical rod has been suggested as a means to reduce the effect of stripes [23]. Rapid oscillatory scanning of the laser sheet across the illuminated plane, achieved with the use of commercial scanners or rotating-mirror galvanometers, has also been utilized to eliminate such non-uniformities. An example of the use of a rotating polygonal mirror to generate a light sheet by sweeping a laser beam is shown in Fig. 5.10(b). Another advantage of the latter method, compared with the conventional light-sheet method, is that the full intensity of the laser beam occurs at some instant at any location on the sheet, rather than being spread over the entire sheet.

An alternative approach to correct for intensity non-uniformity is to calibrate the optical system *in situ*. As an example, consider the laser sheet illumination of dye patterns in a turbulent flow in a water tunnel, for which one wants, in addition to qualitative visualization, also to measure the local concentration of dye. Non-uniformity of the light sheet will produce erroneous values of dye concentration. One can correct this error by applying a correction coefficient that are obtains by calibrating the output of the optical system by using a container filled with a well-mixed dye solution of known concentration, inserted in the water tunnel at the location of interest [24]. Such calibration also can compensate for other imperfections in the optical components. The application of the light sheet to the flow could be continuous or interrupted, with the latter achieved either by use of a pulsed laser or by interruption of the beam of a CW laser by a shutter, rotating perforated disk, or other means (see also Section 5.5).

Three-dimensional illumination can be achieved either by expansion of a laser beam or by the sweeping of a beam or a light sheet. A lens arrangement, consisting of a converging lens to focus the beam, a pinhole on the focal plane to remove unwanted light, and another confocal, converging lens, can be used to expand the original laser beam to a *wide beam* [Fig. 5.10(c)], having a diameter of several centimetres. Obviously this will reduce the light intensity by a factor equal to the beam area ratio, so that this approach may be used with very high power lasers only. Because the intensity of a focussed laser beam is extremely high, one must use a very clean focussing lens. To sweep a light sheet across a volume, one may simply modify the light-sheet arrangement shown in Fig. 5.10(b) by adding a second rotating mirror, with its axis of rotation normal to the rotation axis of the first mirror and parallel to the direction of the original laser beam [25, 26].

5.4 Light scattering

In this section, the term *particle* is used to indicate a small parcel of mass, capable of absorbing and emitting radiation. This includes atoms and molecules as well as *particulates*, namely clusters of molecules suspended in a fluid.

Atomic and molecular spectra: As mentioned in Section 5.1, atomic spectra appear in the form of peaks at characteristic resonant frequencies, flanked by narrow spectral bands, which are due to thermal motion, atomic collisions, and other effects. Molecular spectra are more complex. Molecules have *electronic energy*, associated with the states of electrons in each atom, but also *vibrational* and *rotational energies*, corresponding to vibrations and rotations of the molecule nucleus. All energies are quantized, which means that each molecule is stable at specific energy levels only, which distinguish it from other types of molecules. Each electronic state has a number of vibrational and rotational energy levels associated with it. A representative sketch of the possible energy levels for a diatomic molecule, plotted vs. the separation distance between the nuclei of the two atoms, is shown in Fig. 5.11. Molecular energy transitions involve changes of any or all of these energies. Matters are further complicated by the fact that the three energy modes are coupled to one another: The vibrational energy levels depend

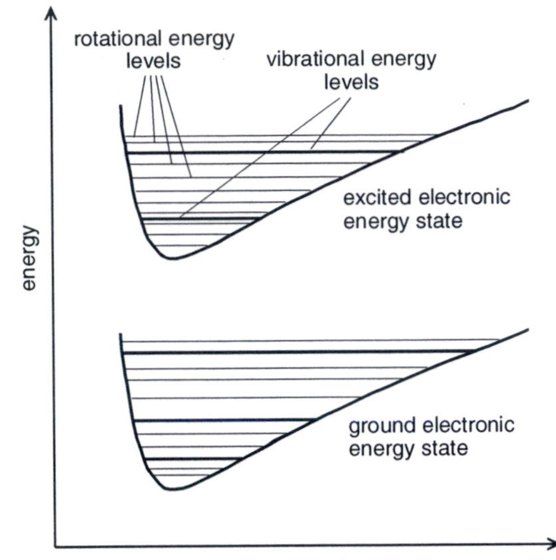

Figure 5.11. Schematic representation of the ground electronic energy state and an excited electronic energy state of a diatomic molecule, together with corresponding vibrational and rotational energy levels.

on the electronic energy state, and the admissible rotational energy levels depend on the vibrational energy of the molecule. Thus molecular spectra have more peaks than atomic ones, broadened by thermal motion, intramolecular interactions, and other effects.

Elastic and inelastic scattering: Consider an atom being approached by a photon, whose energy is associated with a frequency ν, according to Eq. (5.6). The photon will be *absorbed* by the atom if its frequency matches one of the *resonant frequencies* of the atom. Normally, in solids and liquids as well as gases at ordinary pressures, the absorbed energy of the photon will be converted to *thermal energy*, i.e., random motion, without emission of another photon. This process is called *dissipative absorption*. However, if the frequency of the photon is significantly lower than all resonant frequencies of the atom, the absorbed energy will momentarily cause an electron to oscillate, while the atom remains in its ground state. Thus the oscillating electron and the positive nucleus become an *electric dipole* (i.e., a pair of oscillating electric charges), radiating an electromagnetic wave at the same frequency as that of the incident one. This process is called *non-resonant scattering* or *elastic scattering* and takes place essentially instantly, within less than 10^{-15} s. In general, the photons are emitted in random directions, so that, when continuously irradiated, the atom effectively becomes a source of spherical electromagnetic waves. Besides elastic scattering, molecules are also capable of scattering light *inelastically*, namely at frequencies that are higher or lower than the incident frequency. When a molecule is excited by absorbing a photon, it is possible that it will emit a photon with energy that is different from the energy of the absorbed photon, with the (positive or negative) balance being converted to vibrational or rotational energy, or both.

Rayleigh scattering: Light scattering from particles that are smaller than about $\lambda/15$ (λ is the incident-light wavelength) is called *Rayleigh scattering* [1]. The efficiency of light

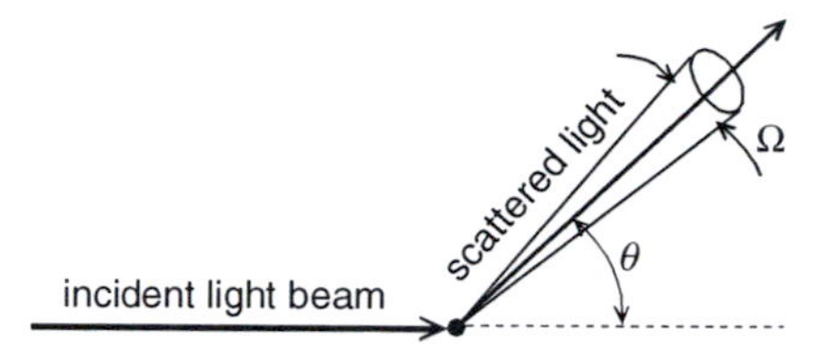

Figure 5.12. Sketch showing light scattered by a spherical particle towards a collecting lens.

scattering from a particle is expressed in terms of its *scattering cross section*, defined as the equivalent area of the incident wave front that has the same power as that emitted by the particle. For a single electron, this is called the *Thomson-scattering cross section*, equal to

$$\sigma_T = 6.65 \times 10^{-29}\ \text{m}^2. \tag{5.38}$$

The classical *Rayleigh-scattering cross section* is

$$\sigma_R = \sigma_T \left(\frac{\lambda_0}{\lambda}\right)^4, \tag{5.39}$$

where λ_0 is the characteristic wavelength of the atom [21]. Therefore the intensity of scattered light is proportional to $1/\lambda^4$, and, consequently, Rayleigh scattering of white light would be much more intense at the violet end of the visible spectrum than at the red end. Scattering from isolated (i.e., spaced by distances equal to λ or more) atoms and ordinary molecules belongs to this type.

Mie scattering: When many atoms and molecules are close to each other, their electromagnetic waves interfere with each other and lateral scattering diminishes, while becoming direction dependent. The intensity of light scattered by spherical particles of arbitrary size can be calculated based on a theory referred to as *Mie-scattering theory* [5,27–30]. Rayleigh scattering is the limit of Mie scattering as the particle diameter diminishes towards zero. Extensions of Mie's theory include predictions of light scattering from coated and optically inhomogeneous spheres and from cylinders [29]. In fact, the term Mie scattering is often used to describe light scattering not only from spherical particles but generally from aggregates of molecules of arbitrary shape. Like Rayleigh scattering, Mie scattering is an elastic process.

Consider a spherical particle with a diameter d_P and a refractive index n_P, immersed in a fluid with a refractive index n_F, and exposed to a collimated beam of randomly polarized, monochromatic light of wavelength λ and intensity (radiant power per unit cross-sectional area) I_0, as shown in Fig. 5.12. According to Mie-scattering theory, the radiant intensity (power per unit solid angle) of light scattered by the particle towards a particular direction forming an angle θ with the incident-light direction would be equal to $I_0\sigma_\lambda$, where the *monochromatic angular scattering cross-section* σ_λ is given by

$$\sigma_\lambda = \frac{\lambda_F^2}{8\pi^2}\left[i_1\left(x, n_P/n_F, \theta\right) + i_2\left(x, n_P/n_F, \theta\right)\right]. \tag{5.40}$$

In this expression, $x = \pi d_P/\lambda_F$, $\lambda_F = \lambda/n_F$, and the *intensity* or *phase functions* i_1 and i_2, in units of inverse steradians, correspond to the two components of the scattered light,

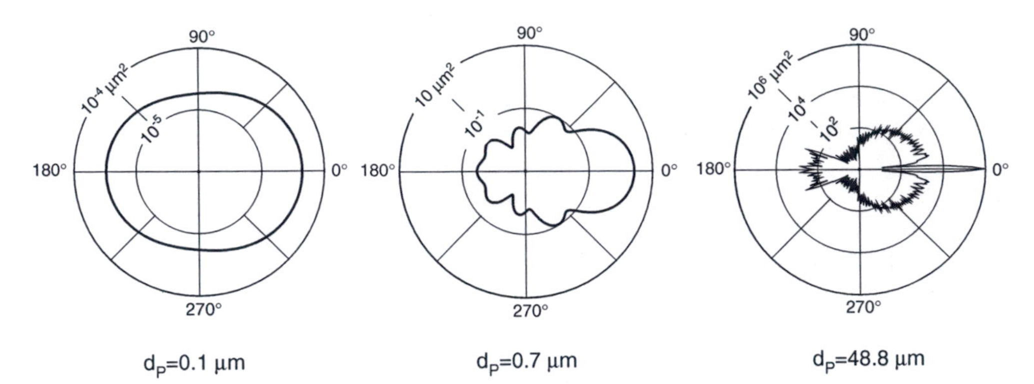

Figure 5.13. Monochromatic scattering cross sections of glass spheres in air, according to Mie-scattering theory; $\lambda = 0.488\ \mu$m. Note that the radial axis is logarithmic, representing changes of orders of magnitude in values.

both of which are plane polarized, with the corresponding waves oscillating on planes perpendicular and parallel to, respectively, the plane formed by the incident and the scattered beams. The intensity functions for various particle sizes and refractive indices can be computed with analytical expressions provided by Mie theory. If the incident light is linearly polarized, the scattered light will be linearly polarized as well, but on a plane perpendicular to the polarization plane of the incident light; its monochromatic angular scattering cross section will be

$$\sigma_\lambda = \frac{\lambda_F^2}{4\pi^2} i_2 (x, n_P/n_F, \theta). \tag{5.41}$$

Whereas the Rayleigh scattering cross section is independent of orientation, the Mie-scattering cross section is a strong function of the scattering angle θ. The maximum amount of scattered light always corresponds to *forward scattering* ($\theta = 0°$), whereas a local maximum also occurs for *backscattering* ($\theta = 180°$). Local maxima and minima may also occur at intermediate angles, the number of which increases with increasing x. As representative of the variation of σ_λ with the scattering angle, three cases are presented in Fig. 5.13. As illustrated in this figure, the directional asymmetry of scattered light increases dramatically as the particle diameter increases, whereas, as the diameter decreases, light scattering tends to become independent of scattering angle.

The total power of scattered light that reaches a certain aperture within a solid angle Ω (Fig. 5.12) would be

$$\Phi_\Omega = I_0 \int_\Omega \sigma_\lambda \mathrm{d}\Omega. \tag{5.42}$$

It is evident that, the larger this aperture is, the smoother the variation of the collected light power with scattering angle θ would be.

Raman scattering: The phenomenon of inelastic light scattering from molecules is called the *spontaneous Raman effect*. This effect is quite rare, typically with an occurrence that is 10^{-5} to 10^{-2} times lower than that for Rayleigh scattering. Thus, the *Raman*

Table 5.5. Wavelengths of Raman-scattered radiation for some common molecules in air at standard atmospheric pressure and a temperature of 295 K; excitation was provided by a ruby laser

Molecule	Anti-Stokes line (nm)	Stokes line (nm)
Incident light	694.30 (Rayleigh line)	
CO_2	638.23	761.17
CO_2	637.92	762.23
CO_2	633.25	768.37
CO_2	632.42	769.61
O_2	626.57	778.44
O_2^+	615.50	796.23
NO	614.26	798.31
CO	604.34	815.73
N_2^+	603.19	817.82
N_2	597.57	828.39
CH_4	577.45	870.46
H_2	538.67	976.41

scattering cross section σ_{Rm} for each species and a specific energy level is several orders of magnitude smaller than the corresponding Rayleigh scattering cross section. If the energy of the emitted photon is higher than that of the absorbed photon, the process is called *Stokes transition*. In addition to Stokes transition, it is possible for a molecule to emit a photon of energy lower than that of the absorbed photon; this process is called *anti-Stokes transition* [1]. The time between photon absorption and emission for Raman scattering is of the order of 10^{-14} s, which is negligible for most practical purposes. Table 5.5 summarizes the Raman-scattered spectral lines of common molecules in air at room temperature and pressure [31]. It can be seen that both Stokes and anti-Stokes lines are present, but it must also be noted that the anti-Stokes lines have significantly higher energy at relatively low temperatures.

Fluorescence and phosphorescence: The Rayleigh and Raman transitions occur essentially instantaneously, thus not allowing other energy conversion phenomena to occur. Certain molecules, however, are capable of emitting radiation after a certain delay following absorption of a photon. When such emission takes place relatively rapidly, typically within the range between 10^{-10} and 10^{-5} s, the phenomenon is called *fluorescence* [4]. When it is relatively slow, following time delays between 10^{-4} s and up to hours, it is called *phosphorescence*. Fluorescence allows sufficient time for collisions of molecules to take place and photon energy to be converted to chemical reaction, dissociation, and ionization energies, before emission takes place. These phenomena, referred to as *quenching*, interfere with the emission process and create measuring uncertainty in fluorescence-based diagnostic and measurement methods. Besides being caused by photon absorption, fluorescence may also be caused by electron bombardment, heating or chemical reaction (*chemiluminescence*). Although *resonant fluorescence*

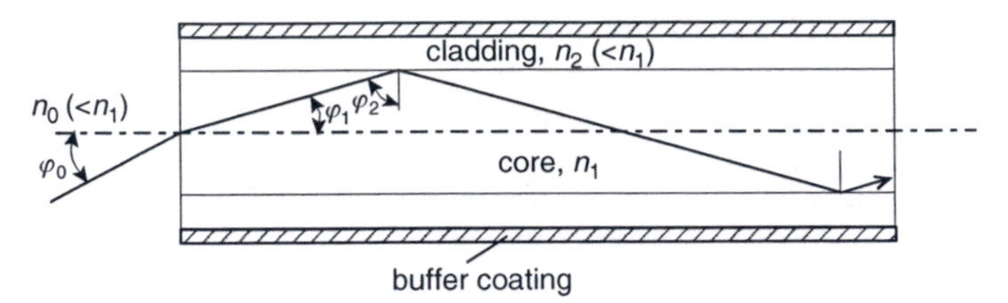

Figure 5.14. Schematic diagram of a step-index optical fibre and the propagation of a light ray along it.

(i.e., emission at the same wavelength as that of the absorbed light) is possible, it is *non-resonant fluorescence* (i.e., emission at wavelength longer than the incident one) that is utilized for flow visualization and measurement, as it permits the separation of fluorescence radiation from the incident light and Mie scattering. This happens when the molecule absorbs a photon that causes transition from the ground energy state to a higher vibrational level of an excited electronic state, but then returns to the ground state from a lower vibrational level, so that the energy of the emitted radiation is lower than the absorbed one. Fluorescence of several species is quite vigorous, exhibiting cross sections that are several orders of magnitude larger than the corresponding Raman-scattering cross sections. The intensity of fluorescence is independent of orientation, and the fluorescent radiation is randomly polarized, even though the incident radiation may be linearly polarized.

5.5 Light transmission, sensing, and recording

Although relatively simple visual and optical experiments may not require more than direct observation of visual patterns, in most cases, it would be necessary to transmit light to a convenient location and then to measure and record its intensity, wavelength, and other properties, which may be further processed later. The present section summarizes commonly used instrumentation and techniques for the transmission, sensing, and recording of light and images.

Fibre optics: *Optical fibres*, or *optical waveguides*, are thin cables capable of transmitting internally light over considerable distances with high efficiency [1,19,32,33]. They consist of a *core*, which is a fine fibre of glass, quartz, or plastic (commonly methyl methacrylate), a *cladding*, which is a thin shroud of glass or plastic surrounding the core, and a *buffer coating*, which protects the fibre (Fig. 5.14). The refractive index of the fibre usually changes abruptly from a value n_1 in the core to a smaller value n_2 in the cladding, although fibres with a gradual decrease of n from the core towards the cladding are also available. Transmission of light along the core is effected through reflections (see Section 5.2) at the interface with the cladding (Fig. 5.14). To ensure that all light is reflected at this interface and none penetrates through the cladding, the incidence angle φ_2 must remain larger than the critical angle $\varphi_c = \arcsin(n_2/n_1)$ for total internal reflection. Application of Snell's law gives the *maximum acceptance angle* for light entering the fibre to be transmitted along the core as

$$\varphi_0 = \varphi_{\max} = \arcsin \frac{\sqrt{n_1^2 - n_2^2}}{n_0}, \tag{5.43}$$

where n_0 is the refractive index of the surrounding medium. For typical values of $n_0 \approx 1.00$ (air), $n_1 \approx 1.62$, and $n_2 \approx 1.52$, one may calculate $\varphi_{\max} \approx 34°$. The parameter

$$\mathrm{NA} = n_0 \sin \varphi_{\max} = \sqrt{n_1^2 - n_2^2} \tag{5.44}$$

is called the *numerical aperture* of the fibre. Commercial fibres are available with NAs in the range between 0.2 and 1.

An important distinction among optical fibres is with respect to the *modes of propagation* of light along them. This arises from the fact that a ray propagating along the axis of the fibre will traverse the fibre faster than a ray that undergoes successive reflections at the cladding. The number of reflections, and thus the delay in propagation, increases with increasing entry angle φ_0. This phenomenon is called *modal dispersion*, which is distinct from *chromatic dispersion*, which is due to variation of refractive index with light wavelength. Thus waves propagating along the fibre accumulate phase shifts; when different waves are in-phase, they interfere constructively, adding their energies; otherwise they interfere destructively and eventually fade away. As a result, only certain modes propagate effectively through the fibre, appearing as distinct light patterns. A parameter that describes the number of these modes is the *V-number*,

$$V = \frac{2\pi}{\lambda} r_1 \mathrm{NA}, \tag{5.45}$$

where r_1 is the radius of the core and λ is the wavelength of propagating light. When $V < 2.405$, only a single mode, called the *fundamental mode*, will exit the fibre, whereas, when $V > 2.405$, multiple modes will. Thus, if the core diameter is maintained sufficiently small, typically less than 10 μm, it is possible to create *single-mode fibres*, which are preferable in most applications in fluid mechanics.

Conventional photography and cinematography: A basic *conventional camera* consists of a *lens*, an *aperture stop*, and a *photographic film*. The lens focusses images on the film, and the aperture stop controls the amount of light that enters the camera. The *clear aperture* is the unobstructed area of the lens that is exposed to light. It is controlled by the aperture stop, which is either the rim of the lens or a diaphragm with an adjustable opening, called the *iris*. The *f-number*, $f^{\#}$, or *focal ratio* of the lens is defined as the ratio of the focal distance of the lens and its clear aperture diameter. The larger the $f^{\#}$, the less the light collected by the lens. The film consists of three main layers: the *emulsion*, which is a suspension of silver-halide grains in gelatin, the *support*, which is made of plastic or glass, and the *dye*, which reduces the blurring effect of scattered light [2]. Grains become 'developable' when exposed to light; when immersed in a *developer*, they are reduced to metallic silver, which is visible; finally, the undeveloped grains are removed by a *fixer* and the photograph becomes permanent. The *exposure* of the film is the amount of light that falls on an area of the film; it is equal to the product of the irradiance and the exposure time. A variety of films are available. The *resolving power* of a film is the number of lines that may be resolved per millimetre; it is obviously related to the average spacing between grains. The *speed* of a film signifies the amount of light required for producing a certain amount of silver after development. Standards, such as those of American Standards Association (ASA) and Deutsche Industrie Norm (DIN) classify films according to their speed. Finally, the *contrast* of a film is its ability to reproduce tone differences on an image; the contrast may be increased by an increase in the development time. *Still photography* [34] produces a single photograph at a time. A main challenge of photography is to capture very fast-changing images. This can be achieved with either the use of a high-speed shutter or by short-duration illumination pulses (see Section 7.3). Mechanical shutters may be used up to a certain speed, whereas *polarization shutters* can reach much higher speeds as they control light by electro-optic or magneto-optic effects (i.e., changes of the polarization plane) that require no moving parts. *Cinematography* produces a series of consecutive images. There is a variety of high-speed cinematography cameras, including mechanical cameras, which use rotating mirrors, image-dissection cameras, image converter and intensifier cameras, and multiple-spark cameras [5]. Despite the convenience of electronic cameras, conventional cameras remain in use in scientific photography and cinematography, often offering advantages in terms of speed and resolution.

Photodetectors: Light may be converted to an electric current with the use of devices called *photodetectors*. Photodetectors utilize the *photoelectric effect*, namely the absorption of photons and the subsequent emission of electrons. Their main characteristic is the quantum efficiency

$$\eta_q = \frac{N_e}{N_p}, \tag{5.46}$$

where N_p is the number of absorbed photons and N_e is the number of emitted electrons. These electrons form an electric current equal to

$$i = \eta_q \frac{\Phi_e q_0}{h\nu}, \tag{5.47}$$

where Φ_e is the collected radiation power, q_0 is the electron charge, h is Planck's constant, and ν is the radiation frequency. As mentioned earlier (see Section 4.1), photodetectors are subject to two types of noise: (a) *shot noise* (or photon noise), which is due to random fluctuations of the rate of photon collection and to background illumination, and (b) *thermal noise*, which is caused by amplification of the current inside the photodetector and by an external amplifier. Even in the absence of a desirable source of light, photodetectors produce a current, called a *dark current*.

The two common types of photodetectors are the *photomultiplier tubes* (*PMT*s) and the *photodiodes* (*PDs*) or *photoelectric cells*. PMTs collect light on a semitransparent *photocathode*, which absorbs photons and emits electrons. The electrons are subjected to electric fields between successive pairs of *grids–dynodes* and increase in numbers until they are captured by an *anode grid*. PMTs have a low quantum efficiency but provide a strong electric output that is due to the internal amplification. PDs are p–n junctions of semiconductors, commonly silicon–silicon type. They have a high quantum efficiency but no internal amplification and so require an external amplifier for producing a current at a usable level. By comparison with PMTs, PDs are much less expensive but they generally provide a lower signal-to-noise ratio, except at very high light levels. A variance of the PD is the *avalanche PD*, which has some internal amplification and so has characteristics intermediate between those of PMTs and common PDs. PDs may operate as isolated transducers (*single-channel photodetectors*) to measure light on a small spot, or in one- and two-dimensional arrays (*multi-channel photodetectors*) to measure light along a line segment or on a plane area, respectively.

Electronic image recording [4,16,35,36]: Until recently, the most widely used means of electronic image recording was the *analogue video camera*, whose operation is based on the *vidicon tube*. The tube front is made of glass, coated on the inside with a thin metallic anode layer and a layer of phosphor. Light entering the tube causes the phosphor to emit free electrons, which affect the local electric conductivity of the layer. An electron beam is created by a cathode at the other end of the tube and scanned by a time-varying electromagnetic field along a line on the tube face and then line-by-line to cover the entire face. The current flowing through the anode depends on the local electric conductivity and thus on the light intensity. This produces an analogue signal, which may be further amplified, digitized by use of a frame grabber or an ADC, and stored electronically. *Silicon-intensified-target* (*SIT*) tubes, also called vidicons, collect the light on a two-dimensional array of silicon PDs, grown on a single crystal. The diodes are first reverse-charged electrically by a fast-scanning electron beam. When exposed to light, they produce a current that depletes the charge. A second scan of the electron beam restores the charge and provides a measure of light intensity. *Intensified SIT* (*ISIT*) tubes first amplify the light by using a phosphor screen, which is fibre-optically coupled to the SIT. The spatial resolution of SITs is determined by the size of the scanning electron beam, which is substantially larger than the size of the individual sensors. In general, vidicons have a high spatial resolution and good sensitivity to low light but are also subject to various distortions. They have been largely superseded by more recent digital camera technologies.

Digital cameras utilize arrays of PDs, each of which measures the light intensity over a small area, called a *pixel*, which is an abbreviation for 'picture element'. The resolution of a digital camera is determined by the number of pixels it has. *Line-scan cameras* consist of a single row of PDs arrayed along a straight line and could be used, for example, for recording light along a laser beam. Line-scan cameras can be used to record two-dimensional images sequentially (i.e., line-by-line) by lateral scanning of the linear image. *Matrix cameras* consist of a two-dimensional array of PDs, so that they can measure two-dimensional images all at once.

Most modern digital cameras consist of arrays of photosensitive electronic circuits, called *charge-coupled devices* (*CCDs*). Each CCD receives the light from a small area of

Table 5.6. Common seeding materials and their refractive indices

Seed material	n_P	ρ_P (kg/m^3)	Fluid
Polystyrene	1.59	1.05×10^3	Gas–liquid
Polyamid	1.5	1.03×10^3	Gas–liquid
NaCl crystals	1.54	2.16×10^3	Gas
Al_2O_3	1.76	3.96×10^3	Gas–liquid
SiC	2.6	3.2×10^3	Gas–liquid
Oil mists	1.4–1.5	$\simeq 0.9 \times 10^3$	Gas
Hollow glass beads	1.52	1.1×10^3	Liquid

the image and converts it to an electric charge, which it then stores or displays. Current scientific matrix cameras have resolutions exceeding 2000×3000 pixels. CCD cameras have a high quantum efficiency, wide dynamic range, and low noise, and, unlike conventional cameras, they retain a good sensitivity at low-intensity light. Even higher sensitivity to low light levels can be achieved with *intensified CCD* (*ICCD*) cameras. Most recently, digital cameras utilizing *CMOS* (complementary metal-oxide semiconductor) technology have become commercially available, offering the advantages of low cost, small size, and operation versatility, although not matching the qualities of CCDs. Although the image acquisition rates of CCDs are restricted between a few and 200 Hz, much higher rates (in the tens of kilohertz) can be achieved with CMOS cameras.

5.6 Characteristics of seeding particles

Visual and optical methods provide visible images of fluids in motion and measure different flow properties by using either variations of the refractive index of the fluid itself or visible *markers*, consisting of foreign material immersed in the fluid. For a correct interpretation of these images and their correlation to flow characteristics, it is necessary to understand the following conditions:

1. The reason for the visibility of a pattern, according to the principles of optics.
2. The physical process of generation or injection of markers.
3. The space–time relationship between marker motion and fluid motion.
4. The physical significance of the observed images (see Chap. 1).

Flow seeding [5, 37, 38]: Untreated air and water flows normally contain a significant concentration of natural foreign particles, such as dust, lint, microorganisms, etc. These may actually be utilized as flow markers by certain visual and optical methods. Because, however, there is no control of the size, density, shape, and optical properties of natural markers, it is often preferable to filter them, either entirely or down to a certain maximum size, and then to introduce deliberately particles whose characteristics are more suitable for the purposes of the study. This process is called *seeding*. Some popular seeding materials and their refractive indices are listed in Table 5.6.

Many visual and optical methods work best when all particles have a uniform size. An indicator of size uniformity is the *monodispersity*

$$\sigma_g = \sqrt{d_2/d_1}, \tag{5.48}$$

where the diameters d_1 and d_2 are defined such that approximately 15.9% of the particles have diameters smaller than d_1 and 15.9% of the particles have diameters larger than d_2. For a perfect monodispersity, $\sigma_g = 1$. If σ_g is close to unity, a particle distribution is called *monodisperse*; otherwise it is called *polydisperse*. Particles may be generated and/or distributed in a flow by several methods, with either homemade or commercial devices. For liquid flows, a usual approach is to mix a powder (e.g., SiC or Al_2O_3) with the same liquid to a suitable concentration of particles, taking care to break down agglomerates, and then to release the suspension into the flow. Particle seeding of gas flows requires more care in order to achieve a relatively uniform particle concentration. Among the methods commonly used, one could mention the following:

Dispersion of powders: This may be achieved by use of a fluidized bed, namely by passing an air stream through a porous screen on which there is a supply of the powder.

Atomization: This consists of creating an aerosol of liquid droplets suspended in a gas stream by feeding the gas into a liquid through special nozzles (*Laskin nozzles*). A variation of the method is to create an aerosol from an aqueous solution of sugar, salt, or other soluble substances, or from a very dilute suspension of solid particles; evaporation of the water in the gas stream leaves a suspension of solid crystals or other solid particles, which act as tracers. Atomization usually produces polydisperse seeds.

Evaporation and condensation: An oil, either natural or mineral, is heated to evaporation and then the vapour is cooled and allowed to condense into droplets (*oil mist*). With proper control, this process produces monodisperse droplets. Water-based fluids can also be used in commercially available *fog generators* to produce non-hazardous *fogs* in small or large quantities.

Optical characteristics of seeding particles [37]: When comparing the visibility of two different types of particles, one should consider not only their actual sizes and refractive indices, but also a number of other factors. An important parameter is the scattering cross section σ_λ. In general, the larger the σ_λ, the more visible the particle. The variation of σ_λ with particle diameter in forward scattering and backscattering for representative particle–fluid combinations is shown in Fig. 5.15. Notice that, for a given particle size, incident-radiation intensity, and scattering angle, the scattered radiation may vary by one or more orders of magnitude, depending on its refractive index. Notice also that σ_λ depends on the refractive index of the surrounding fluid; for example, a particle with $n_p > 1.33$ will appear to be larger in air ($n \simeq 1.00$) than in water ($n \simeq 1.33$). Furthermore, as illustrated in Fig. 5.13, in the Mie-scattering range, σ_λ is a complicated function of orientation, very sensitive to the value of the scattering angle. Forward scattering produces the maximum σ_λ in all cases, and, whenever possible, it

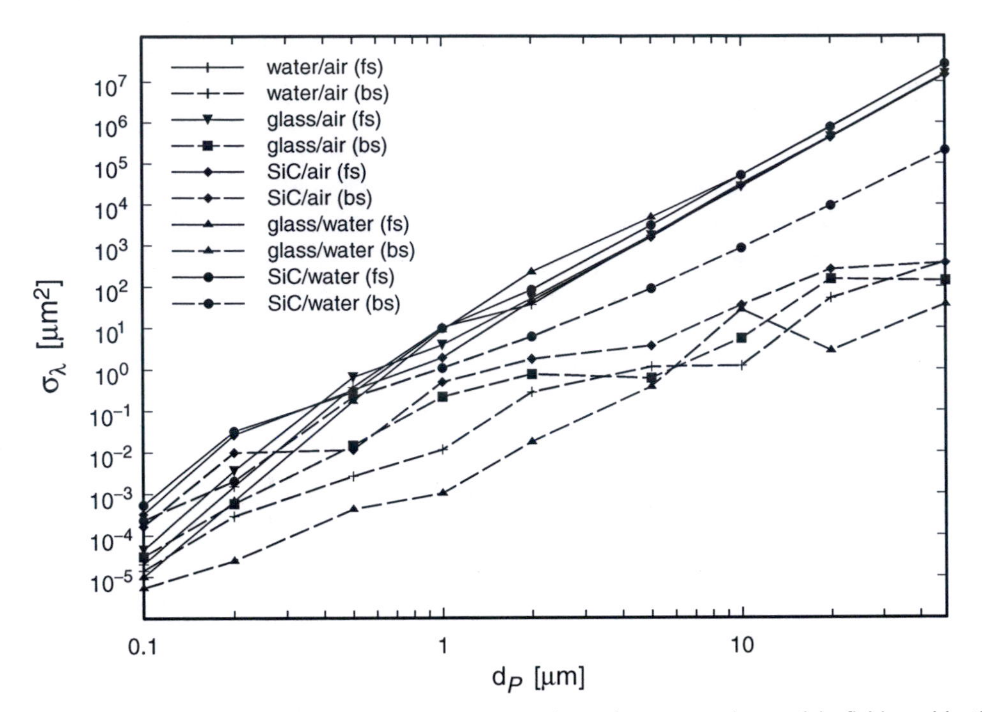

Figure 5.15. Monochromatic Mie-scattering cross sections of representative particle–fluid combinations for forward scatter (fs) and backscatter (bs); $\lambda = 0.488\ \mu$m.

should be the preferable mode of particle observation. However, because of considerations of depth-of-field and convenience in position of transmitting and receiving optical components with respect to the observed flow, backscattering and sidescattering are more common arrangements. Backscattering also permits the use of the same lenses and other components for both transmitting and receiving purposes. The apparent sensitivity of scattered light to the scattering angle is smoothened out by two effects: the averaging effect of the finite aperture of the collecting lens [see Eq. (5.42)] and multiple scattering, namely the successive scattering of light by more than one particle. When the light is not monochromatic, the overall scattering cross section would be a weighted average of all applicable monochromatic scattering cross sections, a process that also tends to smoothen out intensity variations. In the Mie-scattering regime ($d_P \gg \lambda$), the average energy scattered by a particle over a solid angle collected by a lens would approximately increase as $(d_P/\lambda)^2$, whereas, in the Rayleigh-scattering regime ($d_P \ll \lambda$), the same energy would increase as $(d_P/\lambda)^4$ [39].

Various techniques have been developed to improve the visibility of particles. One approach is to coat solid particles with a highly reflective material, such as silver, to increase their refractive index. Another is either to coat them with a fluorescent dye or to embed a fluorescent material in their composition. The particle-laden flow is illuminated by a laser sheet with a spectral peak near the absorption wavelength of the fluorescent substance and the flow is observed through an optical band-pass filter, which removes all radiation except that in the narrow emission band of the material. This way, only the particles are visible, whereas all incident illumination, reflections from the apparatus walls, etc., are removed (see also Section 13.5).

When viewing or recording particle images through a collecting lens, one must also consider the effect of diffraction on particle visibility [39, 40]. Ideally, if the lens *magnification factor* (namely the ratio of image and object diameters) is M, the diameter of the particle image would be Md_P. This is actually the case for relatively large particles; however, the image diameter of small particles would be larger than Md_P because of diffraction on the lens aperture. A length that characterizes the image of a spot viewed through a diffraction-limited lens with an f – number $f^{\#}$ is the diameter of the first dark ring of the Airy disk light intensity distribution (see Sections 5.5 and 5.3):

$$d_s = 2.44\,(M+1)\,f^{\#}\lambda. \tag{5.49}$$

The image diameter of sufficiently small particles would be approximately equal to d_s and independent of d_P. More generally, one may interpolate between the two extreme cases to express the diameter of the particle image as

$$d_e = \sqrt{(Md_P)^2 + d_s^2}. \tag{5.50}$$

As an example, consider light with $\lambda = 488$ nm and a lens with $M = 1$ and $f^{\#} = 5.6$. Then, $d_s = 13.3\ \mu$m and d_e would be within 10% of d_s for particles with $d_P < 6\ \mu$m and within 10% of d_P for $d_P > 30\ \mu$m. A consequence of diffraction is that the images of closely spaced small particles may overlap to the point that they cannot be distinguished from each other. An additional blur of particle images would occur when the particles are outside the lens's *depth-of-field*, which can be calculated as

$$d_f = 4\left(1+\frac{1}{M}\right)^2 f^{\#2}\lambda. \tag{5.51}$$

Finally, when a particle image is recorded on photographic film or on a digital device, it may be further enlarged because of inadequate resolution of the recording medium. It is evident, for instance, that for media with pixels as the recording unit, no image can be smaller than the size of a pixel.

Dynamic response of seeding particles: The motion of particles immersed in a fluid is an extremely difficult problem for which there is no general analytical solution and for which even the mathematical formulation remains a topic of debate. Let us first consider the simplest possible case, for which both the governing equation and its solutions can be treated with confidence. Consider a single, rigid particle with a density ρ_P, mass m_P, and moving at a speed $\overrightarrow{u_P}$, immersed in an unbounded fluid, which has density ρ_F, viscosity μ_F, and uniform far-field velocity $\overrightarrow{u_F}$. For simplicity, neglect body forces, but assume that the particle-to-fluid-density ratio is $\gamma = \rho_P/\rho_F \gg 1$. Then the particle motion will be approximately described by the differential equation

$$\overrightarrow{F_D} = m_P \frac{\mathrm{d}\,\overrightarrow{u_P}}{\mathrm{d}t}, \tag{5.52}$$

where $\overrightarrow{F_D}$ is the drag force by the fluid on the particle. Further assume that the particle is spherical with a diameter d_P and that the relative Reynolds number is

$$\mathrm{Re}_P = \frac{\rho_F \left| \overrightarrow{u_F} - \overrightarrow{u_P} \right| d_P}{\mu_F} \ll 1 \tag{5.53}$$

(*Stokes flow*). Then the drag force is given by the Stokes expression [41],

$$F_D = 3\pi \mu_F \left| \overrightarrow{u_F} - \overrightarrow{u_P} \right| d_P. \tag{5.54}$$

To avoid complications, let us consider that the fluid and the particle move in the same direction. Then the dynamic equation of the particle motion becomes

$$\frac{\rho_P d_P^2}{18\mu_F} \frac{\mathrm{d}u_P}{\mathrm{d}t} + u_P = u_F. \tag{5.55}$$

Thus the motion of this particle may be viewed as a first-order system, in which the undisturbed fluid velocity is the input and the particle velocity is the output. The relative motion of this particle depends on a single parameter, the *characteristic time*

$$\tau_P = \frac{\rho_P d_P^2}{18\mu_F}, \tag{5.56}$$

which, in this case, is equal to the particle's time constant. It can be seen that this time constant is proportional to the cross section and density of the particle and inversely proportional to the viscosity of the fluid. Then the response of the particle to different inputs and for different initial conditions can be easily calculated. For example, the velocity of a particle that is injected with a zero initial velocity at time $t = 0$ into the fluid will be $u_P = u_F \left(1 - e^{-t/\tau_P}\right)$. Similarly, the steady-state frequency response of a heavy particle in oscillatory flow with frequency ω would have an amplitude ratio

$$\eta = \frac{1}{\sqrt{1 + \omega^2 \tau_P^2}}. \tag{5.57}$$

This indicates that such particles would act as first-order low-pass filters of fluid motion with a 3-dB cutoff frequency equal to

$$\omega_c = 1/\tau_P. \tag{5.58}$$

The Stokes drag approximation would be invalid for $\mathrm{Re}_P \gtrsim 1$, however, estimates of particle response based on this assumption would be conservative, because the actual drag would be larger than the Stokes estimate. Among the empirical relations for the drag coefficient of incompressible flow on rigid spheres over the entire range of interest for particle motion, one may use [42]

$$C_D = \frac{F_D}{\frac{1}{2}\rho_F \left| \overrightarrow{u_F} - \overrightarrow{u_P} \right|^2 \pi d_P^2/4} = \frac{24}{\mathrm{Re}_P} \left(1 + 0.15\, \mathrm{Re}_P^{2/3}\right). \tag{5.59}$$

The preceding analysis would describe fairly well the motion of microscopic liquid or solid particles suspended in gases. To avoid particle interference with each other, it is advisable to maintain the distance between particles at greater than $1000d_P$. Concerning non-spherical particles, deviations from the preceding expressions would not be very significant in the Stokes flow regime, as long as the particle dimensions do not vary much in different directions. Such deviations become stronger with increasing Re_P. Expressions for drag coefficients for different object shapes are available in the literature [42].

A similar analysis can be made to predict the motion of particles under the effect of gravity. When the particle density is different from the fluid density, the particle will rise or sink with respect to the surrounding fluid. Consider a single, heavy ($\gamma \gg 1$), rigid particle, released in a fluid moving vertically with a constant speed u_F. Then the particle motion will be described by the differential equation

$$F_D + W_p - B_p = m_p \frac{\mathrm{d}u_P}{\mathrm{d}t}, \tag{5.60}$$

where W_p and B_p are the particle weight and buoyancy, respectively. Further assume that the particle is spherical and that the Stokes drag applies. Then, given an initial condition for the particle velocity, one may solve Eq. (5.60) to find $u_P(t)$. Following a sufficiently long time (e.g., for $t > 6\tau_P$), the relative velocity of the particle will essentially reach its *terminal rise* or *sink relative velocity*, given by the expression

$$(u_P - u_F)_\infty = (1 - 1/\gamma) g\tau_P. \tag{5.61}$$

In flows in which the local velocity changes with time or position, and especially in turbulent flows, the relative vertical velocity of particles will also be variable and may never approach its terminal value. In such cases, the terminal velocity may be considered as an upper bound of the vertical velocity difference between the particle and the fluid. Clearly, to reduce the terminal velocity and generally the vertical relative velocity, one should aim at reducing the density difference and the particle diameter. For liquid flows, it would be of great advantage to match the particle and fluid densities ($\gamma \approx 1$), which would cause the terminal velocity to vanish (*neutrally buoyant* particles).

As the density ratio γ decreases, the preceding analysis would become increasingly inaccurate. The moving particle sets in motion part of the surrounding fluid, whose inertia must be taken into consideration. In simple terms, one must also consider, in addition to the particle acceleration, the acceleration of a mass of fluid, called the *added mass*, which, for spherical particles, is equal to half the particle mass. The unsteady motion of an isolated, rigid, spherical particle in an infinite, incompressible, uniform flow stream, for which the Stokes approximation for the relative velocity applies, is described by the *Basset–Boussinesq–Oseen* (*BBO*) equation. In addition to the terms contained in Eq. (5.55), the BBO equation contains terms representing the added mass and a time integral representing additional resistance that is due to the unsteadiness of the flow. A number of solutions of the BBO equation can be found in the literature [5,37,43,44]. This type of equation would be more appropriate than Eq. (5.55) for solid particles suspended in a liquid medium, for which γ is usually of the order of 1. For liquid particles and bubbles, the analysis would be further complicated by internal motions, which tend to reduce the drag, and possible deformation of the particle, which alters the near field.

The treatment of such effects is beyond the scope of the present review, so that here we shall treat all particles as rigid. A consequence of the added mass and unsteady drag effects is that a gravitating particle would reach its terminal velocity at a slower rate than that predicted by Eq. (5.60). For the case of bubbles rising in a liquid ($\gamma \ll 1$), it has been calculated that the time required for a bubble to reach 90% of its terminal velocity would be many orders of magnitude smaller than τ_P [42,45]. A characteristic time for the particle–fluid interaction that takes into account the added mass and is applicable to arbitrary combinations of spherical rigid particles and fluids in Stokes flow is [46]

$$\tau_{Pa} = \tau_P \left(1 + \frac{1}{2\gamma}\right). \tag{5.62}$$

In the case of bubbles suspended in a liquid, this characteristic time becomes $\tau_P/2\gamma \gg \tau_P$.

A problem of practical interest is the response of suspended particles of arbitrary density in unsteady fluid flow. Solutions of the BBO equation for a few types of unsteady motion are available in the literature. More elaborate momentum equations for rigid-sphere motion in non-uniform flow have also been formulated [47, 48] and solved analytically by various authors for a few sets of simple conditions. As representative, some results are presented that are based on a formulation and solution that applies not only to Stokes flow but also to flows with significantly higher Re_P [49]. Consider the steady-state response of a rigid, spherical particle suspended in a gravity-free flow field that oscillates sinusoidally with a frequency ω. This response can be described by the amplitude ratio η and phase shift φ of the particle motion, relative to the fluid motion. Clearly these parameters would depend on the density ratio γ, the ratio of the characteristic time of the particle τ_P, and the period of fluid motion $1/\omega$. A dimensionless parameter representing this time ratio is the *Stokes number*, here defined as

$$S = \sqrt{\frac{\omega d_P^2}{8\nu_F}}. \tag{5.63}$$

One is warned that different authors use the term Stokes number to describe parameters different from the one just given. The following analytical solution for the amplitude ratio η in terms of S and γ has been derived by asymptotic analysis [49]:

$$\eta = \left\{ \frac{(1+S)^2 + \left(S + \frac{2}{3}S^2\right)^2}{(1+S)^2 + \left[S + \frac{2}{3}S^2 + \frac{4}{9}(\gamma - 1)\, S^2\right]^2} \right\}^{1/2}. \tag{5.64}$$

It is interesting that this solution is accurate in two distinct asymptotic limits:

- for Stokes flow ($\mathrm{Re}_P \to 0$) and any Stokes number S
- for large Stokes numbers ($S \to \infty$) and Re_P not restricted to small values, but significantly exceeding 1.

For intermediate combinations of Re_P and S values, this solution may not be accurate; this, however, is not a serious limitation for the purposes of particle selection, for which we are mainly interested in ensuring that the amplitude ratio remains close to 1.

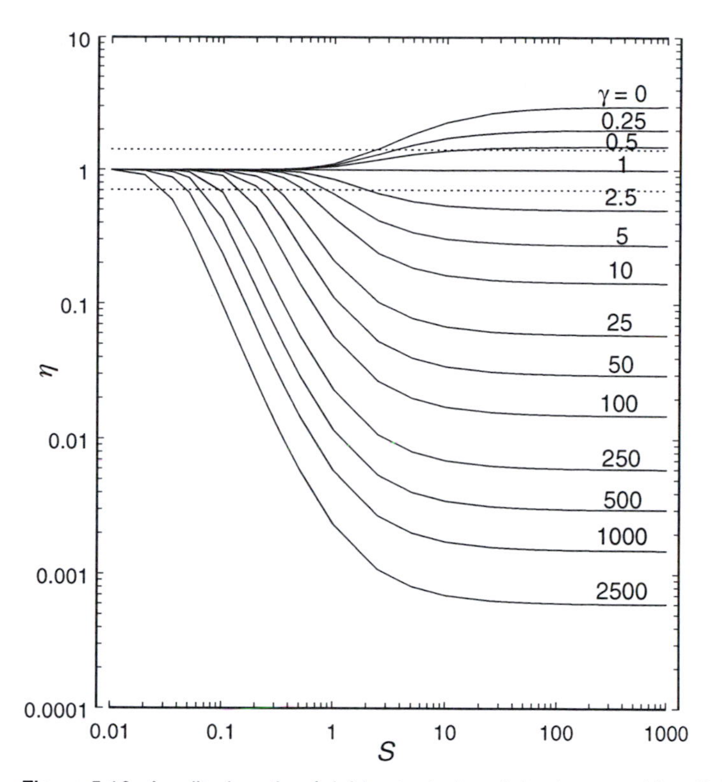

Figure 5.16. Amplitude ratio of rigid spherical particles immersed in a fluid; dotted lines correspond to the 3-dB cutoff frequencies.

Equation (5.64) for representative values of γ has been plotted in Fig. 5.16, from which we can make the following practical observations:

- As $S \to 0$ (steady flow), $\eta \to 1$ in all cases. This means that all particles would have a perfect dynamic response at sufficiently low frequencies.
- Neutrally buoyant particles ($\gamma = 1$) would have a perfect steady-state frequency response, and their steady-state motion would be indistinguishable from that of the surrounding fluid. Even particles with density ratios in the range $0.56 \leq \gamma \leq 1.62$ would be within the $\pm$3-dB tolerance limit for the amplitude ratio at all frequencies and for $0 \leq \mathrm{Re}_P \leq 20$. This confirms the earlier observation that, from the viewpoint of dynamic characteristics, it is preferable to use particles whose density is close to the fluid density. Several such particles are available for liquid flows. For gas flows, the only nearly neutrally buoyant particles are helium-filled soap bubbles; however, because of their relatively large sizes (typically 50–100 μm) and low production rate, these are unsuitable for many applications.
- In general, when $\gamma > 1$, $\eta < 1$ and, when $\gamma < 1$, $\eta > 1$. This means that particles that are heavier than the surrounding fluid would oscillate with lower amplitude than that of the fluid, whereas particles that are lighter than the fluid would oscillate with higher amplitude than that of the fluid.

- In the high-frequency limit of $S \to \infty$, $\eta \to (3/2)/|\gamma - 1|$. This limit for very heavy particles ($\gamma \gg 1$) is an entirely negligible response ($\eta \ll 1$), whereas for very light particles ($\gamma \approx 0$), $\eta \approx 3$, which indicates a motion amplification by a factor of 3.
- The Stokes numbers S_c corresponding to the 3-dB cutoff frequencies of particles with different density ratios can be found by the intersection of the dotted lines in Fig. 5.16 with the corresponding curves. For additional accuracy, one may use the following interpolation relationships:

$$S_c \approx \left[2.380^{0.93} + \left(\frac{0.659}{0.561 - \gamma} - 1.175\right)^{0.93}\right]^{\frac{1}{0.93}}, \quad \gamma < 0.561; \tag{5.65}$$

$$S_c \approx \left[\left(\frac{3}{2\sqrt{\gamma}}\right)^{1.05} + \left(\frac{0.932}{\gamma - 1.621}\right)^{1.05}\right]^{\frac{1}{1.05}}, \quad \gamma > 1.621. \tag{5.66}$$

Given the particle diameter d_P, one can thus estimate the 3-dB cutoff frequency (in hertz) as

$$f_c \approx \frac{4\nu_F}{\pi} \frac{S_c^2}{d_P^2}. \tag{5.67}$$

If, instead, the cutoff frequency is specified, the maximum particle diameter can be calculated as

$$d_{P\,\mathrm{max}} \approx 2S_c \sqrt{\frac{\pi f_c}{\nu_F}}. \tag{5.68}$$

The cutoff frequencies for a few representative particle–fluid combinations have been plotted vs. particle diameter in Fig. 5.17.

Of particular difficulty is the general problem of particle motion in turbulent flows, because of the presence of fluid motions ('eddies') with a wide range of characteristic lengths, times, and amplitudes. The particle would respond differently to different eddies, and its final motion would be the result of a non-linear superposition of the different contributions. One would anticipate that a heavy particle would generally act as a non-linear low-pass filter of fluid motion and that it would not adequately respond to motions whose characteristic frequency is higher than a certain limit. Many attempts have been made to predict particle motion in turbulent flows, without yet reaching a solution of general validity. Considering the available results, it is safer to conclude that the various recommended expressions might be adequate for selecting particles that would be suitable for a particular application (e.g., for selecting a maximum particle size for a given turbulent flow), but rather uncertain as a means of correcting experimental results. A conservative condition for the particle cutoff frequency to ensure that it would follow all motions present in a turbulent flow is

$$f_c > (1/2\pi)(\nu_F/\varepsilon)^{-1/2}, \tag{5.69}$$

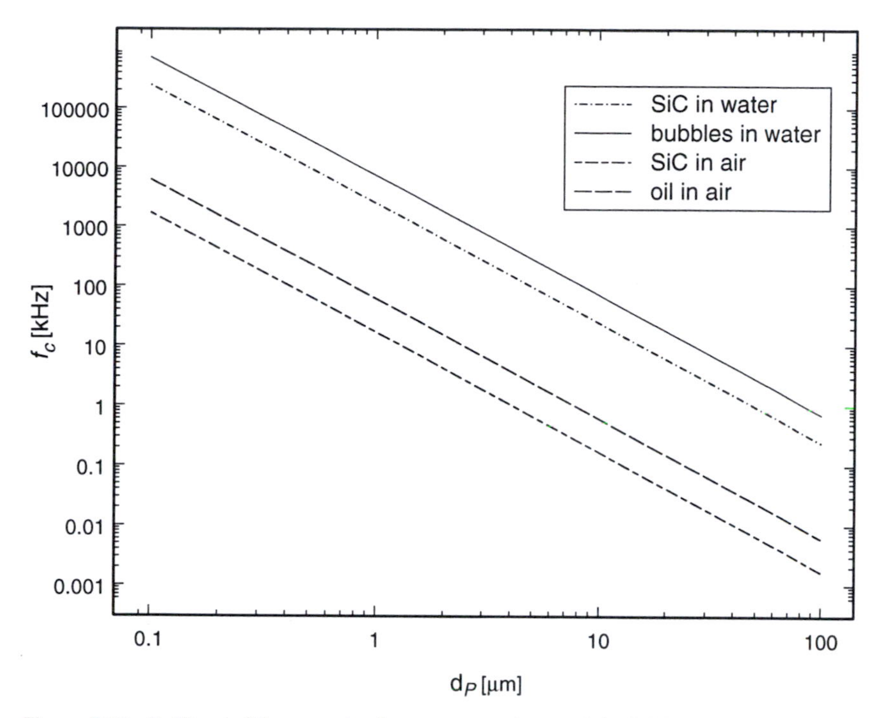

Figure 5.17. 3-dB cutoff frequencies for representative particle–fluid combinations.

where ε is the turbulent kinetic-energy dissipation scale [50–52]. In addition, the size of the particle should be small compared with the length scale of the smallest turbulence eddies (Kolmogorov microscale), namely

$$d_P \ll (\nu_F^3/\varepsilon)^{3/4}. \tag{5.70}$$

Such requirements would be excessive if one is interested in only the large-scale turbulent motions and not in the fine structure.

Besides gravitational forces, a particle may also be subjected to other types of body forces, including electric forces for electrically charged particles in an electric field, magnetic forces for ferromagnetic particles in a magnetic field, centrifugal forces in swirling flows, Coriolis forces in rotating flows, and aerodynamic lift for rotating or non-spherical (e.g., flakes) particles or because of shearing (*Saffman force*). Lift that is due to shearing would be particularly significant near solid walls, tending to move particles away from the wall. Irrespectively of shearing effects, particle concentration near walls would tend to diminish because of deposition, either through gravity or through adherence. In vortical flows, particles heavier than the fluid would tend to centrifuge away from the axis of rotation, whereas lighter particles would tend to concentrate in the core. For this reason, gas bubbles are suitable for flow visualization of vortices in liquids. In flows with a temperature gradient, collisions with fluid molecules in Brownian motion would tend to move small particles from warmer towards cooler regions.

QUESTIONS AND PROBLEMS

1. Compute the Gladstone–Dale constant for monochromatic light with a wavelength of 580 nm passing through the atmosphere of a strange planet, which consists of a homogeneous mixture of the following neutral molecules, with relative number densities given in parentheses: oxygen (2%), nitrogen (10%), carbon dioxide (88%).
2. Consider water in a rectangular tank made of Plexiglas. The beam of a He–Ne laser approaches the wall of the tank at an angle of 30° with respect to the normal. Determine the direction of the laser beam in the water, if its temperature is (a) 20 °C and (b) 34 °C.
3. Water flows in a water tunnel having a rectangular cross section with a width of 0.500 m and a height of 1.000 m and surrounded by glass walls with a thickness of 10 mm. A laser beam enters one vertical wall at an angle of 20° with respect to the normal. Sketch accurately the path of this beam, indicating its inclinations as it crosses the different interfaces between air, glass, and water and exits from the other side of the tunnel. Determine the vertical distance between the entry and exit points of the beam. Furthermore, determine the location of the intersection of the laser beam with the room wall, which is located 3 m away from the water-tunnel back wall. Finally, determine whether there is an entry incidence angle for which the beam will not exit the water tunnel, and, if so, find it.
4. Consider a thin-walled glass tube containing water and surrounded by air. Two parallel beams of a He–Ne laser located on a plane normal to the tube axis approach the tube symmetrically about a plane containing the tube axis. Explain whether it would be possible for the beams to intersect inside the tube. If not, orient and position the beams so that they would intersect at a position on the symmetry plane, half a radius away from the distant end of the tube.
5. Consider water flowing in a cylindrical glass tube with an inner diameter of 100 mm and a wall thickness of 10 mm, surrounded by air. A laser beam is directed parallel to the horizontal plane of symmetry of the tube and at a distance of 25 mm from this plane. Plot the path of the beam through the tube, indicating the values of the different angles. Now consider that the glass tube is immersed in a viewing tank filled with water. Plot again the path of the beam and compare it with the one in the previous case. Finally, consider that the viewing tank and the tube contain a fluid whose refractive index matches the refractive index of glass. Compare the path of the same beam in the latter case with those in the previous two cases.
6. Plot the spectral radiance of a blackbody vs. temperature in the range between 0 and 10000 K, for light wavelengths $\lambda = 300$, 500, 700, and 900 nm. Also plot the spectral radiance of a blackbody vs. light wavelength in the ultraviolet, visible, and infrared ranges, for temperatures $T = 100$, 1000, and 10000 K.
7. Consider light generated by a low-pressure sodium lamp passing through isolated circular apertures with diameters equal to 0.1, 1, and 10 μm. Plot the light intensity exiting the aperture, per unit incident light, vs. the diffraction angle.
8. A laser beam with diameter d_e is passed through a material with an absorption coefficient α. If the intensity of the incident beam has a Gaussian-shaped variation,

determine the thickness variation of the material such that the exiting beam would have a uniform intensity over a cylindrical core with a diameter d_e. In your analysis, allow the minimum possible light absorption.

9. Two identical laser beams are positioned such that their axes intersect perpendicular to each other. For this problem disregard the wave-like character of light and assume that, at each location within a beam, the light intensity is equal to its amplitude I_0. Determine the intensity of light at different positions within the intersection volume. Plot contours of constant light intensity on the plane of symmetry and on other planes parallel to it. Describe the shapes of surfaces of constant intensity.
10. An aerosol consists of water droplets having a Gaussian size distribution, with an average diameter of 85 μm and a standard deviation of 20 μm. The aerosol is injected isokinetically (i.e., at a horizontal speed equal to that of the flow) along the axis of a horizontal wind tunnel having a square cross section with a side of 100 mm. It may be assumed that the flow velocity in the wind tunnel is laminar, uniform, and equal to 5 m/s.
    a. Compute the monodispersity of this aerosol. For the following questions, neglect any droplets of sizes larger or smaller than the average by more than three standard deviations.
    b. Determine the terminal sink velocities of the smallest, the average, and the largest droplets, assuming Stokes drag. Check the validity of the Stokes drag assumption. Without computation, but based on qualitative arguments, discuss whether the actual terminal velocities would be higher or lower than the preceding values.
    c. Determine the downstream distances from the point of injection at which the smallest, the average, and the largest droplets will reach the bottom of the tunnel; assume Stokes drag and neglect boundary-layer effects. Determine also the vertical speeds of these droplets when they reach the bottom.
11. Consider that glass beads with a refractive index of 1.55 and a diameter of 1 μm are injected in air flow and are illuminated by a He–Ne laser. Plot the monochromatic angular scattering cross section of the beads vs. the angle of observation.
12. A sealed tank containing air at 20 °C is mounted on a support oscillating horizontally and sinusoidally with an amplitude of 40 mm and a frequency of 2.0 Hz. A glass bead with a diameter of 50 μm is released isokinetically at the top of the tank while the tank is at the midpoint of its horizontal stroke. Assume that the tank height is sufficiently large for the bead to reach a steady-state response before it touches the bottom. Derive expressions describing the vertical and horizontal velocities of the bead, assuming that the two motions are independent of each other. Describe how you would calculate the position of the bead at different times. Determine the steady-state amplitude of oscillation of the bead.
13. Consider spherical particles with diameters $d_P = 0.5$, 1, 5, and 50 μm, suspended in air. Determine the diameters of their images, if they were viewed through lenses with magnifications $M = 1$, 10, and 100 and $f-$numbers 3.5, 11, and 22, while being illuminated by a He–Ne laser, an Ar-ion laser, a frequency-doubled Nd:YAG laser, or a low-pressure sodium lamp. For the same conditions, determine the ratio

of the depth of field of the lens and the particle diameter. Discuss your results. What would you recommend, if your objective was to measure the particle diameter?

14. A diver breaths through a tube 20 mm in diameter and 300 mm long. The capacity of his lungs is 3 litres and his breathing rate is 20 times/min. Assume that the flow in the tube is uniform and varies sinusoidally with time. Liquid droplets with density of water are released in the air near the inlet of the tube.
    a. Neglecting gravity and diffusion, determine the maximum size of the droplets that will reach the diver. State clearly all assumptions and approximations that you make.
    b. Compute the length of the preceding tube required for the diver to avoid breathing droplets larger than 10 μm in diameter. Do you foresee any other problems with the use of this tube?

REFERENCES

[1] E. Hecht. *Optics* (4th Ed.). Addison-Wesley, Reading, MA, 2002.
[2] M. Young. *Optics and Lasers*. Springer-Verlag, Berlin, 1984.
[3] H. L. Anderson (Editor-in-Chief). *A Physicist's Desk Reference* (2nd Ed.). American Institute of Physics, New York, 1989.
[4] A. C. Eckbreth. *Laser Diagnostics for Combustion Temperature and Species* (2nd Ed.). Gordon & Breach Publishers, Canada, 1996.
[5] R. J. Emrich (Editor). Fluid dynamics. In L. Marton and C. Marton (Editors-in-Chief), *Methods of Experimental Physics,* Vols. 18A and B. Academic, New York, 1981.
[6] H. M. Dobbins and E. R. Peck. Change of refractive index of water as a function of temperature. *J. Opt. Soc. Am.*, 63:318–320, 1973.
[7] H. Fiedler, K. Nottmeyer, P. P. Wegener, and S. Raghu. Schlieren photography of water flow. *Exp. Fluids*, 3:145–151, 1985.
[8] J. B. Pendry and D. R. Smith. Reversing light with negative refraction. *Phys. Today*, June 2004:37–43, 2004.
[9] A. F. Bicen. Refraction correction for LDA measurements in flows with curved optical boundaries. *TSI Quarterly*, 8(2):10–12, 1982.
[10] A. J. Parry, M. J. Lalor, Y. D. Tridimas, and N. H Woolley. Refraction corrections for laser-Doppler anemometry in a pipe bend. *Dantec Information*, No. 09 (September 1990):4–6, 1990.
[11] R. Budwig. Refractive index matching methods for liquid flow investigations. *Exp. Fluids*, 17:350–355, 1994.
[12] M. M. Cui and R. J. Adrian. Refractive index matching and marking methods for highly concentrated solid-liquid flows. *Exp. Fluids*, 22:261–264, 1997.
[13] R. C. Gonzalez and P. Wintz. *Digital Image Processing* (2nd Ed.). Addison-Wesley, Reading, MA, 1987.
[14] G. Sharma (Editor). *Digital Color Imaging Handbook*. CRC Press, Boca Raton, FL, 2003.
[15] CIE. *International Lighting Vocabulary*. CIE Pub. No. 17.4. CIE Vienna, Austria, 1987.
[16] T. Vo-Dinh. Basic instrumentation in photonics. In T. Vo-Dinh, editor, *Biomedical Photonics Handbook*, Chap. 6, pp. 6.1–6.30. CRC Press, Boca Raton, FL, 2003.
[17] F. D. Bennett. Exploding wires. *Sci. Am.*, 206:103–112, 1962.
[18] K. Shimoda. *Introduction to Laser Physics*. Springer, Berlin, 1984.

[19] M. Sayer and A. Mansingh. *Measurement, Instrumentation and Experiment Design in Physics and Engineering*. Prentice-Hall of India, New Delhi, 2000.

[20] F. A. Jenkins and H. E. White. *Fundamentals of Optics* (3rd Ed.). McGraw-Hill, New York, 1957.

[21] E. R. Cohen, D. R. Lide, and G. L. Trigg. *AIP Physics Desk Reference* (3rd Ed.). Springer, New York, 2003.

[22] W. Merzkirch. *Flow Visualization* (2nd Ed.). Academic, New York, 1987.

[23] D. J. Shlien. Inexpensive method of generation of a good quality laser light sheet for flow visualization. *Exp. Fluids*, 5:356–358, 1987.

[24] M. M. Koochesfahani and P. E. Dimotakis. Mixing and chemical reactions in a turbulent liquid mixing layer. *J. Fluid Mech.*, 170:83–112, 1986.

[25] C. Bruecker. Digital-particle-image-velocimetry (DPIV) in a scanning light-sheet: 3d starting flow around a short cylinder. *Exp. Fluids*, 19:339–349, 1995.

[26] A. J. Smits and T. T. Lim (Editors). *Flow Visualization Techniques and Examples*. Imperial College Press, London, 2000.

[27] G. Mie. Beiträge zur optik trüber Medien, speziell kolloidaler Metallösungen. *Ann. Phys.*, 25:377–452, 1908.

[28] H. C. Van de Hulst. *Light Scattering by Small Particles*. Wiley, New York, 1957.

[29] M. Kerker. *The Scattering of Light and Other Electromagnetic Radiation*. Academic, New York, 1969.

[30] C. F. Bohren and D. R. Huffman. *Absorption and Scattering of Light by Small Particles*. Wiley, New York, 1983.

[31] G. F. Widhoff and S. Lederman. Specie concentration measurements utilizing Raman scattering of a laser beam. *AIAA J.*, 9:309–316, 1971.

[32] Anonymous. Fiber optics: Theory and applications. Tech. Rep. Tech. Memo. 100, Burle Technologies, Inc., Lancaster, PA, 2001.

[33] J. E. Anderson. Fiber optics: Multi-mode transmission. Tech. Rep. Tech. Memo. 200, Burle Technologies, Inc., Lancaster, PA, 2001.

[34] A. A. Blaker. *Handbook of Scientific Photography*. Freeman, San Francisco, 1977.

[35] J. C. Russ. *The Image Processing Handbook* (4th Ed.). CRC Press, Boca Raton, FL, 2002.

[36] J. G. Webster. *The Measurement, Instrumentation and Sensors Handbook*. CRC Press, Boca Raton, FL, 1999.

[37] A. Melling. Tracer particles and seeding for particle image velocimetry. *Meas. Sci. Technol.*, 8:1406–1416, 1997.

[38] H.-E. Albrecht, M. Borys, N. Damaschke, and C. Tropea. *Laser Doppler and Phase Doppler Measurement Techniques*. Springer-Verlag, Berlin, 2003.

[39] R. J. Adrian and C.-S. Yao. Pulsed laser techniques application to liquid and gaseous flows and the scattering power of seed materials. *Appl. Opt.*, 24:44–52, 1985.

[40] R. Adrian. Particle-imaging techniques for experimental fluid mechanics. *Annu. Rev. Fluid Mech.*, 23:261–304, 1991.

[41] L. Rosenhead (Editor). *Laminar Boundary Layers*. Oxford University Press, Oxford, UK, 1963.

[42] R. Clift, J. R. Grace, and M. E. Weber. *Bubbles, Drops and Particles*. Academic, New York, 1978.

[43] A. T. Hjemfelt and L. F. Mockros. Motion of discrete particles in a turbulent fluid. *Appl. Sci. Res.*, 16:149–161, 1966.

[44] B. T. Chao. Turbulent transport behaviour of small particles in a turbulent fluid. *Oesterreichisches Ingenieur-Archiv*, 18:7, 1964.

[45] C. F. M. Coimbra and R. H. Rangel. General solution of the particle momentum equation in unsteady Stokes flows. *J. Fluid Mech.*, 370:53–72, 1998.

[46] E. E. Michaelides. Hydrodynamic force and heat/mass transfer from particles, bubbles and drops – the Freeman scholar lecture. *J. Fluids Eng.*, 125:209–238, 2003.

[47] M. R. Maxey and J. J. Riley. Equations of motion for a small rigid sphere in a nonuniform flow. *Phys. Fluids*, 26:883–889, 1983.

[48] E. E. Michaelides and Z.-G. Feng. The equation of motion of a small viscous sphere in an unsteady flow with interface slip. *Int. J. Multiphase Flow*, 21:315–321, 1995.

[49] R. Mei. Velocity fidelity of flow tracer particles. *Exp. Fluids*, 22:1–13, 1996.

[50] J. O. Hinze. *Turbulence* (2nd Ed.). McGraw-Hill, New York, 1975.

[51] S. B. Pope. *Turbulent Flows*. Cambridge University Press, Cambridge, UK, 2000.

[52] S. Tavoularis. Turbulent flow. In J. Saleh, editor, *Fluid Flow Handbook*, Chap. 31, pp. 31.1–31.33. McGraw-Hill, New York, 2002.

6 Fluid mechanical apparatus

Measurement in fluid mechanics encompasses two types of physical flow systems: those in which the flow is set by factors beyond the control of the experimenter, for example environmental flows and flows in existing industrial facilities; and those that are designed and built in the laboratory and over which the experimenter has full, or at least limited, control. This chapter is concerned with laboratory apparatus, particularly the type intended to generate flows ideally matching theoretical concepts (e.g., a uniform stream or a fully developed axisymmetric jet) and, to a lesser degree, those intended to reproduce a complex technological flow. The importance of using properly designed and operating flow apparatus can never be overemphasized. It is obvious that experimental skill and good instrumentation cannot compensate for flaws in the flow itself. Most commonly, experiments in a fluid mechanics laboratory would utilize available apparatus. Even then, understanding of the function of various components would be very useful, in case performance improvements or modifications to suit specific needs become necessary. When building new apparatus is considered, awareness of various options available and careful comparison of the corresponding advantages and disadvantages become absolutely essential. The following sections summarize the basic elements of flow apparatus, directing the reader to selected references for design details.

6.1 Producing the desired flow

The centrepiece of all fluid mechanical apparatus is the *test section*, or *working section*, in which measurements are taken. All others devices and components of the apparatus serve the purpose of producing the desired flow in the test section. This usually requires considerable effort, skill, and expense. A common type of fluid mechanical apparatus is the *general-purpose wind tunnel*, whose test section would ideally contain a uniform, disturbance-free stream. Other types of apparatus may be designed to produce a certain type of turbulent flow or flows in which variations of other properties, such as temperature and concentration, are present. In all cases, it is necessary to include components capable of *flow generation* and *flow management*. Flow generation is normally accomplished with the use of a *fluid mover*, such as a blower or a pump. Fluid movers create significant disturbances, which must be removed before components are deliberately introduced that would generate the desired flow field. An alternative approach of producing a flow is to store a substantial amount of fluid in a *container under pressure* and then release it towards the test section. For water and other liquids this container would be a head-tank, whereas for gases it would be a sealed pressure vessel. A common component of fluid mechanical apparatus is the *settling chamber*, or *plenum*, which is a container upstream of the test section with dimensions much larger than those of the test section itself. As its name indicates, the settling chamber is a relatively low-speed region, in which the fluid is allowed to free itself, at least partially, of upstream disturbances. Then the flow must be accelerated to the test-section speed; this is accomplished with the use of a *contraction*, which is a smooth transition duct that reduces the cross-sectional area of the apparatus. Additional transition pieces, coupling the settling chamber with the flow mover or other intermediate sections, are also required; these are called *diffusers*, as they result in a cross-sectional area increase in the direction of the flow. Specialized components may also be required for particular types of facilities, for example supersonic wind tunnels. Flow management is accomplished with the use of distributed obstructions positioned across the stream in the settling chamber or other parts of the apparatus. These include *flow straighteners* and *screens* of various sorts, whose purpose is to reduce non-uniformity and turbulence in the stream. The same type of devices can be used to generate flows with specified non-uniform characteristics.

An important decision in the early stages of designing flow apparatus is whether to make it open or closed circuit. Open-circuit systems release the test-section fluid into the environment, which, in a broad sense, constitutes the return. At times, open-circuit operation becomes necessary, as in the case of gas flow loops supplied from pressurized tanks. Such an operation has the drawback of wasting fluid mass as well as kinetic energy, which could be a factor if kinetic-energy loss is significant compared with other energy losses. In some cases, open-circuit facilities may be preferable as occupying less laboratory space and generally subject to lower disturbance levels, compared with equivalent closed-circuit facilities. On the one hand, when the flow is supplied with heat or chemical contaminants or contains phase changes or chemical reactions, open-circuit operation removes the burden of restoring the fluid to its original state. On the other hand, closed-circuit facilities are used more commonly as they have the advantage of full control over the entire fluid volume, thus allowing operation of the apparatus independently of its surroundings.

6.2 Changing the flow area

Sudden area expansions and reductions are generally avoided in fluid mechanical apparatus because they cause energy losses as well complications in the flow pattern, including flow separation, recirculation, and, sometimes, unsteadiness. Instead, area decreases and increases are accomplished gradually, with the use of contractions and diffusers, respectively.

Contractions: A *subsonic contraction* [Fig. 6.1(a)] decreases the flow cross-sectional area from A_1 at its entrance to A_2 at its exit, thus accelerating the flow entering the test

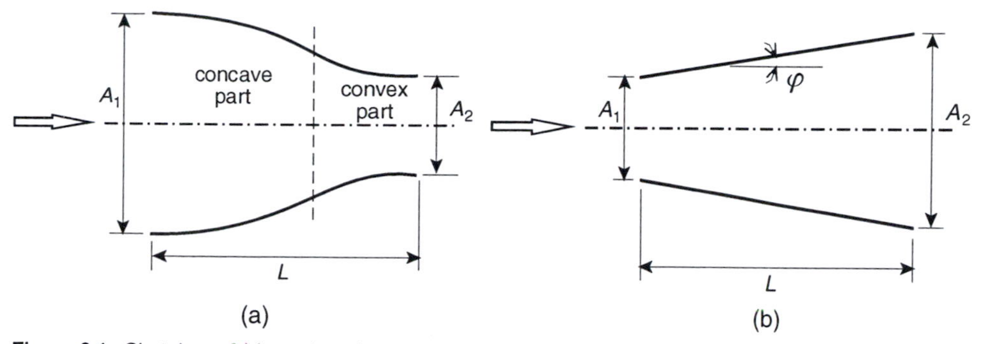

Figure 6.1. Sketches of (a) a subsonic contraction and (b) a subsonic diffuser.

section. The main parameter of a contraction is the *contraction ratio* $c = A_1/A_2 > 1$ Moreover, contractions are characterized as *axisymmetric*, *square*, or *rectangular*, depending on their cross-sectional shape. A contraction serves two important purposes: It tends to make the flow more uniform and it reduces turbulence intensity. This can be demonstrated by simple inviscid flow analysis, originally performed by L. Prandtl [1], as follows. Assume that the entrance and exit streams are parallel to the contraction axis, so that the entrance and exit static pressures are uniform, and neglect friction, gravity, and flow mixing. Consider that the entrance flow velocity is equal to U_1, except over some part in which it is slightly disturbed to the velocity $U_1 + \Delta U_1$, where $|\Delta U_1| \ll U_1$. The exit stream will have a uniform velocity U_2, except in the disturbed part in which the velocity will be $U_2 + \Delta U_2$. Application of continuity and Bernoulli's equations in the two streams and linearization of the corresponding relationships provides the expression

$$\frac{\Delta U_2}{U_2} \approx \frac{1}{c^2}\frac{\Delta U_1}{U_1}, \tag{6.1}$$

which proves that streamwise disturbances, normalized by the local speed, would be reduced by a factor inversely proportional to the square of the contraction ratio. For a modest value of $c = 10$, relative disturbances at the exit would be reduced to a mere 1% of their entrance level. Expression (6.1) would also apply approximately to the streamwise turbulence intensity (defined as the standard deviation of the streamwise fluctuations divided by the local mean velocity), which would thus be significantly dampened by the contraction. A more detailed study [2] predicts that the ratio of entrance-to-exit streamwise turbulence intensities would be equal to

$$\left(1/c^2\right)\sqrt{(3/4)(\log 4c^3 - 1)} \tag{6.2}$$

rather than $\left(1/c^2\right)$, as predicted by expression (6.1). The difference between the two expressions is about 40% for $c = 10$ and 60% for $c = 20$, which is not too much considering the rough purposes for which such expressions are used. On the other hand, contractions may actually introduce or amplify other types of disturbances. Transverse velocity fluctuations will actually increase in magnitude along a contraction, whereas the transverse turbulence intensity at the exit will be reduced only to $\sqrt{3/(4c)}$ of its entrance level [2]. Furthermore, density non-uniformities in the entrance stream would generate velocity non-uniformity at the exit. For a rough estimate, assume that the entrance flow has uniform velocity U_1 and consists of a main part with uniform density ρ and a disturbed part with density $\rho + \Delta\rho$. Making simplifying approximations as previously, one can calculate that the exit flow will contain a velocity non-uniformity ΔU_2 given by [1]

$$\frac{\Delta U_2}{U_2} \approx -\frac{1}{2}\frac{\Delta\rho}{\rho}\left(1 - \frac{1}{c^2}\right). \tag{6.3}$$

For example, a positive temperature disturbance of 5 K at the entrance would generate a positive velocity disturbance of the order of 1% at the exit.

Contraction design is an art rather than a science, and, whenever possible, it is advisable to copy designs that are known to perform well under comparable conditions. The contraction ratio c should be selected to be as large as possible, within the constraints of available laboratory space, fabrication capabilities, and budget. A value of $c \geq 16$ would generally be sufficient, although values near 10 are also commonplace. The cross-sectional shape is usually dictated by the shape of the test section or other practical needs, and it is not based on fluid mechanical considerations. Axisymmetric shapes are the choice for contractions leading to circular test sections, whereas rectangular shapes are easier to fabricate and should be preferred when the test section is rectangular. Note that the aspect ratio (height-to-width ratio) of rectangular contractions may be varied along the flow direction to match the dimensions of adjacent sections. The most important factor for contraction performance is the shape of its walls, as it determines the wall-pressure variation and thus the state of the boundary layers. All contractions consist of a concave section towards the entrance, followed by a convex section towards the exit. Monotonically decreasing wall pressure may be achieved only with infinitely long contractions, whereas finite-length contractions are subject to local adverse pressure gradients in both the concave and convex sections, particularly in the concave one; these correspond to the well-known phenomena of 'overshoot' and 'undershoot' for the potential flow wall velocity. Contraction profiles, optimized to eliminate boundary-layer separation as predicted by the Stratford separation criterion [3], have been presented for both axisymmetric [4] and rectangular [5] contractions, whereas shapes optimized to produce the shortest possible contraction that meets the separation criterion have also been proposed [6]. These profiles usually consist of two or more circular, elliptical, cubic, or other polynomial-type curves, matched tangentially to each other and to the axial direction at the entrance and exit. As practical guidelines, one is advised to maintain the concave section considerably longer and more gradual than the convex one and to select the contraction length L approximately 50% larger than the average entrance dimension. In fact, the exact shape of the profile is not particularly important, and the experienced designer may exercise judiciousness in respecting these rules while making a totally empirical choice. To be on the safe side, however, when novel designs are used, it seems worthwhile to compute the wall-pressure distribution by use of potential flow simulation and then to verify that a boundary-layer separation criterion is respected.

Diffusers: A *subsonic diffuser* [Fig 6.1(b)] is a duct that gradually leads to a larger flow area, thus allowing static-pressure recovery by reduction of the fluid kinetic energy and increasing the efficiency of operation of the apparatus. Flow facilities may contain more than one diffuser. Diffusers are commonly placed between the flow mover and the settling chamber but also following the test section. In open-circuit facilities, a diffuser placed at the test-section exit increases the test-section speed, acting as suction. In blowdown wind tunnels, an exit diffuser also decreases kinetic-energy losses and laboratory drafts. An optimal diffuser would accomplish the required area change by using the shortest possible length to minimize frictional energy losses and space usage. On the other hand, diffusers generate an adverse pressure gradient, which induces boundary-layer separation when sufficiently large. Although shape optimization would lead to non-linear diffuser wall profiles, convenience in fabrication almost always dictates a straight wall, which essentially limits diffusers to conical ones or those with rectangular cross sections. A number of studies have dealt with flows in diffusers, particularly addressing the phenomenon of *stall*, namely boundary-layer separation from the diffuser wall [7–9]. Depending on the diffuser geometry and flow conditions, stall may occur either intermittently, resulting in the formation and detachment of large-scale vortices that would be transported downstream, creating unsteady disturbances to the flow, or steadily, in part of or in the entire wall. To prevent stall of low-speed diffusers, the angle φ between the wall and the diffuser axis must be kept below about 5°, and even less if the boundary layers at the diffuser entrance are particularly thick or if the diffuser length is more than 10 times the inlet height. The area increase that can be achieved by single small-angle diffusers is limited to about 3–4. In cases of highly non-uniform or turbulent flow entering a diffuser leading to a settling chamber, it is preferable to divide the diffuser into sections with flow management devices (see Section 6.3) inserted between sections. When a large area increase is required or the available length is limited, a *wide-angle diffuser*, with a half-angle as large as 70°, may be used. This diffuser will definitely separate unless fitted with a number of pressure-reducing screens, which would counteract the adverse pressure gradient [10]. In extreme cases, the flow discharges directly into the settling chamber, in the form of a *sudden expansion*; this situation is to be avoided, if possible, as it leads to total loss of kinetic energy and to a jet-like flow, which would require substantial effort for its management.

6.3 Flow management

An important parameter of the performance of a flow management device is the *pressure-loss coefficient*

$$K = \frac{\Delta p}{\frac{1}{2}\rho V^2}, \tag{6.4}$$

where Δp is the pressure drop across the device and V is the flow velocity through it. The higher the value of this coefficient, the stronger the influence of the device on the flow. On the other hand, this coefficient is proportional to the energy loss caused by the device, and so higher values of K correspond to lower energy efficiency of the apparatus. To reduce the associated pressure losses, flow management devices are usually positioned in the settling chamber or other wide sections of the apparatus, in which the flow speed is relatively low. In incompressible flows, the pressure-loss coefficient depends on the shape of the device and the Reynolds number, whereas the Mach number would also be a factor in compressible flows. Flow management devices usually consist of regularly arrayed identical geometrical elements (e.g., cylinders, disks), which can be characterized by two parameters:

- the *spacing* of elements; for square elements, this spacing is called the *mesh size* M; for simplicity, this term is used to characterize all shapes of flow management devices, with the understanding that it denotes an average spacing of obstruction elements
- the *solidity* σ of the device, defined as the ratio of the projected blocked area to the total projected area; alternatively, one may use the *porosity*, which is equal to $1 - \sigma$

Flow straighteners: When a wind tunnel or other flow apparatus has corners or other sections in which the flow direction changes, it is a good practice to guide the flow to the new direction with the use of *turning vanes* [11, 12]. Properly designed vanes eliminate flow separation and secondary flows that are due to streamline curvature, at the expense of some pressure drop. Rolled or bent plates, with ends aligned with the desired flow directions, are the simplest design of turning vanes and have been used extensively in closed-circuit wind tunnels, whereas the use of optimized airfoils could result in reduced pressure losses [13].

The most common device for straightening a non-parallel flow within a duct is the *honeycomb*, which consists of an array of cellular channels, usually of hexagonal cross-sectional shape (other shapes are also available), through which the flow is forced to pass. Honeycomb materials include aluminium and various alloys as well as paper and plastics. Cell sizes range typically between 0.5 and 20 mm and lengths vary from a few millimetres to considerably longer [14]. Wall material is quite thin, resulting in low solidity, in the range between 1% and 5%. For small facilities, flow straighteners can be made by packing plastic drinking straws or other thin-walled tubes. In general, honeycombs pose small pressure losses to the flow; as a representative example, the pressure-loss coefficient for an aluminium honeycomb with a cell size of 6.4 mm, a length of 25.4 mm, and a solidity of 4% is $K \sim 0.4$. The main function of flow straighteners is to obstruct transverse velocity components, including swirl, thus generating a flow that is nearly parallel to their walls. Honeycombs are usually positioned near the entrance of suction tunnels or closely after the flow mover in order to straighten non-parallel flows. As long as they have a low K value, they are ineffective in removing streamwise velocity non-uniformity present in the stream. The effect of honeycombs on flow turbulence depends on both the honeycomb and the turbulence characteristics; they are most effective when used in combination with screens [15–19].

Screens and similar devices: A *screen* per se is a thin, fine mesh, made of metal, plastic, or other material and fabricated by weaving, welding, etching, or another method; such a screen is also referred to as a *gauze*. More generally, the term screen may also be used to describe *grids* of parallel rods, *perforated plates*, honeycombs, foam and fibre sheets, loosely woven fabrics, and other relatively thin flow obstructions with a flow resistance and mounted across the flow stream. Usually screens are plane and inserted normal to the flow direction, but they may also be used in inclined or curved configurations. This subsection is concerned with the use of screens to improve flow uniformity and to reduce turbulence; some other uses are described in Section 6.6. For more details, the reader is advised to consult the extensive literature that is available on screens and turbulence management [1,9,18,20,21].

The pressure-loss coefficient K depends on the geometrical solidity σ of the screen, the Reynolds number (with a diminishing effect as this number increases), and the shape of the screen's elements. For example, a grid of parallel square rods would have a K larger than that of a grid of circular rods with the same solidity, because the wakes of the square rods are wider than those of circular ones. Thus an *effective solidity*, taking into account the geometrical shape and even material of the screen, seems to be more appropriate to characterize its dynamic performance. Among the various relationships that have been suggested for square-mesh, woven gauzes, one may use as representative the following one [22]:

$$K = \left(0.52 + \frac{17}{\mathrm{Re}_d}\right) \frac{\sigma\,(2-\sigma)}{(1-\sigma)^2}, \tag{6.5}$$

where $\mathrm{Re}_d = Vd/\nu$ is the Reynolds number based on the thickness d of the wire.

A uniform screen normal to a flow with a spatially varying velocity whose scale of non-uniformity is larger than the mesh size tends to reduce the peaks of velocity, while increasing the valleys, thus making the flow more uniform. This can be easily explained as follows. The drag force on one mesh of the screen will be

$$F_D = \frac{1}{2}\rho K M^2 V^2. \tag{6.6}$$

Thus a mesh immersed in higher speed will have a higher drag force than a mesh immersed in lower speed and will tend to be decelerated more. Because continuity requires that the velocity averaged over the entire screen remain constant, the low-speed regions of the flow will actually be accelerated across the screen. As an example, the mean velocity gradient of incompressible flow was found to decrease to about 60% of its upstream value when crossing a fine screen with solidity $\sigma = 0.3$ and to about 40% of its upstream value when the solidity increased to $\sigma = 0.4$ [22]. One must also be aware that two or more screens would have a stronger reduction of non-uniformity than a single screen with a pressure-loss coefficient equal to the sum of those of the individual screens used separately. Thus multiple screens of relatively low solidity would be preferable to a lower number of high-solidity screens. At any rate, it is recommended that the solidity not exceed the value of about 0.40 to 0.45, because, at high solidities, screens tend to produce large-scale flow non-uniformity [1].

A more thorough understanding of screen–flow interaction is required for using screens effectively for free-stream turbulence reduction. The flow closely downstream of screens consists of a cluster of jets and wakes, which, within the Reynolds number range of most fluid mechanics experiments, are unstable and break down to turbulence that tends to be nearly homogeneous and isotropic. Thus a screen necessarily generates turbulence; however, this turbulence receives no energy away from the screen and decays with distance from the screen. The screen-generated turbulence kinetic energy k, normalized by the mean velocity $\overline{V}$, decays following the empirical law

$$\frac{k}{\overline{V}^2} \sim a \left(\frac{x_1 - x_0}{M} \right)^n, \quad \frac{x_1}{M} > 10, \tag{6.7}$$

where the coefficient a depends on the geometry of the screen, x_1 is the distance from the screen, x_0 is an empirical effective origin, typically equal to $3M$, and the exponent n is in the range -1.2 to -1.4. It is generally accepted that, for $x/M > 500$, screen-generated turbulence is negligible. A more tolerant limit for flow facilities that have significant disturbances from other sources would be $x/M = 100$.

Why then are screens used to reduce turbulence? Screens are usually positioned in parts of the apparatus in which the turbulence of oncoming flow has dominant eddies with relatively large size, comparable, for example, with the dimensions of the blades of the fluid mover. As these eddies cross a screen with a substantially smaller mesh size, they are broken down to smaller sizes. The smaller the eddies, the more they are subjected to viscous dissipation (i.e., reduction of local velocity differences by friction) and the faster they decay. The optimal range of screen mesh sizes that produce effective eddy breakdown extends to about 1 or 2 orders of magnitude less than the scale of oncoming turbulence, whereas much finer screens of the same solidity would have a lesser effect. Thus a strategy emerges on how to reduce flow turbulence with the use of screens. To reduce pressure losses, screens must be positioned in low-speed sections of the apparatus, preferably the settling chamber or the downstream half of a diffuser. Typically four or more screens should be selected, arranged in order of diminishing mesh size and with sufficient distance from each other for the turbulence generated by each screen to decay before the next screen is approached. To avoid generating flow non-uniformity, the screens must be of good quality and free of defects, kinks, and bulges, which could have occurred for example, when the screen was rolled for shipping. They must be maintained very taught, with uniform tension applied on all sides. During operation of the apparatus, the screens must be maintained clean and rust free; this may require periodic inspection, cleaning, and maintenance.

6.4 Wind tunnels

A wind tunnel is a duct containing flow of air or another gas. Wind tunnels are commonly used for the study of air flow past models of aircraft, vehicles, and structures as well as for fundamental fluid mechanical experiments, such as studies of vortices, the structure of turbulent flows, and turbulent diffusion. There are a great number and variety of wind tunnels around the world, including many large-scale and specialized national facilities [9, 23–27]. This section is concerned with a few general-purpose wind-tunnel designs, of the kinds that can be designed and built with a moderate amount of effort and resources [11, 28–30] and likely to be found in an academic institution or a small industry. All types subsequently discussed are classified as *low-speed wind tunnels*, which means

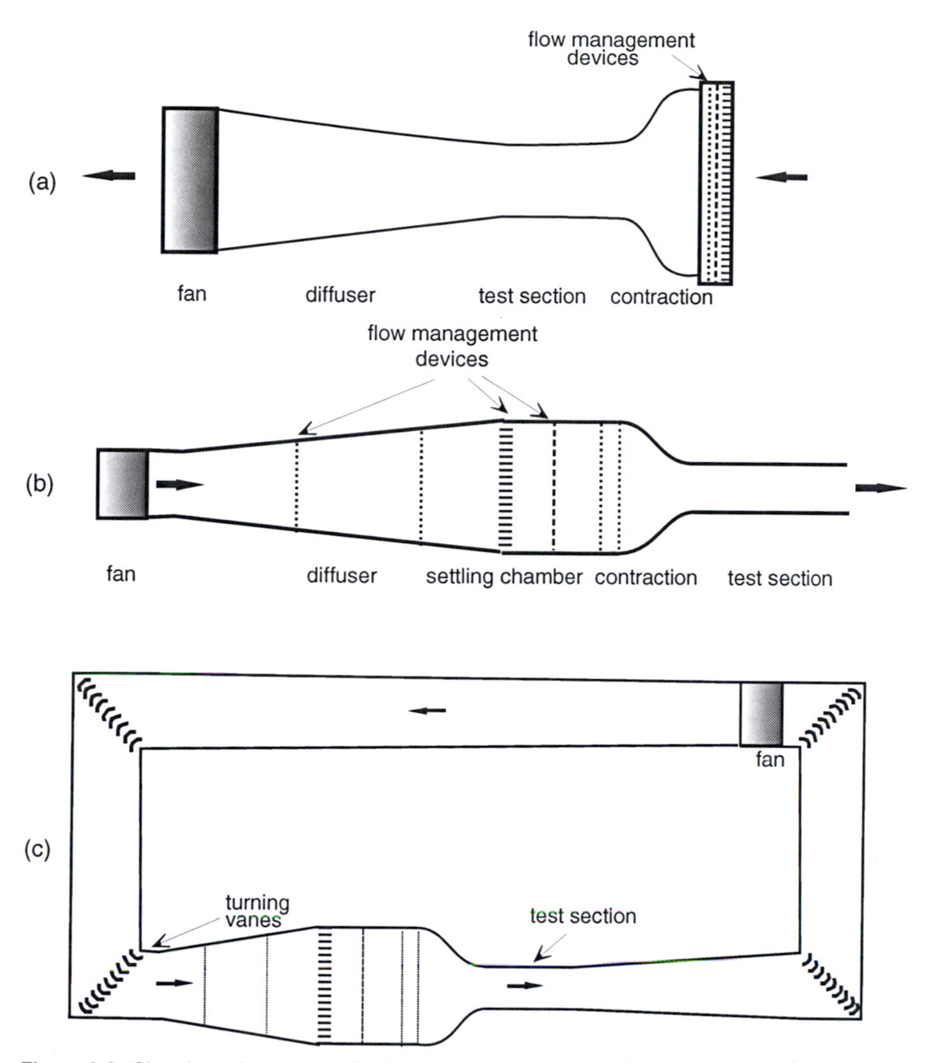

Figure 6.2. Sketches of representative low-speed wind tunnels: (a) suction tunnel, (b) blowing tunnel, and (c) closed-circuit tunnel.

that their maximum speed would be lower (or, in many cases, considerably lower) than about 100 m/s, so that compressibility effects would be negligible or very small. Such facilities can also be classified as small or medium size, having test-section cross-sectional areas in the range between a fraction of to a few square metres. Representative designs of such wind tunnels are subsequently listed.

Suction wind tunnels: These consist of a straight-line, open-circuit arrangement, in which air is drawn from the surroundings into the test section through suction generated by a fan located downstream of the test section [Fig. 6.2(a)]. The flow usually passes through an air filter, a honeycomb, and screens before entering the contraction that leads to the test section. The latter is connected to the blower section through a diffuser in order to reduce kinetic-energy losses. Compared with other types, suction tunnels are easier to design, less expensive to build, and occupy less laboratory space. Although free of disturbances that are caused by an upstream fan, they are vulnerable to disturbances that are due to external obstructions and large-scale recirculation in the room, as there is little one can do to control the effects of such problems when they occur.

Blowing wind tunnels: Like suction tunnels, blowing tunnels are also arranged in a straight-line, open-circuit configuration, with the difference that the fan is positioned upstream of the test section [Fig. 6.2(b)]. Although this arrangement may appear to result in more disturbances than the previous case, it creates the possibility for better flow control, with the use of a settling chamber and numerous flow management devices. On the other hand, it requires additional power for covering the significant pressure losses that occur in such devices.

Because open-circuit wind tunnels freely communicate with their surrounding space, whether it is the open atmosphere or an enclosed laboratory, they are subjected to contamination by dust and other impurities. For the same reason, they are generally unsuitable for studies requiring injection of smoke, seeding particles, or other contaminants. On the positive side, the use of a large volume of air as a return nearly eliminates the temperature increase that is due to accumulation of self-heating effects, thus eliminating the need for a cooling system.

Closed-circuit wind tunnels: Compared with open-circuit tunnels, closed-circuit tunnels feature higher power efficiency, reduced noise levels, flow containment, and better flow management. For these reasons, most large wind tunnels are of the closed-circuit type [Fig. 6.2(c)]. Many variants of this design are available, some suitable mainly for conventional aerodynamic studies, others offering special capabilities. Many contain a refrigeration unit to remove the self-heating load, which would otherwise introduce temperature differences, leading also to flow non-uniformity.

Specialized facilities: As an indication of the variety of specialized wind tunnels and related facilities, the following examples are listed:

Open-test-section wind tunnels, also called *Eiffel-type* wind tunnels. The test section in these tunnels is not bounded by solid walls but is in the form of a free jet, in which the models are inserted. Compared with wind tunnels with enclosed test sections, they offer the advantage of being free of wall effects (see Section 6.7). Open test sections can be used in closed-circuit configuration by being surrounded by a much larger enclosure leading to the return. To eliminate the need for separate blowers, modern large-scale wind tunnels of this type are usually combined with conventional types by use of flow gates and diverters.

High-speed wind tunnels, providing compressible air flow; these include transonic, supersonic, and hypersonic tunnels and require the addition of a converging–diverging nozzle to produce the desired Mach number in the test section.

Pressurized wind tunnels, which maintain the test-section pressure at levels substantially above the atmospheric pressure, by either releasing the exhaust air through a throttling valve or by being entirely enclosed within a pressurized vessel; the advantage of pressurization is the increase of the flow Reynolds number that is due to increasing air density.

Cryogenic wind tunnels, which maintain a gas stream at an extremely low temperature by injection of the same gas in liquid form; this also results in an increased Reynolds number that is due to the reduced viscosity, while pressurization keeps the density at high values.

Meteorological wind tunnels, which simulate the characteristics of the lower atmospheric boundary layer and are suitable for wind engineering studies.

Low-temperature wind tunnels, specialized for the study of low-temperature flow phenomena, such as icing on aircraft wing and power cables.

Combustion wind tunnels, with wall materials suitable for the study of flames and other high-temperature or corrosive phenomena.

Vertical or short take-off and landing (V/STOL) wind tunnels, which require a particularly large test section and are usually combined with conventional wind tunnels.

Shock tubes, for studies of compressible flows and the effects of shock waves on various flow phenomena [31].

A variety of devices in which relative motion is produced by moving the model and instrumentation through still air, rather than placing them in an air stream; these include *whirling arms*, *towing systems* mounted on rails, *rotating frames*, and *swinging arms* (pendulum-like devices).

Free jets, issuing from nozzles and supplied with air from a fan or a compressed air tank; these are particularly suitable as calibration rigs for velocity and temperature transducers.

6.5 Water tunnels and towing tanks

These are facilities for which water is used as a medium, although other liquids have also been used for specific purposes.

Water tunnels: With the possible exception of some very small flow rigs, water tunnels are normally closed circuit (see example in Fig. 6.3). They can be subdivided into entirely enclosed facilities, which can be properly called water tunnels, and facilities with a free surface, often called *water channels* or *flumes*. Water tunnels consist, in principle, of the same types of components as wind tunnels, namely a flow mover (pump), a settling tank, a contraction, diffusers, and flow management devices, adapted to the use of water rather than air [24,30,32–34]. Besides general-purpose facilities used for aerodynamic–hydrodynamic and fundamental fluid mechanical studies in single-phase homogeneous water flow, the following specialized types are also instructive to mention.

Stratified flow channels, usually with stable density stratification (i.e., heavier fluid layers below lighter layers), produced by varying the salinity or temperature of different layers; closed-loop operation requires prevention of vertical mixing;

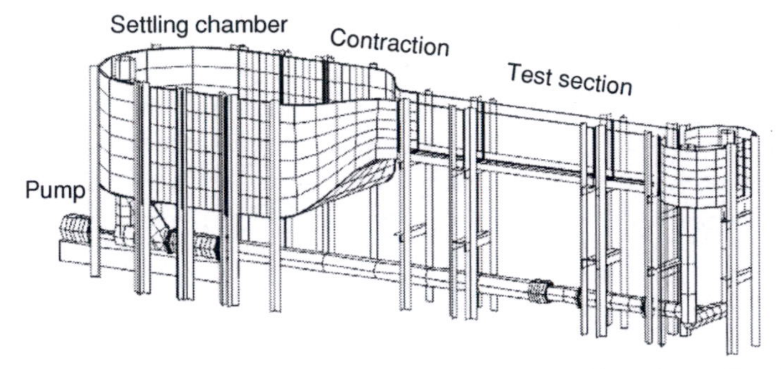

Figure 6.3. Sketch of a recirculating-flow water tunnel (drawing by Ben Kislich-Lemyre).

examples of such facilities include a design in which horizontal flow is produced through viscous actions with the use of two sets of meshing, rotating disks [35] and a design in which different horizontal layers are skimmed separately and recirculated through separate pumps [36].

Low-Reynolds-number facilities, containing a liquid with very high viscosity, such as glycerol or silicon oil; temperature control may be required in these facilities because of the relatively large viscous dissipation of kinetic energy and the strong temperature sensitivity of viscosity.

Matching refractive-index facilities, containing a liquid whose refractive index matches that of plastics, thus enabling the visual and optical studies of flows past and through complex boundaries.

Cavitation tunnels, which include a mechanism capable of reducing the static pressure in the enclosed test section below atmospheric level to facilitate cavitation.

Towing tanks: Compared with water tunnels of the same size and operating at the same relative velocity between fluid and towed object, towing tanks [24, 37] have the advantages of lower power consumption, lower free-stream disturbances, and absence of boundary-layer effects. Because of their low background turbulence, towing tanks are particularly suitable for studies of hydrodynamic stability and transition to turbulence. Moreover, stratified and multilayered inhomogeneous liquids can be most conveniently produced when still. In combination with two- or three-dimensional traversing and/or rotation of towed objects, towing tanks offer the possibility of simulating complex object paths and unsteady flows, as it is much easier to control the motion of towed objects and instrumentation than that of a flowing fluid. The main disadvantage of towing tanks is their finite length, which limits the testing time for each experimental run. They also impose some 'dead time' between runs in order to allow the liquid to acquiesce. Inevitably, towing also generates mixing, which is a limiting factor for studies in inhomogeneous and stratified fluids. The carriage could be simply made to slide on a smooth track, driven by a falling weight through a pulley. Most commonly, it employs an electric motor, preferably one with accurate speed control, which turns a rubbing wheel or a pinion gear. A point of concern in many experiments is to minimize disturbances of the motion that are due to carriage vibration; the use of continuous lubrication and backlash-free gears would certainly help in this respect. *Ripple tanks* are shallow tanks dedicated to the study of surface waves, as models of other types of wave motion.

Surface waves: A common problem with all facilities that contain a liquid with a free surface is the generation of travelling surface waves, which get reflected at the end of the facility and travel back and forth along the test section, thus creating a periodic disturbance to the flow. The travelling speed of surface waves is [38]

$$c = f\lambda = \sqrt{\frac{g\lambda}{2\pi} \tanh \frac{2\pi h}{\lambda}}, \tag{6.8}$$

where f is their frequency, λ is their wavelength, and h is the depth of the channel, assumed to be uniform. For relatively shallow channels ($h < 0.07\lambda$) one gets a simplified expression for the wave speed, as

$$c \approx \sqrt{gh}, \tag{6.9}$$

which is independent of wavelength. Given an average depth and the length l of the facility, one may estimate the period of disturbance as

$$T \approx \frac{2l}{\sqrt{gh}} \tag{6.10}$$

Shallow-water surface waves are difficult to eliminate, but one may neglect their effects if their amplitude is very small and their period is much larger than the time scales of interest in a particular experiment.

6.6 Turbulence and shear generation

Flow management devices described in the previous section may also be used to generate flows with a desired variation in the mean velocity and/or a desired turbulence structure. Only a few relatively simple examples are presented, as it would be impossible to foresee the great variety of possible configurations that one may wish to recreate in the laboratory.

Nearly isotropic turbulence: The statistically simplest type of turbulence is homogeneous and isotropic, namely a flow whose statistical moments and other averaged properties are independent of location and orientation. This is a flow that has been studied extensively on its own merit, but also a good environment for testing the effect of turbulence on a great variety of flow-related phenomena, for example on separation from a wing or on combustion efficiency [39, 40]. It can be approximately realized in wind and water tunnels by the passing of a stream through a normal array of periodically spaced obstructions, such as a grid, perforated plate, or honeycomb, although the most commonly used devices are grids of parallel rods or biplanar square-mesh grids. The two important macroscopic parameters of isotropic turbulence are its kinetic energy k (or, equivalently, its *intensity*, usually defined as the standard deviation of the streamwise velocity component normalized by the mean velocity) and its integral length scale.

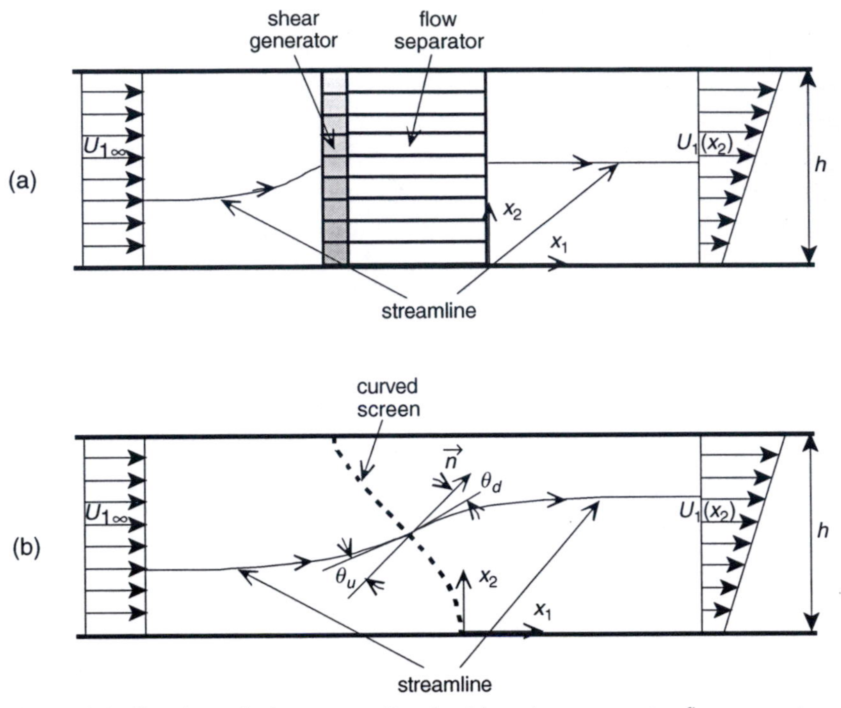

Figure 6.4. Sketches of shear generation by (a) a shear generator–flow separator combination and (b) a curved screen; boundary layers have been disregarded, for simplicity.

The kinetic energy of grid-generated turbulence decays according to law (6.7); however, the coefficient a increases with increasing solidity and also depends on the design of the grid. A preferable solidity range is 0.30–0.38, which should not be exceeded substantially to avoid flow unsteadiness and large-scale non-uniformity. The mesh size should be maintained sufficiently small for wall effects to be negligible in the core of the stream; it is recommended to maintain the mesh size at less than about 1/10 of the smallest transverse dimension of the test section, considering that the integral length scale of isotropic turbulence starts from a value comparable in order of magnitude with the mesh size and grows downstream following a power law similar to law (6.7) with an exponent between 0.4 and 0.5.

Uniformly sheared flow: A rectilinear flow with a constant transverse velocity gradient (*shear*) is the simplest type of shear flow and is often used as an idealized environment to study the effects of shearing on a variety of phenomena. Also known as *Couette flow*, it is the paradigm used to illustrate the concept of viscosity, representing the laminar flow in the narrow gap between two infinite parallel walls, one of which is fixed and the other of which moves parallel to itself with a constant velocity [41]. By definition, the velocity of such a flow in a channel with a rectangular cross section having a height h would be (see Fig. 6.4)

$$U_1 = U_{1c} + \frac{\mathrm{d}U_1}{\mathrm{d}x_2}\left(x_2 - \frac{h}{2}\right), \tag{6.11}$$

where $dU_1/dx_2 = \text{const}$ and U_{1c} is the centreline speed. The strength of shearing is measured by the value of the *shear parameter*

$$\beta = \frac{h}{U_{1c}} \frac{dU_1}{dx_2}, \tag{6.12}$$

which, in the absence of flow reversal, is limited to the range between 0 and 2.

The generation of Couette flow in the laboratory by moving plane walls in the form of belts is possible but rather impractical, whereas coaxial cylinders, at least one of which rotates (*circular Couette flow*), are mostly utilized for the measurement of viscosity or the study of rotation effects. The most popular way of generating shear is by the passing of a parallel stream through a non-uniform transverse obstruction, called the *shear generator* [Fig. 6.4(a)]. The great variety of such obstructions that have been used in different laboratories includes planar grids, screens, perforated plates, and airfoil cascades with a varying solidity, produced by variable spacing of identical elements, variable thickness of equally spaced elements, or variable mesh size [22]. Such elements would introduce significant turbulence, which would be sustained and even amplified by production that is due to shear. Mesh size non-uniformity also results in length-scale non-uniformity, which is an undesirable feature if the apparatus is meant to generate homogeneous turbulence. For this reason, it is advisable to utilize a *flow separator*, placed immediately following the shear generator and consisting of a set of evenly spaced plates parallel to the flow, thus separating the cross section into parallel channels. The flow separator helps straighten the flow, which otherwise may start with curved streamlines. Moreover, it imposes its own spacing as an initial scale for the energy containing eddies of device-generated turbulence. The use of a flow separator also introduces the possibility of generating mean shear by supplying each channel with fluid from a separate fluid mover or head-tank. If one achieves variable obstruction by varying the solidity of a perforated plate, one may achieve an approximately uniform shear by utilizing a linearly varying solidity, and then one may improve the shear uniformity by trial-and-error minor adjustments.

As previously mentioned, a shear generator–flow separator type of device would introduce substantial turbulence, much stronger than typical free-stream levels in wind tunnels and unacceptably high for certain types of experiments in which turbulence might obscure shearing effects. A means of producing low-turbulence shear, typically at intensities of about 0.5%–1%, is the use of curved screens [Fig. 6.4(b)]. The determination of the shape of a uniform fine screen that would produce uniformly sheared flow with a desired value of shear parameter β can be achieved by use of idealized theory [20, 42, 43]. According to this analysis, the screen is modelled as a discontinuity in the tangential velocity component, which generates circulation about the screen elements (wires) and thus lift. At the same time, the screen also produces a pressure drop, corresponding to drag and characterized by the pressure-loss coefficient K, referenced to the normal velocity V_n, which, by continuity, remains unchanged across the screen. Thus the screen effectively deflects the velocity vector towards its normal. This is indicated in Fig. 6.4(b) by the reduction of the upstream incidence angle θ_u to the lower downstream value θ_d. For small incidences, the ratio of these angles reaches a finite limit [20,44],

$$\alpha = \lim_{\theta_u \to 0} \frac{\theta_d}{\theta_u}, \quad 0 \leq \alpha \leq 1, \tag{6.13}$$

which depends on the pressure-loss coefficient. Among the available empirical expressions that describe the relationship between α and K, one may use [45]

$$\alpha = \left[\left(\frac{K}{4} \right)^2 + 1 \right]^{1/2} - \frac{K}{4}, \quad 0.7 < K < 5.2. \tag{6.14}$$

Let V_{tu} and V_{td} be the velocity components tangential to the screen just upstream and just downstream of it. Then, one may define the *deflection coefficient*

$$B = 1 - \frac{V_{td}}{V_{tu}}, \tag{6.15}$$

which would normally be in the range between 0 and 1. The following expression was obtained by polynomial least-squares fitting to a theoretical estimate, also including an empirical correction [43], of the shape of a screen that would produce uniform shear with a shear parameter β:

$$\frac{BK}{(2 + K - B)\beta} \frac{x_1}{h} = -0.738 \left(\frac{x_2}{h} \right)^6 + 2.812 \left(\frac{x_2}{h} \right)^5 - 3.839 \left(\frac{x_2}{h} \right)^4 + 2.687 \left(\frac{x_2}{h} \right)^3 - 1.224 \left(\frac{x_2}{h} \right)^2 - 0.0054 \frac{x_2}{h}, \tag{6.16}$$

where the deflection coefficient was estimated as

$$B = 1 - \frac{1}{\sqrt{1 + \sqrt{K}}} \tag{6.17}$$

The value of K in Eq. (6.17) could be based on direct measurement or estimated from Eq. (6.5). This procedure has been applied successfully by several investigators for the generation of shear in wind and water tunnels, with typical values of β of about 0.4 or lower.

Turbulent boundary layers: Experimental realizations of two-dimensional turbulent boundary layers in wind and water tunnels are very useful configurations, both for fundamental turbulence studies as well as for simulations of flows in the lower atmosphere in a variety of applications. In such studies, it is desirable for the turbulent region to be as thick as possible, to maximize the spatial resolution of instrumentation and to permit the use of relatively large models. The generation of naturally thick boundary layers requires a long test section. This has been achieved in specially designed facilities, with the initial growth further augmented with the use of large arrays of roughness elements. Turbulent boundary layers, roughly simulating wind in the lower atmosphere, have also been generated in short test sections by the passing of the flow through large *spires* or strakes, in combination with distributed wall roughness elements [46–50]. A growing boundary layer in a duct with a fixed cross-sectional area would result in free-stream acceleration, thus generating a favourable streamwise pressure gradient. If it is desired to maintain the flow mean pressure uniform, it is necessary to increase the cross-sectional area gradually or to bleed part of the free stream. A common practice in wind-tunnel design is to use *corner fillets* to reduce secondary flows. An increasing cross-sectional area can be achieved by tapering of the fillets or by use of slightly divergent walls. Boundary layers with favourable or adverse pressure gradients can be generated by similar means, most commonly by making adjustments to a flexible wall opposing the one on which the boundary layer grows.

Other shear flows: Variable-solidity shear generators and curved screens can be also used to generate flows with non-uniform mean shear, making it possible, in principle, to simulate any desired two-dimensional velocity profile. A splitter plate separating two streams with different speeds is normally employed to initiate mixing layers, whereas obstructions of various shapes are used to generate wake flows. Considering the large variety of possible configurations, the reader is advised to consult with the available specialized experimental literature before attempting to generate a complex flow field.

6.7 Model testing

The laboratory study of flows through and past various objects often employs the use of geometrically scaled models. In most cases, the model is scaled down, compared with the object of interest (e.g., models of aircraft and hydraulic turbines), but full-size or scaled-up models are also used to permit convenient flow visualization and measurement (e.g., models of insects and micromachines). Scaling normally preserves geometric similarity, although the use of different scales in different directions can be employed in specific studies, as, for example, in hydraulic models of large terrains, in which the vertical system-to-model scale ratio is taken to be larger than the horizontal one.

For an exact correspondence between model studies and actual system properties, it is necessary to respect *dynamic similarity*, which requires that the values of corresponding relevant dimensionless parameters (see Section 1.6) be identical in the model and actual systems. For incompressible single-phase flows of homogeneous fluids, it is normally the Reynolds number value that has to be matched. In the majority of cases, the fluid in the model studies is, by necessity, air or water, namely the same fluid as that used in most actual systems. This, in combination with the use of small-scale models, would necessitate higher-than-actual speeds for the tests, in order to maintain Reynolds number similarity. The generation of adequately high speeds may not be feasible in the available facilities and, even when feasible, it may introduce compressibility effects. In many internal and external flow configurations, Reynolds number effects on properly non-dimensionalized forces, pressures, etc., as well as on flow patterns (e.g., flow separation from sharp edges, vortex shedding from bluff objects), are known to be weak for large Reynolds numbers, and so it may be acceptable to conduct model tests at lower Reynolds numbers. For improved estimates of actual system properties, it may be possible to apply empirical corrections (e.g., corrections for hydraulic turbines and pumps, based on size or Reynolds number) or extrapolate trends to higher Reynolds numbers.

When, in addition to viscous effects, other physical phenomena are important in a flow, one has also to match additional dimensionless parameters. For example, in studies involving a free surface of a liquid, the Froude number must be matched, in addition to the Reynolds number; in high-speed gas flows, it is also the Mach number that has to be matched; when cavitation, surface tension, buoyancy, rotation, or other effects are significant, additional parameters (see Section 1.6) also have to be matched. The matching of more than one parameter may not be possible for scaled models, especially if additional constraints such as use of the same fluid are imposed. In such cases, one is forced to compromise similarity to a weak or incomplete form, and it is best to follow established practices in the particular field of application.

An important effect that distorts the correspondence between model-test measurements in wind and water tunnels and actual systems is the confinement of the flow by solid walls. In this discussion, the case of open-test-section wind tunnels is excluded; information on these may be found in the references subsequently mentioned.

When a solid model is inserted in a duct flow, it blocks part of the stream, which is diverted around the model. By application of the continuity principle, it is easy to see that the flow speed at cross sections intersecting the model or near it should be higher than the *reference velocity* V_∞ in the duct, defined as the flow velocity (assumed to be uniform) in the absence of the model. This effect is referred to as *solid blockage*. Defining an appropriate velocity increase as ΔV_∞, one can also deduce that the various dimensionless parameters should be evaluated by using $V_\infty + \Delta V_\infty$, rather than V_∞, as the velocity scale. Thus, compared with unconfined flow, solid blockage effectively increases all dimensionless pressure differences, forces, and moments, if these are referenced to V_∞. Blockage does not cease at the downstream end of the model, as its wake (which is a low-speed region) also produces a velocity increase in the free stream, an effect known as *wake blockage*. Associated with the velocity increase in the free stream is a static-pressure drop across the model, which increases the drag force by an amount called *longitudinal buoyancy drag* or simply *buoyancy drag*. This additional drag occurs in all confined flows, irrespectively of their orientation with respect to the gravitational direction. A similar drag is produced by flow acceleration that is due to boundary-layer growth on the duct walls; this drag can be removed by equalization of the pressure through deliberately introduced test-section wall divergence.

Confinement also affects flow motion around models, as it tends to restrict streamline displacement away from the model walls, as well as to direct streamlines towards directions parallel to the wall planes. Moreover, it may alter the nature of the flow in ways that would not occur in the absence of walls. For example, the increased effective velocity and resulting increased effective Reynolds number may dramatically affect the flow separation pattern from bluff objects with rounded edges operating under ‘critical’ or near-critical conditions, at which the drag coefficient presents a well-known sudden ‘dip’ [41, 51]. On the other hand, separation from bluff objects with sharp edges is known to be insensitive to Reynolds number changes, as such objects have no critical regime. Pressure drop that is due to blockage in liquid streams may introduce or affect cavitation, whereas, in high-speed gas flows, it may affect shock-wave patterns.

Several types of corrections for blockage effects are available, based on theoretical, numerical, and experimental studies of flows around both streamlined and bluff objects, which usually provide estimates of the velocity ratio $\Delta V_\infty / V_\infty$ [23, 52–55]. For incompressible single-phase flows, it is generally accepted that such effects may be neglected when the projected frontal area of the model is less than 1% of the test-section area, whereas such simple corrections might become inappropriate when the blocked area ratio exceeds 10%–15%.

QUESTIONS AND PROBLEMS

1. Derive expressions (6.1) and (6.3).
2. Consider a wind-tunnel contraction with a contraction ratio c. Two parallel streams of air enter the contraction, the first one with speed U_1 and density ρ, and the second one with speed $U_1 + \Delta U_1$ and density $\rho + \Delta\rho$, where $|\Delta U_1| \ll U_1$. Determine the density difference $\Delta\rho$ required for the flow at the exit of the contraction to have uniform velocity. Neglect, friction, heat transfer, and mixing.
3. Using procedures presented in Ref. 5, design a contraction with a contraction ratio of 10:1, a rectangular cross-sectional shape that maintains a width-to-height ratio of 1.5, and a length-to-initial-width ratio of 1.5. Make any reasonable assumption for the values of other required parameters, considering that the contraction will be used in a medium-size, low-speed wind tunnel. Sketch the shapes of the two wall profiles. Also, assuming that the contraction will be made by rolling contoured cutouts of sheet metal and welding them along the edges, draw plane patterns of the cutouts.
4. Consider the right-angle bend in the return duct of a closed-circuit wind tunnel. To prevent secondary flows, consider stretching inclined plane screens at appropriate positions. The air speed in the duct is 10 m/s and the duct cross section is rectangular. Explain the rationale for this design. Give an example of a proper screen arrangement, specifying the mesh sizes, solidities, locations, and orientations of the screens. Explain a procedure by which you would optimize this arrangement so that the combined pressure loss would be minimized.
5. Consider flow in a water tunnel with a water depth of 1 m and a desired centreline speed of 0.2 m/s. Determine the shape of a curved, woven-wire screen with a solidity of 0.40 and a mesh size of 2.5 mm that would produce a uniformly sheared flow with a velocity gradient of 0.06 s^{-1}.

REFERENCES

[1] S. Corrsin. Turbulence: Experimental methods. In S. Flügge and C. Truesdell, editors, *Handbuch der Physik* [*Encyclopedia of Physics*], Vol. 8(2), pp. 524–590. Springer, Berlin, 1963.

[2] G. K. Batchelor. *The Theory of Homogeneous Turbulence*. Cambridge University Press, Cambridge, UK, 1953.

[3] S. Tavoularis. Flow past immersed objects. In J. M. Saleh, editor, *Fluid Flow Handbook*, Chap. 20, pp. 20.1–20.44. McGraw-Hill, New York, 2002.
[4] T. Morel. Comprehensive design of axisymmetric wind tunnel contractions. *J. Fluids Eng.*, 97:225–233, 1975.
[5] J. H. Downie, R. Jordinson, and F. H. Barnes. On the design of three-dimensional wind tunnel contractions. *Aeronaut. J.*, 88:287–295, 1984.
[6] M. N. Mikhail and W. J. Rainbird. Optimum design of wind tunnel contractions. Tech. Rep. AIAA Paper 78–819, American Institute of Aeronautics and Astronautics, New York, 1978.
[7] G. Sovran and E. D. Klomp. Experimentally determined optimum geometries for rectilinear diffusers with rectangular, conical or annular cross-section. In G. Sovran, editor, *Fluid Mechanics of Internal Flow*, pp. 270–319. Elsevier, New York, 1967.
[8] J. P. Johnston. Diffuser design and performance analysis by a unified integral method. *J. Fluids Eng.*, 120:6–18, 1998.
[9] P. Bradshaw. http://vonkarman.stanford.edu/tsd/pbstuff/tunnel/index.html
[10] M. M. Seltsam. Experimental and theoretical study of wide-angle diffuser flow with screens. *AIAA J.*, 33:2092–2100, 1995.
[11] P. Bradshaw. *Experimental Fluid Mechanics*. Pergamon, Oxford, UK, 1970.
[12] T. F. Gelder, R. D. Moore, J. M. Sanz, and E. R. McFarland. Wind tunnel turning vanes of modern design. Tech. Rep. AIAA-86-0044, American Institute of Aeronautics and Astronautics, New York, 1986.
[13] A. Sahlin and A.V. Johansson. Design of guide vanes for minimizing the pressure loss in sharp bends. *Phys. Fluids A*, 3:1934–1940, 1991.
[14] Hexcel. Honeycomb in air directionalizing applications. Tech. Rep. TSB 102, Hexcel International, Arlington, TX, 1986.
[15] J. L. Lumley. Passage of a turbulent stream through honeycomb of large length-to-diameter ratio. *J. Basic Eng.*, 86:218–220, 1964.
[16] J. L. Lumley and J. F. McMahon. Reducing water tunnel turbulence by means of a honeycomb. *J. Basic Eng.*, 89:764–770, 1967.
[17] R. I. Loehrke and H. M. Nagib. Experiments on management of free-stream turbulence. Tech. Rep. AGARD Rep. No. 598, Advisory Group for Aerospace Research and Development, Neuilly sur Seine, France, 1972.
[18] R. I. Loehrke and H. M. Nagib. Control of free stream turbulence by means of honeycombs. *J. Fluids Eng.*, 98:342–353, 1976.
[19] C. Farell and S. Youssef. Experiments on turbulence management using screens and honeycombs. *J. Fluids Eng.*, 118:26–32, 1996.
[20] E. M. Laws and J. L. Livesey. Flow through screens. *Annu. Rev. Fluid Mech.*, 10:247–266, 1978.
[21] J. Tan-Atichat, H. M. Nagib, and R. I. Loehrke. Interaction of free-stream turbulence with screens and grids: A balance between turbulence scales. *J. Fluid Mech.*, 114:501–528, 1982.
[22] U. Karnik and S. Tavoularis. Generation and manipulation of uniform shear with the use of screens. *Exp. Fluids*, 5:247–254, 1987.
[23] W. H. Rae and A. Pope. *Low-Speed Wind Tunnel Testing*. Wiley, New York, 1984.
[24] S. P. Parker (Editor-in-Chief). *Fluid Mechanics Source Book*. McGraw-Hill, New York, 1987.
[25] National Research Council Canada. Aerodynamics laboratory, 2002, available on-line: http://iar-ira.nrc-cnrc.gc.ca

[26] NASA, Ames Research Center. Wind tunnels, 2003, available on-line: http://windtunnels.arc.nasa.gov/

[27] The Worthey Connection. The wind tunnel connection, 2001, available on-line: http://www.worthey.net/windtunnels/

[28] P. Bradshaw and R. C. Pankhurst. The design of low-speed wind tunnels. In D. Kuechemann and L. H. G. Sterne, editors, *Progress in Aeronautical Science*, Vol. 5, Chap. 1, pp. 1–69. Pergamon, Press, Oxford, UK, 1964.

[29] R. D. Mehta and P. Bradshaw. Design rules for small low speed wind tunnels. *Aeronaut. J.*, 73:443–449, 1979.

[30] R. Gordon and M. S. Imbabi. CFD simulation and experimental validation of a new closed circuit wind/water tunnel design. *J. Fluids Eng.*, 120:311–318, 1998.

[31] G. Briassulis, J. H. Agui, J. Andreopoulos, and C. B. Watkins. A shock tube research facility for high-resolution measurements of compressible turbulence. *Exp. Therm. Fluid Sci.*, 13:430–446, 1996.

[32] G. E. Erickson, D. J. Peake, J. Del Frate, A. M. Skow, and G. N. Malcolm. Water facilities in retrospect and prospect – an illuminating tool for vehicle design. Tech. Rep. NASA Tech. Memo. 89409, NASA, November 1986.

[33] Eidetics International. Flow visualization water tunnels. Tech. Rep., Eidetics International, Inc. Torrance, California.

[34] T. M. Ward. The hydrodynamics laboratory at the California Institute of Technology – 1976. *J. Fluids Eng.*, 1976:740–748, 1976.

[35] G. M. Odell and L. S. G. Kovasznay. A new type of water channel with density stratification. *J. Fluid Mech.*, 50:535–543, 1971.

[36] D. C. Stillinger, M. J. Head, K. N. Helland, and C. W. Van Atta. A closed loop gravity driven water channel for density-stratified shear flows. *J. Fluid Mech.*, 131:73–90, 1983.

[37] M. Gad-el-Hak. The water towing tank as an experimental facility. *Exp. Fluids*, 5:289–297, 1987.

[38] R. L. Panton. *Incompressible Flow*. Wiley, New York, 1984.

[39] G. Comte-Bellot and S. Corrsin. The use of a contraction to improve the isotropy of grid-generated turbulence. *J. Fluid Mech.*, 25:657–682, 1966.

[40] P. E. Roach. The generation of nearly isotropic turbulence by means of grids. *J. Heat Fluid Flow*, 8:82–92, 1987.

[41] R. W. Fox, A. T. McDonald, and P. J. Pritchard. *Introduction to Fluid Mechanics* (6th Ed.). Wiley, New York, 2004.

[42] J. W. Elder. Steady flow through non-uniform gauzes of arbitrary shape. *J. Fluid Mech.*, 5:355–368, 1959.

[43] D. J. Maull. The wake characteristics of a bluff body in a shear flow. In *The Aerodynamics of Atmospheric Shear Flow*, pp. 16.1–16.13. AGARD C.P. 48, Paper 16, Advisory Group for Aerospace Research and Development, Neuilly sur Seine, France, 1970.

[44] G. I. Taylor and G. K. Batchelor. The effect of a gauze on small disturbances in a uniform stream. *Q. J. Mech. Appl. Math.*, 2:1–29, 1949.

[45] J. C. Gibbings. The pyramid gauze diffuser. *Ing. Arch.*, 42:225–233, 1973.

[46] H. P. A. H. Irwin. The design of spires for wind simulation. *J. Wind Eng. Industr. Aerodyn.*, 7:361–366, 1981.

[47] J. Counihan. An improved method of simulating an atmospheric boundary layer in a wind tunnel. *Atmos. Environ.*, 3:197–214, 1969.

[48] J. Counihan. Simulation of an adiabatic urban boundary layer in a wind tunnel. *Atmos. Environ.*, 7:673–689, 1973.

[49] N. M. Standen. A spire array for generating thick turbulent shear layers for natural wind simulation in wind tunnels. Tech. Rep. LTR-LA-94, National Aeronautical Establishment, Ottawa, Canada, 1972.

[50] J. E. Cermak. Wind tunnel design for modeling of atmospheric boundary layers. *J. Eng. Mech. Div. ASCE*, 107:623–642, 1981.

[51] F. M. White. *Fluid Mechanics* (4th Ed.). McGraw-Hill, New York, 1999.

[52] Engineering Sciences Data Unit. Blockage corrections for bluff bodies in confined flows. Tech. Rep. Item No. 80024, Institution of Structural Engineers, London, 1980.

[53] Engineering Sciences Data Unit. Lift interference and blockage corrections for two-dimensional subsonic flow in ventilated and closed wind-tunnels. Tech. Rep. Item No. 76028, Institution of Structural Engineers, London, 1976.

[54] H. C. Garner, E. W. E. Rogers, W. E. A. Acum, and E. C. Maskell. Subsonic wind tunnel data corrections. Tech. Rep. AGARDograph 109, Advisory Group for Aerospace Research and Development, NATO, Paris, October 1966.

[55] V. J. Modi and S. El-Sherbiny. Effect of wall confinement on aerodynamics of stationary circular cylinders. In *Proceedings of the Third International Conference on Wind Effects on Buildings and Structures*, pp. 365–375. Saikon Co., Tokyo, Japan, 1971.

7 Towards a sound experiment

The success of an experiment relies on the skill, experience, and background preparation of the experimenter, as much as it depends on the suitability, quality, and condition of the apparatus and instrumentation. Personal enthusiasm and interest are propitious, but not sufficient, qualities of the good fluid mechanics experimenter. In addition, one must consider time-tested sound practices to be followed while preparing for and performing an experiment. This chapter summarizes such considerations.

7.1 Planning the experiment

When it comes to fluid mechanics experiments, there is no substitute for hard and arduous work. Nevertheless, the task can be lightened significantly by proper preparation and planning. A non-exclusive list of preliminary actions and considerations is as follows:

- Understand the physical problem at hand and study its theoretical aspects.
- Establish the need for experimental work and identify the information that would be desirable to obtain experimentally.
- Clearly define the objectives and the scope of the experiment.
- Identify the ideal apparatus and equipment that would provide the desired information; list and compare alternatives.
- Identify all resources available for your experiment and those that can be used by arrangement with other projects or laboratories. Besides hardware, this list could include space, utilities (e.g., compressed air, water, etc.), technical services, computational resources, and financial support. In cases in which there is more than one suitable item (e.g., wind tunnels or similar pieces of equipment), compare them and rank them in terms of desirability. As a rule, use the simplest, least expensive, and least time-consuming approach that would produce the desired results within an acceptable uncertainty.
- From the previous list, chose the most suitable items and identify what is missing for carrying out your experiment.
- Explore the possibilities of borrowing, leasing, purchasing, or constructing needed apparatus, equipment, and software. Identify and compare suppliers and sources,

obtain quotations and cost estimates, and establish delivery and construction times. Prepare a budget for the project. Consider the need for writing proposals or requests and obtaining approvals.
- Before using any piece of equipment for actual measurement, become familiar with its principle of operation, specifications, range, and general capabilities and limitations.

7.2 Safety

Any laboratory user must, at all times, safeguard without compromise the health of laboratory personnel and the public at large. Laboratory usage is subject to safety legislation and regulations at national, provincial/state, municipal, and institutional levels, and many organizations require safety training before granting permission to work in a laboratory. Besides been familiar with regulations, one must also exercise common sense to prevent accidents and mishaps. Ideally, a laboratory user should be in good physical condition and free of stress and should not act under extreme time pressure. If at all possible, avoid working in the laboratory alone, especially during late hours or holiday periods. Identify alternative nearby locations of room exits, fire extinguishers, fire alarms, and first aid kits. Avoid wearing loose clothing and footwear and unrestricted long hair, which can easily be entangled in moving components or suction ports. Secure gas cylinders and other unstable objects. Particular attention should be given to rotating and reciprocating components, high electric voltages, ungrounded equipment, electric cables, high-pressure releases from valves, hot surfaces and fluids, unpleasant odours, toxic and combustible substances, and sharp or pointed objects. When dealing with lasers or other sources of radiation, follow meticulously all regulations and instructions of use. Wear protective glasses and avoid exposure of skin to radiation, even at low levels. Besides immediate hazards, such as damage to the cornea by exposure to a laser beam, there are long-term effects of low-level radiation, which may have not yet been understood. Release of harmful substances to the environment is strictly prohibited, and their disposal should be made according to the existing regulations. Be aware that older or specialized apparatus and instruments may contain currently regulated substances such as mercury, chlorofluorocarbons, or asbestos. It goes without saying that one must report immediately any accident or hazardous situation.

7.3 Qualitative assessment

As emphasized in Section 7.1, a principle of the experimentalist should be that all required information ought to be collected by the simplest and most economical means available. To illustrate this principle, we consider a few examples of simple and inexpensive experimental methods, which should always precede measurement in any experiment. More often than not, preliminary qualitative assessment has revealed unexpected features and led the experimenter to radically rethink the experiment, in some cases identifying flaws in the objectives and critical limitations of the earlier planning process.

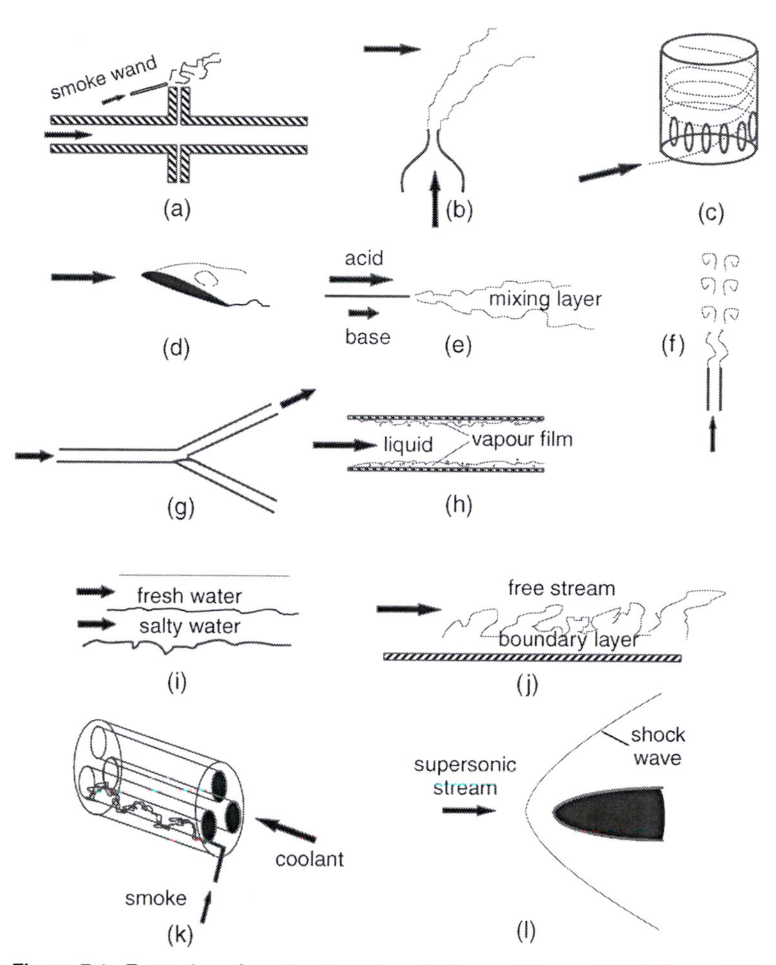

Figure 7.1. Examples of application of qualitative assessment techniques: (a) leakage detection, (b) detection of a draft or cross stream, (c) swirling flow pattern in a Diesel engine, (d) separation bubble over an airfoil, (e) mixing layer of reacting streams resulting in color change, (f) jet instability and break up, (g) flow direction in a fluidic switch, (h) steam–water flow in a heated tube, (i) density stratified flow, (j) turbulent boundary layer, (k) flow pulsations and coherent structures in gap regions of rod bundles, and (l) shock wave in supersonic flow over a bluff object.

Why perform a qualitative assessment? The general objective of a qualitative assessment is to provide some preliminary insight into the overall appearance and characteristics of a flow before engaging in possibly cumbersome, expensive, and lengthy experimentation. It provides an opportunity to reassess the need for detailed measurement, to evaluate the appropriateness of the setup, and to select the most suitable measuring instrumentation and procedures. Among the many flow features that may be conveniently detected by qualitative assessment, one may identify the following ones:

- Possible defects (e.g., leakage) and undesirable influences (e.g., motor-induced vibrations) in the operation of the flow apparatus [Figs. 7.1(a) and 7.1(b)].
- The flow direction, particularly separated and recirculating flow regions [Figs. 7.1(c) and 7.1(d)].

- Flow boundaries and material interfaces.
- The degree of mixedness of different streams [Fig. 7.1(e)].
- Unsteadiness and instability [Figs. 7.1(f) and 7.1(g)].
- The presence of impurities or a second phase [Fig. 7.1(h)].
- Regions with chemical reactions or combustion [Fig. 7.1(e)].
- Differences in density or temperature, density stratification, and natural convection [Fig. 7.1(i)].
- Turbulence and typical eddy structure; interface between turbulent and non-turbulent flow regions; orders of magnitude of the turbulence intensity and length scale [Fig. 7.1(j)].
- Special flow patterns and coherent structures [Fig. 7.1(k)].
- Shock waves, expansion waves, and compression waves [Fig. 7.1(l)].
- Interfacial and internal waves.

Techniques of qualitative assessment: A qualitative assessment technique is usually a simple version of a more sophisticated qualitative or quantitative method used in scientific studies. However, at the preliminary stage, one should resist the temptation to refine the technique at the expense of time, effort, and cost. For the present purposes, an experimental procedure may be considered as qualitative assessment of a flow if it meets the following conditions:

- It is relatively easy and fast to implement.
- It utilizes already available or inexpensive equipment and materials and does not require the design or purchase of sophisticated instrumentation.
- It is non-destructive and non-contaminating, such that the facility could be easily returned to its original condition at the end of the assessment.
- Its results are relatively easy and clear to interpret.

The vast majority of qualitative assessment techniques are visual, because of the relative ease by which many flow phenomena can be made visible. Some flows have inherently visible characteristics, which can be observed without any special effort, as, for example, liquid jets in air, bubbly flows of liquids, flames, and flows carrying impurities. However, in most situations, fluids are optically homogeneous and isotropic and thus their internal motions are invisible to the naked eye. In such cases, visualization may be achieved by the addition of *flow markers* or the use of suitable illumination and recording methods.

Easily visible flow markers are relatively large, typically with dimensions between 50 and 300 μm. These can be released either from an orifice on a surface or from a hypodermic tube in the stream [Fig. 7.2(a)].

Among the common flow markers that are suitable for a qualitative assessment of flows are these:

- Natural markers, such as dirt, lint, or dust.
- For gas flows, smoke, powders, or aerosols.
- For liquid flows, air bubbles, powders, metal filings, plastic beads, or dyes.

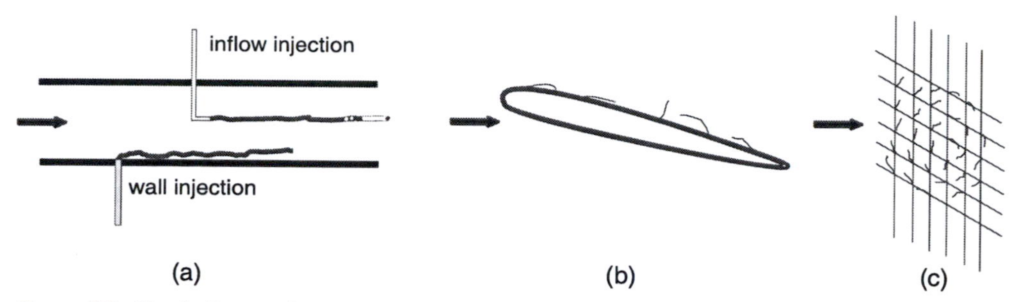

Figure 7.2. Simple flow-marker techniques: (a) dye injection through hypodermic tubes and wall taps; (b) use of tufts to identify flow separation over an airfoil; and (c) use of a tufts screen to visualize a wing-tip vortex.

- Tufts, attached to a surface [Fig. 7.2(b)] or a thin-wire screen across a stream [Fig. 7.2(c)].

Illumination can be conveniently and easily achieved by room light, or other flood-type light, optimized in position and intensity by trial and error. For better results, one may utilize the nearly collimated white light produced by a slide projector, a stroboscope, or a camera flash. One may also consider the use of an inexpensive, low-power laser, which could be of the CW type or pulsed; laser beams may be turned into a light sheet with the use of a cylindrical glass or acrylic rod [Fig. 7.3(a)]. In flows with density variations, a simple version of a shadowgraph, consisting of a collimated light beam and a projection screen, may be appropriate [Fig. 7.3(b)]. More details on natural markers and illumination methods can be found in Chap. 5.

The human eye is capable of discriminating with a fairly good resolution among a great variety of images. In certain situations, however, light recording may be necessary to improve the visibility of an image or to produce a permanent record of it. For example, high-speed flows, even if visible, would appear to be blurred to the eye; then, clear images could be obtained with the use of a still, movie, or video camera, in combination with a high-speed shutter or light source, such as a flash, a stroboscope, or a pulsed laser. The judicious use of close-up lenses, telelenses, or other attachments could improve the view, whereas infrared or special films may reveal images that cannot be seen by naked eye. Digital cameras, in combination with the digital image processing capabilities provided by many popular software packages, may be employed as an inexpensive and fast way to enhance our visual capability. One may also consider the instant replay,

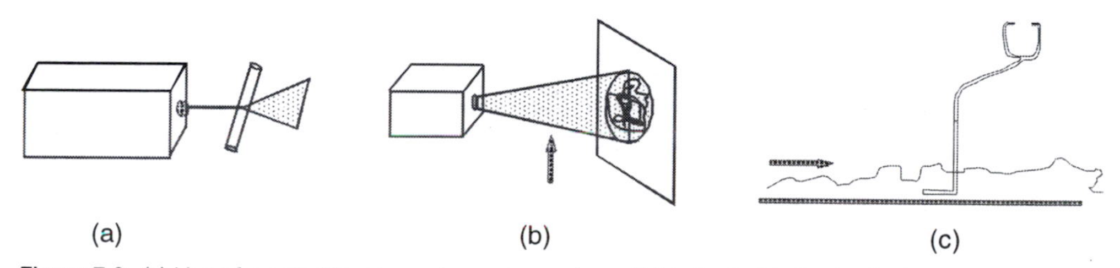
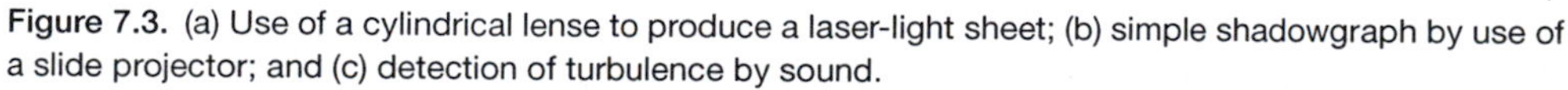

Figure 7.3. (a) Use of a cylindrical lense to produce a laser-light sheet; (b) simple shadowgraph by use of a slide projector; and (c) detection of turbulence by sound.

reverse playback, and frame-by-frame observation features provided by analogue and digital video cameras.

In addition to visual methods, qualitative assessment of flows may be achieved by the use of any other sense. By touching the apparatus or feeling the flow, one may identify vibrations, flow direction, temperature differences, or leakage. By listening to a flow through a small tube connected to a stethoscope, one may detect transition and turbulence [Fig. 7.3(c)]. Carefully listening to the flow sound may reveal cavitation, flow-induced vibrations, or other phenomena accompanied by characteristic noise. Needless to say, such approaches should only complement the experimenter's intuition and experience, and they cannot replace proper measurement.

Interpretation of preliminary observations: The interpretation of the results of a qualitative assessment method has to made with caution, particularly because such methods are, by definition, relatively crude and of moderate sensitivity. For example, separation on the surface of a wing may not be detectable by heavy or stiff tufts, although the same separation may be clearly indicated by finer ones. Injection of dye or smoke in excessive amounts or non-isokinetically (namely at speeds substantially different from the local flow speed) may generate patterns unrelated to the flow of interest. Under certain conditions, non-isokinetic marker injection or the presence of the injection tube may result in prolonged memory effects, trigger flow instability and premature transition to turbulence, or change the turbulence structure. Finally, one is warned against the perils of misinterpretation of visual patterns (e.g., pathlines, streaklines, and streamlines; see Chap. 1). In unsteady flows, even carefully conducted flow visualization may sometimes lead to erroneous or misleading results [1].

7.4 Record keeping

When performing any experiment, one must be able to retrace one's steps in order to document the experimental conditions, explain unexpected results, and even reconstruct the experiment under conditions identical to the previous ones. For these reasons, it is a sound practice to maintain a detailed, chronologically arranged, scientific journal of all conditions, actions, and results obtained. The most suitable format for such journal is a hardcover logbook, of the type available in university bookstores, with pages lined and numbered and preferably containing some pages with graph grids. It is advisable to leave the first few pages blank, to be filled as the experiment progresses in the format of a table of contents. The logbook should be updated on a continuing basis, with the date and time of day of all actions clearly marked. Typical information to be recorded includes the following:

- The title and objective of the experiment or activity, with a short title included that can be used in case more than one experiment is recorded in the same logbook.
- Any references and personal communications related to the project.
- A list of all equipment used with brief technical details (e.g., dimensions, output units, resolution of readings), sketches of connections among different pieces of equipment, and availability of technical manuals or instructions for use.

- A list of all acquired measurements recorded during each experiment, the settings for different instruments (e.g., positions of switches), and reference values (e.g., room temperature, barometric pressure, and humidity).
- Personal observations concerning the experiment, including possible difficulties, malfunctions of equipment, notes for the interpretation of the results, etc.
- All potential sources of error and uncertainty for the measured quantities.
- Simple plots of various parameters to verify that the experiment is progressing as planned and to detect spurious phenomena that may affect the results.
- Brief summaries of postprocessing progress, comparisons with previous experiments, theories, or numerical studies.

All information should be recorded neatly and coherently, so that it can be reviewed easily not only by the author but also by other members of the group. In general, information should be recorded by hand directly on the logbook pages. If any additional material is attached to the logbook, it must be pasted firmly on blank pages, should not be oversize, and should not interrupt the page numbering or the continuity of the logbook.

7.5 Scientific ethics

While performing experiments and when reporting results one must comply with appropriate scientific and professional ethics and practices and exercise honesty and collegiality. All ideas that contributed measurably to the project and any non-trivial assistance provided during the various stages of the work (e.g., technical services, help with collecting and analysing the measurements, and preparation of graphics or a manuscript) should be duly acknowledged. Verbatim copying of material from unacknowledged sources or intentional presentation of previously published material as the author's creation constitutes *plagiarism*, and the fabrication, falsification, or 'doctoring' of experimental results constitutes *fraud*; both are punishable at different levels. Trademarks, copyright, and intellectual property in general are protected by legislation and specific regulations and guidelines. The duplication of copyrighted material in a thesis or publication requires written permission, usually of the publisher of the journal or book in which the material first appeared or, for unpublished material, of the author. An important aspect of all scientific reports and publications is *documentation*. This means that all presented material or ideas that have been borrowed from other sources must be accompanied by *citation* of the proper references. Thus it is understood that any undocumented material contained in a report is either common knowledge (e.g., the First Law of Thermodynamics) or original.

A contentious issue in scientific research is coauthorship in joint publications and public presentations. This refers to both the names to be included as coauthors and the order of names. Practices vary widely among different fields, and one should not venture to refer to general rules. It is generally accepted, however, that a coauthor is a person who has made a substantial intellectual contribution to the project and/or the report. When one perceives that coauthorship is not entirely clear, it would be desirable to obtain consensus of all persons concerned well in advance of the completion of a project.

Issues concerning intellectual property can be very complex, as evidenced by the large number of legal contests and the ongoing updating of laws and regulations. When uncertain how to approach such an issue, it is best to consult with an experienced person or an authority on the subject.

QUESTIONS AND PROBLEMS

1. A hydraulics laboratory has just completed the construction of a scale model of a river and its surrounding flood planes. It is desired to assess, quickly and inexpensively, the flow quality in the model for different water levels, and particularly to determine whether there is flow separation at the bends, whether the flow is laminar or turbulent, and whether there is a hydraulic jump. Describe the type of tests that you recommend, and list the instrumentation and materials required. Estimate the time and cost of these tests, if they have to be repeated for three flow rates, including the maximum flow rate that can be provided by the available pumping system.
2. A student team is modifying the exterior shape of a classic car in the hope to achieve a better aerodynamic performance. They have no access to wind-tunnel facilities, and their budget is very limited. Describe the type of tests that you would recommend, including materials and a cost estimate.
3. A pressure tube is used to measure flow properties in a transonic stream. Discuss whether there is a simple type of test that can be used to determine whether there are shock waves forming around the tube.

REFERENCE

[1] M. Gad-el-Hak. Splendor of fluids in motion. *Prog. Aerosp. Sci.*, 29:81–123, 1992.

PART TWO

MEASUREMENT TECHNIQUES

8 Measurement of flow pressure

Pressure-measuring instrumentation and techniques have been described in numerous sources, both of general [1–8] and specific [9,10] scope. The following sections represent a digest of relevant material that would be useful for the measurement of pressure in the fluid dynamics laboratory.

8.1 What exactly is pressure?

The terms *pressure* and *static pressure* are sometimes used casually in experimental fluid mechanics (e.g., Ref. 2) to describe the normal force per unit area, namely the normal stress. This definition is unambiguous when one deals with fluid pressure on an immersed solid surface; however, it is inappropriate for a point in a moving fluid, because normal stresses generally depend on direction, even though such dependence is often disregarded. The appropriate mechanical definition of pressure is as the average normal stress [11, 12],

$$p = -\frac{1}{3}\left(\sigma_{11} + \sigma_{22} + \sigma_{33}\right), \tag{8.1}$$

which is an invariant under coordinate system rotations and reflections. The negative sign accounts for the fact that, by convention, pressure is positive when compressive, whereas a normal stress is considered positive when tensile. In static fluids, the normal stress is independent of orientation, and therefore the pressure is equal to any normal stress at a given position. In static fluids subjected to a gravitational field, the *hydrostatic pressure difference* between two locations A and B can be found as

$$p_A - p_B = -\int_{z_A}^{z_B} \rho g \mathrm{d}z, \tag{8.2}$$

where z is a vertical upwards direction, g is the gravitational acceleration, and ρ is the fluid density; both g and ρ could be functions of position.

Besides its mechanical definition, which relies on the continuum assumption, pressure has also been defined through thermodynamic concepts [12, 13]. In static fluids, the thermodynamic and mechanical pressures coincide, whereas in incompressible fluids, it is only the mechanical pressure that can be defined. To simplify matters, one may generally employ the *Stokes assumption*, by which the thermodynamic pressure may be taken as equal to the mechanical pressure; this assumption is accurate for Newtonian fluids, provided that their rate of expansion is not exceedingly high.

Pressure as just defined is also referred to as the *absolute pressure*. Sometimes, however, a pressure-measuring instrument may indicate not the absolute pressure, but rather its difference from the reference pressure in the surrounding environment, usually the atmospheric pressure p_{atm}; the pressure difference $p_g = p - p_{atm}$ is called the *gauge pressure*. Another parameter of importance in fluid dynamics is the *total pressure*, or *stagnation pressure*, p_0 defined as the static pressure that a fluid particle would achieve if it were decelerated isentropically (namely adiabatically and reversibly) to zero speed (stagnation). In incompressible flows,

$$p_0 = p + \frac{1}{2}\rho V^2 \tag{8.3}$$

where the second term on the right-hand side is called the *dynamic pressure*. In compressible flows, however, static and total pressures are related as

$$p_0 = p\left(1 + \frac{\gamma - 1}{2}\mathrm{M}^2\right)^{\frac{\gamma}{\gamma-1}} \tag{8.4}$$

where γ is the ratio of specific speeds and M is the Mach number (note that this expression applies to ideal gases with constant ratio of specific heats γ). Particular care must be taken when a shock wave forms upstream of an instrument. Assuming that the shock wave is normal, one may compute the ratios of static and total pressures across the shock as, respectively [14],

$$\frac{p}{p'} = \left(\frac{2\gamma}{\gamma + 1}\mathrm{M}^2 - \frac{\gamma - 1}{\gamma + 1}\right)^{-1}, \tag{8.5}$$

$$\frac{p_0}{p_0'} = \left(\frac{2\gamma}{\gamma + 1}\mathrm{M}^2 - \frac{\gamma - 1}{\gamma + 1}\right)^{\frac{1}{\gamma-1}} \left[\frac{(\gamma - 1)\mathrm{M}^2 + 2}{(\gamma + 1)\mathrm{M}^2}\right]^{\frac{\gamma}{\gamma-1}}, \tag{8.6}$$

where non-primed and primed properties refer, respectively, to positions just upstream and just downstream of the shock.

8.2 Pressure-measuring instrumentation

Liquid-in-glass manometers

Liquid-filled, U-shaped manometers are a simple and effective means of measuring constant or slowly varying pressures. For the basic configuration, shown in Fig. 8.1(a), the difference between the pressures at positions A and B is given by

$$p_A - p_B = -\rho_1 g\,(z_A - z_C) + \rho_2 g\,(z_D - z_C) + \rho_3 g\,(z_B - z_D)\,. \tag{8.7}$$

In this configuration, fluid 2 must be a liquid, whereas fluids 1 and 3 may be either gases or liquids immiscible with liquid 2. If fluids 1 and 3 are gases and the elevations of

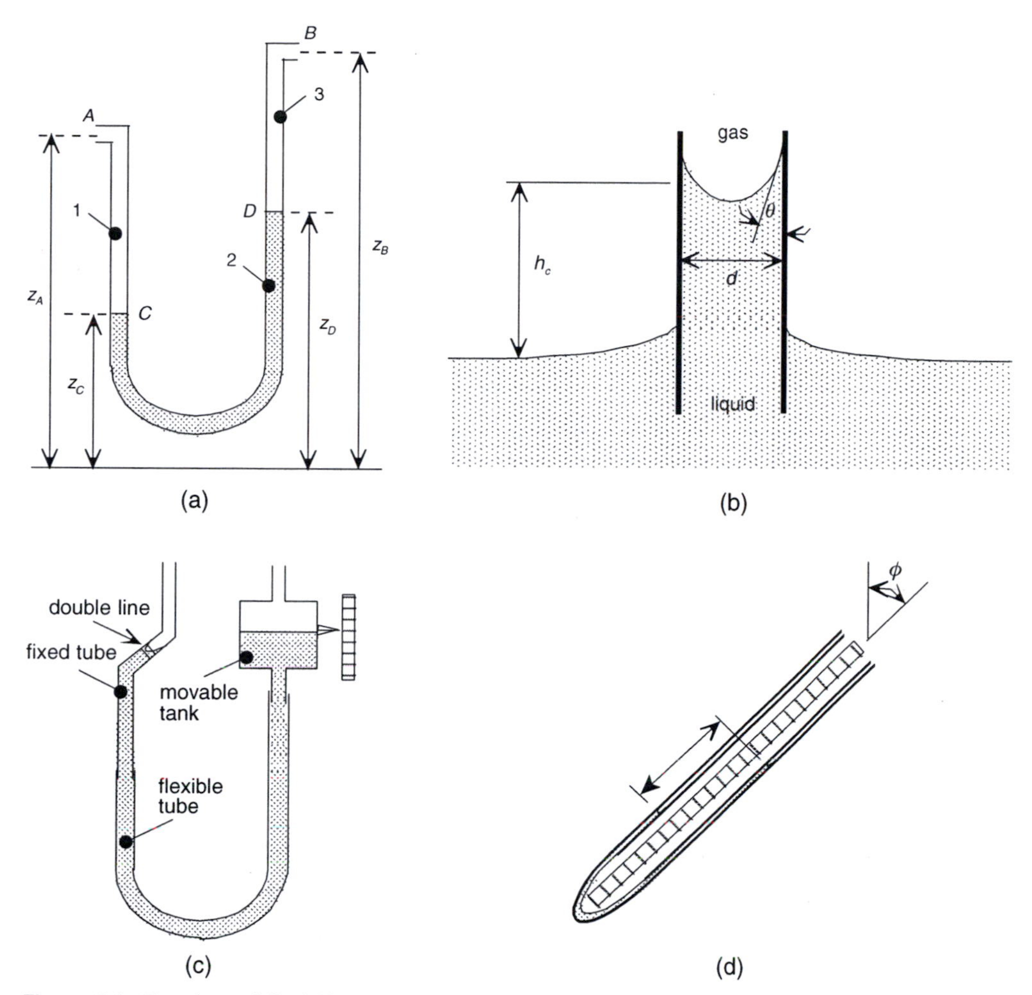

Figure 8.1. Sketches of liquid-in-glass manometers: (a) U-tube manometer, (b) the capillary effect, (c) Prandtl-type micromanometer, and (d) inclined manometer.

points A and B are not much higher than that of the manometer, the Eq. (8.7) may be simplified to an approximate one,

$$p_A - p_B \simeq \rho_2 g (z_D - z_C). \tag{8.8}$$

Considering the pressure difference as the input and the column of liquid as the output, one gets the static sensitivity of the U-tube manometer, when used in gases, as $1/(\rho_2 g)$; clearly the manometer sensitivity increases with decreasing liquid density. When used to measure pressure differences in liquids, the static sensitivity would be $1/[(\rho_2 - \rho_1) g]$, which increases with decreasing density difference between the manometric and the experimental liquids. On the other hand, for a given height of the U-tube, the range of the manometer would be proportional to $\rho_2 g$, or $(\rho_2 - \rho_1) g$, so that the choice of manometric fluid density should be a compromise between the conflicting requirements of high sensitivity and wide range. Manometric fluids of different densities and immiscible with water are available. Mercury is the liquid with the highest density

Table 8.1. Surface-tension values and contact angles for some representative manometric material combinations

Material combination	σ (N/m)	θ (°)
Water–air–glass	72.8×10^{-3}	0
Mercury–air–glass	470×10^{-3}	140
Mercury–vacuum–glass	480×10^{-3}	140
Mercury–water–glass	380×10^{-3}	140

under usual laboratory conditions (specific weight of 13.55 at 20 °C), but can no longer be used in open systems. The heaviest allowable liquids are various types of oils, none of which, however, exceeds 3.0 in specific weight.

The U-tube manometer response equations clearly indicate that errors in the measurement of pressure may be introduced by variations of the fluid densities (e.g., as a result of temperature variations) or the gravitational acceleration (e.g., by use of the manometer at different elevations); such errors are relatively easy to correct [2]. Other errors could be introduced by non-vertical positioning of the manometer and by the mounting of the manometer on an accelerating object. When the manometer is used to measure gauge pressure, for example by opening side B in Fig. 8.1(a) to the atmosphere, which would subsequently be used to estimate the absolute pressure, care must be taken to use the correct value of *barometric pressure*, which can be measured with a *barometer*.

A different source of error is *capillarity*, namely the rise or fall of the free surface of a liquid inside vertical or inclined tubes. This error, indicated as h_c for the case of the vertical tube shown in Fig. 8.1(b), is [11]

$$h_c = \frac{4\sigma \cos\theta}{\rho g d}. \tag{8.9}$$

In this expression, the surface tension σ depends on the liquid–gas combination, whereas the contact angle θ depends on the combination of liquid–gas–solid-wall materials. A few representative values are presented in Table 8.1 [2]. Clearly, the larger the tube diameter is, the smaller the capillarity error would be.

As an example, consider an open glass tube with a diameter $d = 10$ mm, dipped vertically into water whose free surface is exposed to air, as shown in Fig. 8.1(b). Using values from Table 8.1, one can compute a rise of the water in the tube by $h_c \simeq 3$ mm. If mercury were used instead of water, similar analysis gives $h_c \simeq -1.1$ mm, which represents a drop below the free surface of the surrounding tank.

Errors in manometer readings could also be caused by inconsistent readings of the meniscus (i.e., the free surface inside the tube) or changes of the meniscus shape that are due to dirt, variations in the tube size, and shape and inclination of the tube. Capillarity and meniscus-reading errors can be minimized if the meniscus is always positioned at the same reference position, marked clearly by a line. This is the basis of the *Prandtl micromanometer*, which permits the vertical traversing of the liquid containing tank

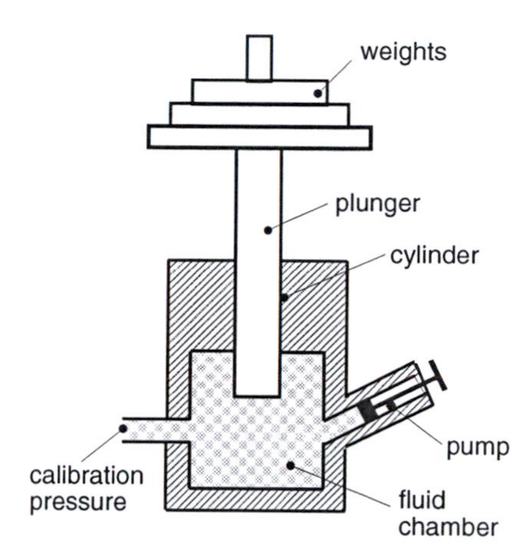

Figure 8.2. Sketch of an idealized deadweight gauge.

to restore the meniscus at the reference position [Fig. 8.1(c)]. The precision of this instrument is increased with the use of a precision lead screw and a rotary scale [2].

A number of variants of the basic U-tube manometer have been proposed to enhance its operation for specific purposes. A useful configuration is the *inclined manometer* [Fig. 8.1(d)], whose static sensitivity (if used in gases and if one considers the reading of the meniscus position along the inclined tube as the output) is $1/(\rho_2 g \cos\varphi)$; for an inclination of $\varphi = 80°$, the sensitivity of the inclined manometer would be nearly six times higher than that of the vertical manometer with the same liquid used.

Mechanical pressure gauges

Deadweight gauges: *Deadweight gauges* are highly accurate devices, but fairly cumbersome in their use, so that they find application mainly as standards for the calibration of other pressure gauges, rather than for the measurement of pressure in the laboratory [2, 15]. An idealized deadweight gauge is shown in Fig. 8.2. Both hydraulic and pneumatic versions are available. Pressure is built up inside the fluid chamber with the use of a pump, and weights are added on a platform on top of the plunger until the plunger stops moving. Then the gauge pressure inside the fluid chamber (omitting hydrostatic pressure) should ideally be equal to the weight (excluding the *tare*, namely the weight of the plunger and other accessories) divided by the plunger cross-sectional area. For a higher accuracy and to account for the small clearance that must be provided between the plunger and the cylinder, an average area between the plunger and the cylinder cross-sectional areas should be used instead. Additional corrections can be made for buoyancy, altitude, and temperature-variation effects. Improved versions of deadweight gauges are available, with rotation or oscillation of the cylinder or the plunger used to reduce friction and various compensation methods used to account for leakage. Typical ranges of deadweight gauges are between 10^2 and 10^8 Pa within an uncertainty of 0.01%–0.05% of the reading.

Elastic-element gauges: These widely used, general-purpose pressure gauges contain an elastic element that deforms under pressure and creates a linear or angular

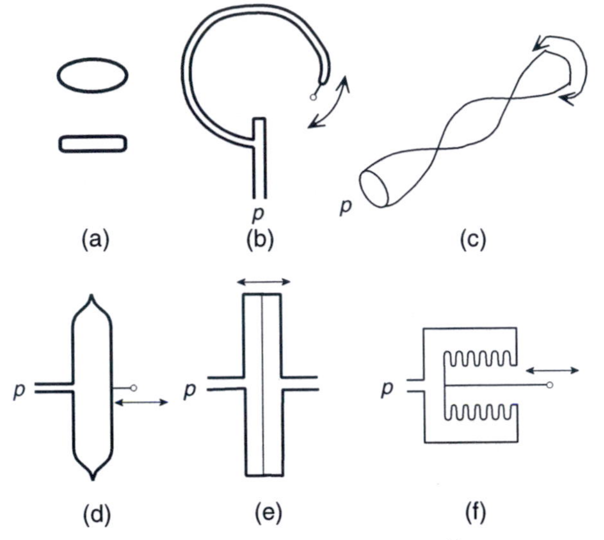

Figure 8.3. Sketches of elastic elements used in pressure gauges: (a) cross-sectional shapes, (b) curved Bourdon tube, (c) twisted Bourdon tube, (d) capsule, (e) diaphragm, and (f) bellows.

displacement of a component that is either displayed on a dial by means of purely mechanical linkages or transformed to an electric signal that can be displayed or recorded at will. Although the operation of these devices can be explained from first principles, their output cannot be predicted accurately and so they require calibration, although calibration need not be performed for precisely manufactured commercial gauges. A common type of elastic element is the *Bourdon tube*, namely a sealed metallic tube with an oval or flattened cross-sectional shape [Fig. 8.3(a)] that tends to approach the circular shape as the difference between the inner and the outer pressures increases. Thus curved tubes [Fig. 8.3(b)] tend to become straight, causing a linear displacement, and twisted tubes [Fig. 8.3(c)] tend to unwind, causing an angular displacement. Other types of elastic elements include capsules, diaphragms, and bellows [Figs. 8.3(d)–8.3(f)]. Elastic deformation gauges are available for both absolute and gauge pressures. As might be expected, they have a relatively slow response and at best a moderate level of accuracy, especially in the lower 10%–20% of their ranges. They find extensive application in the monitoring of industrial flows, but not as much as laboratory instruments, other than for the purpose of monitoring supply pressures and the like.

Electrical pressure transducers

These devices provide an electric output signal that is linearly or non-linearly dependent on the absolute pressure or a pressure difference. There is a variety of physical phenomena that have been utilized for this purpose. The only truly passive devices are piezoelectric crystals, which generate an electric voltage when deformed, but even these require external signal amplification. The operation of all other devices is based on a change of an electric property (resistance, capacitance, or inductance) as a result of pressure-induced displacement or deformation; such devices are active, namely they require some excitation power for their operation.

Electrical pressure transducers usually require calibration when they are first employed and frequent calibration checks thereafter. They are generally susceptible to temperature and humidity effects, which could seriously contaminate the measurements. Temperature effects often manifest themselves in the form of a zero drift, although they may not significantly influence the static sensitivity. In such cases, an adjustment of the zero or subtraction of the zero-offset value from the results may be sufficient to account for this effect. Humidity effects can be more unpredictable, particularly for variable-capacitance transducers. An easy way to remove humidity from the chambers of pressure transducers used in air is to leave them connected overnight to a flask containing hygroscopic crystals. When pressure transducers in liquids are used, care must be taken to remove gas pockets and bubbles from the transducer chambers and all connecting lines.

The great advantage of electrical transducers over liquid-in-glass manometers and mechanical gauges is their superior frequency response, which makes them suitable, under certain conditions, for the measurement of unsteady and turbulent pressures. In fact, several of the electrical pressure transducers (piezoelectric, variable-capacitance, and strain-gauge types) originated as *microphones*, namely devices used for acoustical measurements. The disadvantage of some of these devices is that they do not measure the absolute or gauge pressure itself, but only pressure fluctuations.

Variable-capacitance transducers: A pressure transducer that is quite common in fluid mechanics laboratories is the *barocell* type, available for both absolute and gauge pressures. It contains a chamber with a metallic diaphragm–electrode, mounted parallel to a fixed backplate that acts as the second electrode of a capacitor. A *polarizing voltage* must be provided to the capacitor by a power supply. Pressure is measured by means of changes in capacitance between the two electrodes, caused by changes in the gap between them as a result of diaphragm deformation.

The common-type *condenser microphones* or *capacitor microphones* [16] also operate on the same principle. The elastic diaphragm is usually made of nickel or plated Mylar or glass. The microphones are also available in miniature forms, with diameters as small as a few millimetres, and have a wide frequency response, ranging from a few hertz to nearly 100 kHz, which makes them ideal as audible-sound transducers, although they cannot resolve steady pressure. In addition to externally polarized microphones, a permanently polarized variation is available, called the *electret* microphone. Electret microphones contain a polymer diaphragm with embedded electric charges. Their advantages over capacitor microphones are lower cost and lower sensitivity to humidity.

Piezoelectric transducers: The most commonly used piezoelectric materials are lead zirconate titanate and barium titanate, although several other crystalline materials have also been utilized as the passive elements [16, 17]. A limitation of piezoelectric transducers is their relatively high sensitivity to vibrations and acceleration; to reduce such sensitivities, various compensation methods (e.g., the use of two crystals in tandem) have been proposed [7].

Strain-gauge transducers: These transducers utilize one or more strain gauges (i.e., components whose electric resistance is proportional to axial strain) attached to an elastically deformable element. The strain gauges are connected in a Wheatstone bridge and supplied with an excitation voltage to produce a pressure-dependent electric signal. Because their outputs require substantial amplification, they may be subject to considerable drift and temperature sensitivity. An advantage of strain-gauge transducers is that they can be made quite thin, with a thickness as low as about 1 mm, and so are suitable as wall-pressure sensors [1].

Variable-reluctance transducers: *Reluctance* is the ratio of magnetic 'force' to magnetic flux in a magnetic circuit; its reciprocal is called *permeance* [18]. These transducers contain a magnetically permeable diaphragm mounted between two symmetrically located magnetic coils. If the diaphragm is deflected towards one of the coils, as a result of applied pressure, its reluctance with respect to the magnetic field of one coil would increase and with respect to that of the other would decrease. Thus the inductances of the two coils would be changed. The coils are connected in a bridge configuration such that the inductance ratio is measured as an electric voltage. Both dc and ac outputs can be provided.

Linear-variable differential transformers: In these transducers, pressure-induced deformation of an elastic element is transmitted to the core of a transformer consisting of a central primary coil, flanked by two secondary coils at either of its ends. When no pressure is applied, the core is in a symmetric position and the two secondary coils are in balance. When pressure is applied, imbalance in the circuitry produces an output voltage proportional to pressure.

Semiconductor and microelectromechanical pressure transducers: During the past two decades, semiconductor fabrication technology has been applied to the manufacturing of silicon-based pressure transducers, which can be made to sizes much smaller than the corresponding conventional transducers. Transducers with sizes of 1 mm or larger are usually referred to as *miniature-type*, whereas the term *microelectromechanical systems* (*MEMS*) applies to devices with sizes between 1 μm and 1 mm [19–22]. Such transducers are usually manufactured by the use of photolithographic methods and are distinguished into three categories:

Piezoelectric transducers: These have relatively low sensitivity and high noise level.

Piezoresistive transducers: These have a relatively high sensitivity to pressure variations, but are also sensitive to temperature variations and stresses; they are usually connected in a Wheatstone bridge configuration and are compensated for temperature variation.

Capacitive transducers: These operate similarly to capacitor microphones and require electronic preamplification; their advantage is a high sensitivity to pressure variations but low sensitivity to temperature variations; electret-type MEMS transducers are also available.

MEMS transducers can be manufactured at low cost and can also be produced in the form of two-dimensional arrays, which makes them suitable for the simultaneous measurement of pressure fluctuations over a surface. Their frequency response extends to several tens of kilohertz, which is adequate for the measurement of pressure fluctuations in many turbulent flows.

Gauges for extreme pressures

The measurement of pressure in ranges that are several orders of magnitude higher or lower than the usual, near-atmospheric, laboratory range can be made only with special instrumentation.

Vacuum gauges: Liquid-in-glass micromanometers, including special designs that 'amplify' pressure, can be used effectively to measure pressures as low as about 0.1 Pa. *Vacuum*, namely the extremely low-pressure range, can be measured with devices relating pressure to various other fluid properties, including viscosity, capacitance, and thermal conductivity [10,15,23–26]. Among the available instruments one may mention the following, keeping in mind that the indicated ranges may vary significantly from one model to another:

Device	*Pressure range* (Pa)
McLeod gauge	10^{-1}–10
Pirani (resistance) gauges	10^{-3}–10^{2}
Thermistor and thermocouple gauges	10^{-2}–10^{2}
Philips cold-cathode gauges	10^{-3}–1
Penning cold-cathode gauges	10^{-6}–10^{-1}
Capacitance gauges	10^{-5}–10^{-2}
Ionization gauges	10^{-11}–10^{2}
Mass spectrometers	to less than 10^{-11}

High-pressure gauges: The high-pressure range may be conventionally defined as exceeding 10^8 Pa. Strain-gauge and elastic-element transducers can be made to measure pressures at most an order of magnitude larger than that value. For even higher pressures, up to a limit of about 10^{12} Pa, the only transducers available are of the variable-resistance type [1,9,15,27], but of a design quite different from that of strain-gauge transducers. The high-pressure gauges contain a coiled wire of a gold-chrome (2.1%) alloy or manganin, sealed inside metallic bellows filled with kerosene. The pressure is transmitted by the bellows to the kerosene, which compresses the wire. The gauge's output is proportional to the electric resistance R of the wire, which is related to pressure as [1]

$$\frac{\mathrm{d}R/R}{p} = \frac{2}{E} + \frac{\mathrm{d}\rho_e/\rho_e}{p},$$

where E and ρ_e are, respectively, Young's modulus of elasticity and the electric resistivity of the wire.

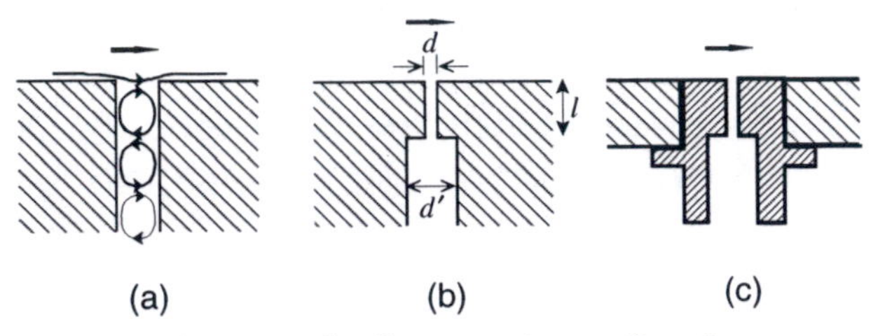

Figure 8.4. Sketches of wall-pressure tap configurations.

8.3 Wall-pressure measurement

Static-pressure taps: The simplest method of sensing steady or slowly varying static pressure at a solid wall is by machining a small orifice (*tap*) facing the flow and connecting it to a manometer or pressure transducer. This approach is used widely; however, it may introduce appreciable systematic errors when not implemented carefully. To begin with, flow over a cavity generally induces motions of the fluid contained in it in the form of a sequence of counter-rotating vortices [Fig. 8.4(a)], each having a strength that weakens with increasing distance from the orifice exit. Such motions entrain high-speed fluid from regions at some distance from the wall and tend to create a pressure inside the cavity that is higher than the true wall static pressure. Thus wall pressure will be measured accurately only by an infinitesimally small tap. The use of extremely small taps, however, is impractical for several reasons. Besides the technical difficulty of machining small holes that are clean and perpendicular to a surface, such holes would be amenable to blockage by flow impurities and would have an unacceptably slow response. Practical hole sizes usually range between 0.5 and 3 mm and, even in the ideal case of a perfect hole on a smooth wall, they introduce a systematic error $\Delta p = p_m - p$, where p_m is the measured pressure (usually $\Delta p > 0$). This error depends on several geometric and dynamic parameters. To combine the benefits of a small tap diameter and a reasonably fast response time, it is a common approach to connect the tap to a larger-diameter cavity [Fig 8.4(b)], which may contain a pressure transducer or a tube leading to remotely located instrumentation. An important length scale characterizing the dynamics of near-wall flow is the *viscous length* ν/u_τ, where ν is the kinematic viscosity of the fluid and $u_\tau = \sqrt{\tau_w/\rho}$ is the friction velocity (τ_w is the wall shear stress and ρ is the fluid density). Thus an appropriate dimensionless tap diameter can be defined as $d^+ = d/(\nu/u_\tau)$, where d is the tap diameter (d^+ may also be viewed as the friction Reynolds number of the tap). The error in pressure measurement can be normalized either by the local wall shear stress as $\Delta p/\tau_w$ or by the free-stream dynamic pressure $q = \frac{1}{2}\rho V^2$ as $\Delta p/q$. Figure 8.5 summarizes some of the available error measurements, plotted vs. d^+. Early measurements [30] have demonstrated that the error increases with increasing tap length-to-diameter ratio l/d, but becomes insensitive to this parameter for $l/d \geqslant 1.5$. It may be noted that, in a different set of experiments [31], this error was found to assume negative values for $l/d \leqslant 3$. To avoid a variable error, and considering that very large values of l/d would lead to slow response, it seems sensible to maintain l/d in the range 5–15. It has further been assessed [28] that, for sufficiently large l/d, the error $\Delta p/\tau_w$ increases with increasing d^+ to an asymptotic value of around 3.5 for

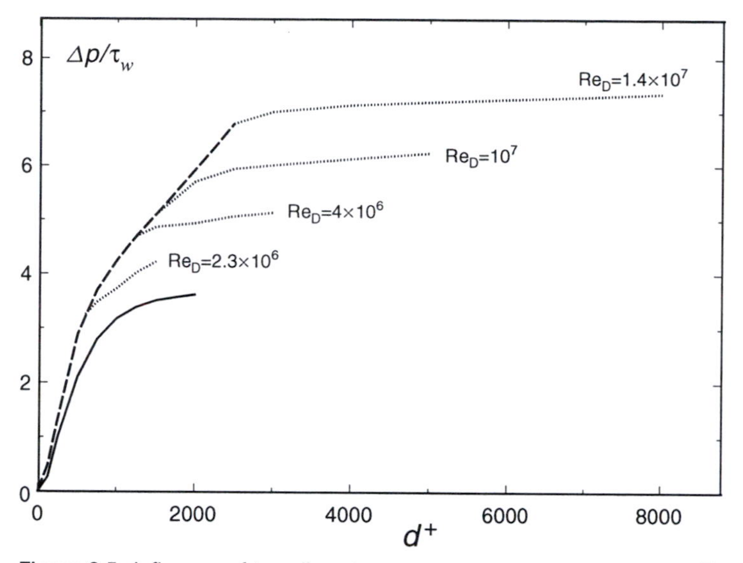

Figure 8.5. Influence of tap diameter on pressure measurement error. The solid curve corresponds to measurements on a flat plate (free-stream Reynolds number was not specified [28]). The dashed curve is the envelope of measurements in pipe flow at different pipe Reynolds numbers, Eq. (8.10) [29]. The dotted curves represent measurements in the pipe flow at the specified pipe Reynolds numbers.

$d^+ > 2000$. In addition to these effects, however, it has been realized that the free-stream Reynolds number also has an effect on the error [2]. Recent measurements in pressurized pipe flow [29] have demonstrated that the asymptotic value of $\Delta p/\tau_w$ increases with increasing pipe Reynolds number $\mathrm{Re}_D = DU_b/\nu$ (D is the pipe diameter and U_b is the bulk velocity), reaching values near 7 for $d^+ > 2500$ and $\mathrm{Re}_D = 1.4 \times 10^7$ (Fig. 8.5). A polynomial fit to the envelope of these measurements for $d^+ < 2500$ gives

$$\frac{\Delta p}{\tau_w} = 0.8306\left(\frac{d^+}{1000}\right)^6 - 6.623\left(\frac{d^+}{1000}\right)^5 + 19.83\left(\frac{d^+}{1000}\right)^4 - 26.68\left(\frac{d^+}{1000}\right)^3 + 13.49\left(\frac{d^+}{1000}\right)^2 + 3.481\left(\frac{d^+}{1000}\right). \quad (8.10)$$

The preceding error estimates apply to smooth pipes and very precisely machined holes. Even slight imperfections in the shape or orientation of the hole are likely to introduce additional systematic errors, which may be positive (as in the case of rounded hole edges) or negative (as in the case of holes with chamfer) and typically up to $1\%q$ in magnitude [2, 32, 33]. Thus geometrical imperfections may introduce unpredictable errors that are much larger than the finite hole error. When it is difficult to machine a precise tap through the flow side of a surface, it may be preferable to do so on a removable plug [Fig. 8.4(c)], which can then be inserted in a much larger hole through the wall, thus reducing the effects of geometrical distortions. Tap errors become quite unpredictable when one is dealing with rough walls. Flow distortions that are due to the roughness elements tend to introduce additional errors, whereas roughness-induced turbulence

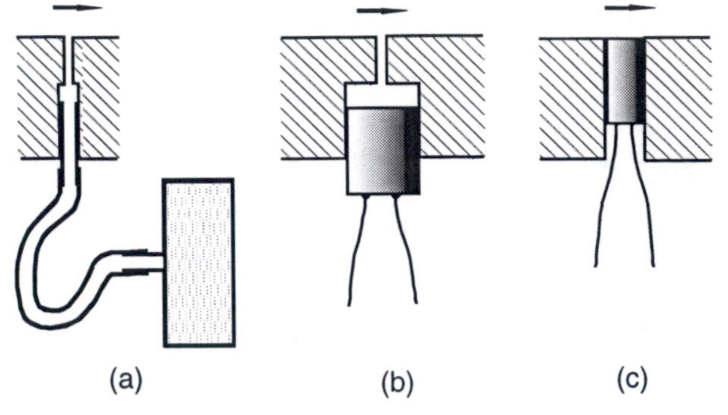

Figure 8.6. Various ways of connecting pressure transducers for the measurement of wall pressure: (a) remote connection, (b) cavity mounting, and (c) fluch mounting.

tends to mix the flow near the tap, possibly reducing error. When wall roughness is considerable, wall tap measurements may be totally unreliable.

Additional errors are due to compressibility [2, 34]. In the case of transonic and supersonic flows, wall-pressure measurement might be exceedingly complicated by the presence of unsteady shock waves and shock–boundary-layer interactions. For subsonic flows, the error that is due to finite hole size tends to increase with increasing Mach number M. For example, the asymptotic (i.e., large d^+) error $\Delta p/q$ increases by 13% for $M = 0.4$ and 45% for $M = 0.8$, with larger increases for smaller hole diameters [2, 32, 33].

Several authors have addressed the issue of turbulence effects on wall-pressure measurement, without, however, reaching a satisfactory conclusion. It appears that turbulence tends to decrease the finite hole error and so, in the absence of a better approach, turbulence effects may be disregarded [2, 35].

Alternative approaches are needed in experimental settings in which it is not possible or convenient to install pressure taps. An example is the surface-mounted disk probes, which, with proper corrections applied, may provide reasonably accurate wall static-pressure measurements and mountable externally on immersed surfaces [36].

Transducer connection: Three common configurations of pressure transducers used in the measurement of wall pressure are illustrated schematically in Fig. 8.6. *Remote mounting* [Fig. 8.6(a)] through flexible or metallic tubing is the simplest one and can utilize any pressure transducer or manometer, irrespective of size or operation principle. This approach can be applied to *multi-port measurement* through the use of a manual or automated *pressure-scanning valve*, which sequentially connects a single transducer to a series of pressure taps. An alternative device for multi-port measurement, also utilizing remote mounting, is the *electronic pressure scanner*, which contains a number of small transducers, fabricated by use of semiconductor technology and mounted together on a block, each with a separate pressure port. An obvious disadvantage of remote mounting is the deterioration of the dynamic response of the pressure-measuring system that is due to the interference of tubing, connectors, valves, etc., which sometimes dictate the frequency response of the system, while the transducer characteristics become secondary or irrelevant (see Section 8.5). To better utilize its capabilities, one must mount the transducer close to or, if possible, on the immersed surface. Surface mounting is further distinguished to *cavity mounting* [Fig. 8.6(b)] and *flush mounting* [Fig. 8.6(c)]. Cavity mounting permits the transmission of wall pressure, sensed by a small tap, to a nearby transducer and improves the dynamic response of the system by the elimination of tubing. This configuration, however, is unable to exploit the full capabilities of the transducer, as one would have to consider the dynamic response of the tap–cavity–transducer combination (see Section 8.5). Compared with flush mounting, cavity mounting has the advantages of improved spatial resolution and utilization of a wider range of transducers, without strict requirements on size or quality of sensing surface. On the other hand, flush-mounted transducers are the only ones that offer their full dynamic response. This technique, however, is subjected to two restrictions. First, to avoid extensive spatial averaging of the measured pressure, the transducer dimensions must be very small, which effectively excludes all transducers except the miniature and MEMS types. Second, the transducer interface with the flow must match closely the shape of the surface in order to avoid local pressure distortion. Small transducers with flat tips are mainly of the piezoelectric and piezoresistive types. Such transducers may be mounted only on plane surfaces or when the radius of curvature of the surface is several orders of magnitude larger than the transducer diameter.

One important consideration in choosing a pressure-measuring configuration is cost. When temporal response requirements are not essential, it would be preferable to use a multi-port connection to a single, general-purpose transducer, or a small number of such transducers. Surface mounting of a large number of transducers may prove to be expensive, cumbersome to implement (need for room under the solid wall and between transducers), and possibly subjected to electronic interference and cross-talk among the transducers.

Pressure-sensitive paints [37–40]: Surface coatings of *pressure-sensitive paints* (*PSPs*) are used for flow visualization purposes (e.g., to indicate locations of flow separation or shock waves), as well as to map local surface pressure, which may then be integrated to provide forces and moments. This method has found application mostly in high-speed air flows ($M > 0.3$), but has also been used in low-speed air studies.

The paint consists of an active material whose luminescence depends, in an inversely proportional fashion, on pressure, and a polymer binder. The principle of operation is a photophysical process, which may be summarized as follows. Light produced by an illumination source having a suitable wavelength range (usually an ultraviolet or blue radiation lamp) is absorbed by the photosensitive molecules in the paint, which undergo transition to an unstable state. Some of these molecules return to their original state by losing energy through radiation emission at a higher wavelength (yellow or red), namely by fluorescence. Other molecules convert the absorbed energy to vibrational energy, namely heat, while others yet lose energy by collisions with oxygen molecules, a process called *oxygen quenching*. By *Henry's law*, the oxygen concentration in a material in contact with air is proportional to the partial pressure of oxygen in the air, which is proportional to the surface static pressure. Therefore the coating at a location on the surface exposed to relatively high pressure will have a relatively high concentration of absorbed oxygen, which would result in more intense oxygen quenching, lower luminescence, and therefore a darker appearance compared with locations of lower pressure. The image of the surface is monitored with a CCD camera, through an optical filter that removes the incident light and allows only the fluorescent radiation to pass.

To recover the local pressure values, the method needs to be calibrated, for example vs. readings from static-pressure taps distributed on the surface. A relationship between luminescence I and local pressure is given by the *Stern–Volmer equation*

$$\frac{I_{\text{ref}}}{I} = A(T) + B(T)\frac{p}{p_{\text{ref}}}, \tag{8.11}$$

where the subscript 'ref' refers to reference conditions, and the Stern–Volmer coefficients A and B depend on the local temperature. This relationship shows that the use of PSP maps for pressure measurement requires knowledge of the local temperature and that temperature variations would result in an increased pressure uncertainty. PSP calibration may be done by enclosure of the model, or a sample coated with the same paint, in a sealed chamber whose pressure and temperature can be controlled. Calibration for varying pressure must be repeated for different values of constant temperature, so that the coefficients $A(T)$ and $B(T)$ can be established.

The accuracy of the preceding technique also depends on the exact knowledge of the intensity of incident light I_{ref}. In addition to difficulties in illuminating complex surfaces uniformly, light variations may be caused by a fluctuating power of the light source or deformation of the model when exposed to different air speeds. Correction methods for some of these effects have been devised. An alternative technique that is insensitive to incident-light variations is the *lifetime-mode* method. The fluorescent light intensity emitted by a surface decays exponentially with time as

$$I = I_0 e^{-kt}, \tag{8.12}$$

where I_0 is the initial intensity and the decay rate (or inverse time constant) is

$$k = k_R + k_{nR} + k_Q, \tag{8.13}$$

with k_R, k_{nR}, and k_Q representing the decay rates that are due to radiation, non-radiative energy transfer, and quenching, respectively. The technique consists of using two light detectors to integrate the fluorescent light intensity over two contiguous time intervals and then compute the ratio

$$\frac{\int_T^{2T} I \mathrm{d}t}{\int_0^T I \mathrm{d}t} = e^{-kT}, \tag{8.14}$$

which is sensitive to pressure and temperature, but not to the illumination intensity. The radiative decay rate k_R is relatively insensitive to temperature, whereas the non-radiative decay rate k_{nR} depends on absolute temperature T as

$$k_{nR} \sim e^{-\frac{\Delta E}{RT}}, \tag{8.15}$$

where ΔE is the activation energy of the process and R is the molecular constant. Materials with relatively high ΔE would have a relatively low temperature sensitivity, which would make them suitable as PSPs. Nevertheless, the dependence of the quenching rate k_Q on temperature as well as on binder properties further complicates the method.

PSP luminophores and binders are quite complex, and, although possible to assemble from generic products, they would be best acquired in ready-to-use form from specialized suppliers.

8.4 In-flow pressure measurement

The exact measurement of the local static pressure in a flowing fluid is extremely difficult, if not impossible, as there is no non-intrusive pressure-measuring method available, and the pressure field would inevitably be distorted by any intrusive method. Relatively simple methods, using inserted tubes and other objects, are available for the measurement of constant and time-averaged static pressures, and a few fast-response probes have been developed for the measurement of unsteady and turbulent in-flow pressures [7]. In-flow fluctuating pressure has also been measured indirectly as the difference between total pressure, measured with a fast-response total-pressure probe, and dynamic pressure, inferred from measurements of flow velocity slightly upstream of the total-pressure probe tip obtained by hot-wires or LDV [41].

The measurement of pressure in a flowing stream can be achieved only by intrusive means, as there is no available non-intrusive method, except for estimating pressure from other measured properties. Although it is preferable to measure pressure locally, in most cases it is transmitted through tubing to a location outside the flow, where there is no concern about using relatively large-size instrumentation. The in-stream pressure is monitored with the use of thin tubes of various designs, collectively known as *pressure tubes* [2,3,42–45], which can be classified into the following categories, depending on their function:

Static tubes, which monitor the local static pressure;
Pitot tubes, also known as *impact tubes* and *total pressure tubes*, which measure the local total pressure;
Pitot–static tubes, which are combinations of the two preceding tubes and provide the local dynamic pressure $\frac{1}{2}\rho V^2$, from which one can easily calculate the local flow velocity;
Multi-hole probes, which consist of combinations of several tubes arranged such that they measure the local static and total pressures as well as the local velocity in both magnitude and direction.

Pressure tubes have a relatively low spatial and temporal resolution. Fast-response probes have been designed by the incorporation of miniature and micro-pressure transducers within the probe body, in the vicinity of the pressure-sensing orifices [20,22,46].

Static-pressure tubes: *Static-pressure tubes* [2,3,42–44] are thin hollow tubes, which are sealed at the tip facing the flow, and have holes or slits on their side. When a

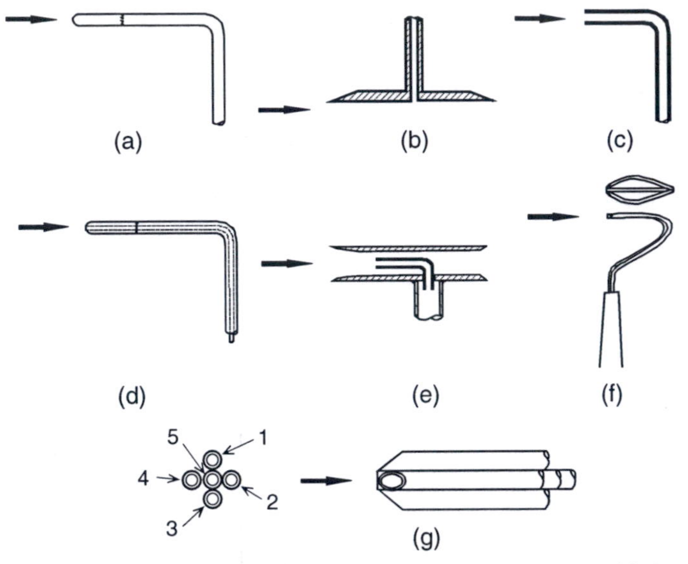

Figure 8.7. Sketches of pressure tubes: (a) static-pressure tube, (b) sharp-edge static tube, (c) Pitot tube, (d) Pitot-static tube, (e) Kiel tube, (f) cobra probe, and (g) five-hole probe.

long static-pressure tube is inserted in uniform flow with its axis aligned with the flow direction, the pressure inside it, sensed through the side openings, would be somewhat lower than the free-stream pressure, because the flow would have to accelerate around the nose. Instead of straight tubes, however, a more practical configuration for measurements in pipes, ducts, wind tunnels, etc., is the bent tube, in which the tip forms a right angle with the stem [Fig. 8.7(a)]. Then flow deceleration that is due to stem blockage would cause an increase in static pressure along the tube, which partly or totally counteracts the nose effect and makes it possible, at least in principle, to design static tubes free of error by proper positioning of the side holes. The response of static-pressure tubes depends on the shape of the nose, the number, size, and location of holes, and the location of the stem, but tube design is also dictated by ease and precision of construction and consideration to sensitivity to misalignment with the flow direction, possible damage during usage, and possible hole blockage that is due to the presence of impurities in the stream. A variety of designs have been proposed and tested. Common designs available commercially have hemispherical or ellipsoidal noses and six or eight evenly spaced holes located at some distance between the tip and the stem. Some caution is required as the designation 'standard' may vary among different suppliers. Moreover, it cannot safely be assumed that the nose and stem effects are perfectly balanced. Typically, even well-aligned, right-angle static-pressure tubes could have an error (usually negative) in the static-pressure reading of up to 1% q, where $q = \frac{1}{2}\rho V^2$ is the dynamic pressure. When higher accuracy is required, it would be advisable to calibrate each tube separately vs. a well-constructed static tap. Static-pressure tubes are much more sensitive to flow direction than total-pressure tubes are. As a result, even a slight misalignment may introduce a significant error to the static-pressure measurement. For example, the static-pressure reading would typically drop by 1% at 5° yaw (i.e., rotation about the stem axis),

3% at 10°, and 5% at 15°. Therefore a sizeable error might occur when exact alignment of the tube cannot be achieved or when the flow direction varies, as in the case of streamline divergence that is due to the presence of a model in a wind tunnel. In such cases, an on-site evaluation of the static tube response would be advisable. For wind tunnels and smooth pipes, one might also consider using wall-pressure readings, rather than in-flow pressure. Other factors affecting the response of static tubes are flow turbulence, internal motions (see Section 8.3), cavitation, and viscous and compressibility effects. Because of its dependence on tube geometry and operating conditions, it would be difficult to predict accurately the error of a specific tube when such effects are present. The literature on such topics (e.g., Ref. 42) is contradictory and incomplete. For example, a reading of the few publications that have addressed the effects of turbulence would conclude that static-pressure errors would be of the order of 1%q (usually positive) or lower for turbulence intensities lower than 10%. However, correction methods for turbulence effects are not of general validity, as the error would depend not only on the turbulence intensity, but also on the length scale of turbulence, the turbulence structure, and the presence of organized motions (coherent structures). For turbulence levels greater than 15%, the errors are likely to be significant. Vibrations of the probe that are due to turbulence or vortex shedding from the stem would also tend to produce readings that are higher than the true static pressure.

Besides cylindrical tubes, a variety of objects with different shapes can be used to measure or estimate in-flow static pressure. A useful type of device for industrial flows containing significant amounts of suspended solids is the *disk-static probe*, available in sharp-edge [Fig. 8.7(b)] and rounded-edge variations. Their main advantage over static tubes is that they can be produced with relatively large orifice openings (reportedly up to 50 mm), which would not be easily blocked by impurities. Such devices are sensitive to flow direction and may have systematic errors even when aligned, so some testing of their response is advisable before use. Finally, static-pressure values can also be recovered from multi-hole probes of various shapes; these probes are discussed later in this section.

Pitot and Pitot-static tubes: A *Pitot tube* is an open-ended, hollow cylindrical tube facing the flow [Fig. 8.7(c)], thus forming a stagnation region on its face, where the pressure would be equal to the total pressure p_0. At relatively large Reynolds numbers and relatively low Mach numbers, the flow upstream of the tube may be accurately described by Bernoulli's equation, which determines the ideal pressure coefficient as

$$C_P = \frac{p_0 - p}{\frac{1}{2}\rho V^2} = 1. \tag{8.16}$$

Then one may use the measured p_0, together with a local measurement of the static pressure p, to compute the flow velocity as

$$V = \sqrt{\frac{2\,(p_0 - p)}{\rho}}. \tag{8.17}$$

Arrays of parallel Pitot tubes (*Pitot rakes*) are used to measure velocity profiles in wakes and other two-dimensional flows, whereas coaxial *Pitot-static tubes* of optimized designs [Fig. 8.7(d)] are routinely used to measure local flow velocity from the directly indicated pressure difference $p_0 - p$.

Pitot tubes are simple, versatile instruments with the additional advantage of not requiring calibration, as their response is based on a basic principle. Nevertheless, their applicability is limited by a number of factors, the most common of which are discussed in the next six subsections.

Misalignment effect: Pitot tubes do not have to be precisely aligned with the flow direction, as the indicated total-pressure reading is insensitive to orientation for small misalignment angles. However, as soon as such angle exceeds a critical value, the indicated C_P would decrease rapidly from the ideal value. For thin-walled, square-ended cylindrical tubes, the critical angle is about 20°, whereas, for similar tubes with the commonly used inner-to-outer diameter ratio $d_i/d_o = 0.6$, it decreases to about 12°. Tubes with rounded or conical faces are significantly more sensitive than square-ended tubes and would require more careful alignment. Specially designed probes consisting of tubes shielded by a surrounding cylinder and sometimes referred to as *Kiel probes* [Fig. 8.7(e)] are particularly insensitive to flow direction, with typical critical angles of about 45° [47]. As mentioned earlier, static-pressure tubes are quite sensitive to misalignment, much more than total-pressure tubes are. For this reason, in three-dimensional flows, it would be preferable to measure static pressure separately and not with the use of a Pitot-static tube.

Shear effect: When inserted in a shear flow with a mean velocity gradient $\partial V/\partial n$ normal to its axis, a Pitot tube will introduce an obstruction, displacing the streamlines towards lower velocities. Let δ be the displacement of the streamline that passes through the tube axis. Then the total pressure indicated by the tube would not correspond to the free-stream velocity V at the extension of its axis but to the higher value $|V| + |\partial V/\partial n|\,\delta$. A secondary source of error, also tending to increase the total pressure reading, is the averaging of pressure oven the tube's orifice. Among the expressions that have been suggested to estimate δ, a reasonably accurate one is [48,49]

$$\frac{\delta}{d_o} = 0.13 + 0.08\frac{d_i}{d_o} \tag{8.18}$$

For the common value $d_i/d_o = 0.6$, one gets $\delta \approx 0.18 d_o$. This error is known as the *displacement effect* and is particularly significant in the inner boundary layer at large Reynolds numbers. To reduce this effect, one may use thinner tubes, thin-walled tubes, and tubes with conically shaped tips. There is a limit, however, on tube diameter, as a decrease in cross-sectional area would slow down the tube's response. For this reason, in highly sheared flows, tubes with nearly rectangular cross sections, obtained by pressing the tips of circular tubes, would be more usable than circular ones.

Wall-proximity effect: When a pressure tube is positioned near a wall, it blocks the flow in its vicinity, displacing streamlines away from the wall, which is the opposite of the displacement effect previously discussed. Thus the tube reading would correspond to a velocity lower than that at the extension of its axis and would need to be corrected upwards. Circular tubes in a moderate-Reynolds-number range would be free of wall effects as long as their axis is positioned more than about $2d_o$ away from the wall [50]; for flattened tubes, it seems that a similar bound is valid provided that the width (largest dimension) of the tube is used instead of d_o [51]. The velocity correction for circular tubes in contact with a wall is about 1.5%.

Turbulence and vibration effects: Although there has been considerable controversy concerning the effect of turbulence on Pitot and Pitot-static tube responses, it is clear that this effect would depend not only on the turbulence intensity $u'/\overline{V}$ (u' is the rms velocity and $\overline{V}$ is the time-averaged velocity, both in the direction parallel to the tube axis), but also on its length scale and its structure, as indicated by the relative magnitudes of the various normal and shear turbulent stresses. For turbulence intensities lower than about 10%, one may neglect turbulence effects, whereas for intensities greater than 20%, such effect are likely to be significant and not easy to correct for. Because of the non-linear relationship between velocity and pressure, turbulent fluctuations always tend to increase the tube reading above the time-averaged value. As a rough estimate of the associated error of Pitot-static tubes, one may use the expression

$$\overline{C}_P = \frac{\overline{p}_0 - \overline{p}}{\frac{1}{2}\rho \overline{U}^2} \approx 1 + \alpha \left(\frac{u'}{\overline{U}}\right)^2, \quad 1 < \alpha < 5, \tag{8.19}$$

where the value $\alpha = 1$ corresponds to a turbulence length scale that is small compared with the distance between total- and static-pressure orifices, whereas the value $\alpha = 5$ corresponds one that is relatively large. Similar to turbulence effects are the effects of vibration, as they result in the tube's being exposed to pressure fluctuations. A rough estimate of such effects for axial vibrations with frequency f (in hertz) and amplitude a gives

$$\overline{C}_P = \frac{\overline{p}_0 - \overline{p}}{\frac{1}{2}\rho V^2} \approx 1 + \left(\frac{2\pi a f}{V}\right)^2. \tag{8.20}$$

Viscous effect: When the Reynolds number Re_{d_i} based on the internal tube diameter d_i is less than about 50, the tube response no longer follows the inviscid flow relationship and the pressure coefficient $(p_0 - p)/\frac{1}{2}\rho V^2$ is greater than unity. As an estimate of circular tube response at low Reynolds numbers, one may use the following expressions [45]:

$$\frac{p_0 - p}{\frac{1}{2}\rho V^2} = \begin{cases} 4.1/\mathrm{Re}_{d_i}, & \mathrm{Re}_{d_i} < 0.7 \\ 1 + 2.8/\left(\mathrm{Re}_{d_i}\right)^{1.6}, & 0.7 < \mathrm{Re}_{d_i} \end{cases}. \tag{8.21}$$

Viscous effects appear when thin tubes are used in highly viscous fluids or in low-speed flows, including wall regions.

Compressibility effect: In compressible flow, Eq. (8.16) is no longer valid and one may instead use the also idealized, isentropic expression

$$C_P = \frac{p_0 - p}{\frac{1}{2}\rho V^2} = \left[\left(1 - \frac{\gamma - 1}{2}\mathrm{M}^2\right)^{\frac{\gamma}{\gamma-1}} - 1\right]\frac{2}{\gamma \mathrm{M}^2}, \tag{8.22}$$

which is valid along a streamline. In the low subsonic regime, expression (8.22) may be approximated through a binomial expansion by

$$C_P \approx 1 + \frac{M^2}{4} + \frac{(2-\gamma)M^4}{24}. \tag{8.23}$$

Compared with the more accurate expression (8.22), incompressible expression (8.16) would give an error of less than about 0.5% for $M < 0.15$, whereas the use of the first two terms in expression (8.23) would give a less than 5% error for $M < 0.7$. Clearly, Eq. (8.22) would not hold in transonic and supersonic flows, in which the presence of the tube may introduce shock waves and other discontinuities. Keep in mind that the total pressure decreases across a normal shock, according to the *Rankine–Hugoniot relationship*, Eq. (8.6). Conical-nose tubes are preferable to blunt-nose tubes in supersonic flows, particularly for static-pressure measurement. Compressibility effects are coupled with other sources of error, such as displacement and misalignment effects, introducing a large uncertainty into any correction procedure.

Multi-hole probes: Multi-hole probes are arrangements of two or more orifices at the tip of a probe body with faces inclined with respect to the probe axis so that they can provide information about the flow direction. In combination with total- and static-pressure orifices, they may also measure the flow velocity in magnitude and direction. The simplest type of such probes, called *yawmeters*, consists of two symmetrically inclined thin tubes. They normally operate in the *nulling mode*, which means that they are rotated about their axes until the pressure readings of both tubes become equal, in which case the flow direction should be normal to the line connecting the two orifice centres; for plane flows, this would determine the flow direction. The addition of an upstream-facing orifice between the two inclined ones would provide the total pressure, from which the flow velocity magnitude can be determined; an example of such probes is the *cobra probe* [Fig. 8.7(f)], in which the tip centre is aligned with the probe axis but the plane of the three orifices is normal to this axis for convenient insertion and rotation through the apparatus wall. Probes with two pairs of inclined orifices with faces located in a pyramidal arrangement, like the *claw-type* probe, have also been developed for use in three-dimensional flows, in which it would be necessary to determine both the *yaw* and the *pitch angles* of the flow with respect to a fixed axis. In many experimental settings, nulling of probes is inconvenient and time consuming, and it is preferable to recover flow information from a fixed probe or one that is traversed parallel to a fixed direction. There is a variety of multiple-hole probes that provide flow direction and velocity based on analytical expressions or calibration curves. Probes with spherical or hemispherical tips may utilize the potential flow solution that provides the pressure coefficient on the surface of a sphere as

$$C_P = \frac{p - p_\infty}{\frac{1}{2}\rho V^2} = 1 - \frac{9}{4}\sin^2\theta, \tag{8.24}$$

where P_∞ is the undisturbed stream pressure and θ is the angle formed by radial axes passing through the surface point and the stagnation point. Because of imperfections in probe manufacturing, the response of a specific probe may deviate from the theoretical expression and it would be preferable to compute C_P from calibration data. Probe designs with wedge-like, conical, pyramidal, and other shapes have been developed; however, the most popular design, known as the *five-hole probe*, consists of a set of five parallel tubes, a central one leading to an orifice normal to the tube axis at the apex of a pyramid formed by the inclined orifices of four peripheral tubes, arranged symmetrically [Fig. 8.7(g)]. Such probes may be constructed by a skilled machinist on-site and be made at sizes as small as nearly 1 mm in cross-sectional diameter [52, 53]. For their calibration [44], these probes are inserted in a uniform stream at different combinations of pitch and yaw angles φ and ψ, respectively, and the five pressures p_1, p_3 ('pitch' pressure pair), p_2, p_4 ('yaw' pressure pair), and p_5 ('total' pressure) are recorded. The actual total pressure p_0 of the stream is also recorded and an average side-orifice pressure is computed as

$$p_m = \frac{1}{4}(p_1 + p_2 + p_3 + p_4). \tag{8.25}$$

Then, the values of the pitch, yaw, total-pressure, and dynamic pressure coefficients are computed, respectively, as

$$C_\varphi = \frac{p_3 - p_1}{p_5 - p_m}, \tag{8.26}$$

$$C_\psi = \frac{p_4 - p_2}{p_5 - p_m}, \tag{8.27}$$

$$C_0 = \frac{p_0 - p_5}{p_5 - p_m}, \tag{8.28}$$

$$C_q = \frac{p_5 - p_m}{\frac{1}{2}\rho V^2}, \tag{8.29}$$

and plotted in iso-contour form vs. the two angles φ and ψ. During an experiment, one measures the five pressures, which can be used to compute C_φ and C_ψ. Assuming that each pair of values of C_φ and C_ψ corresponds to a unique pair of φ and ψ, one may then find the orientation of the flow velocity with respect to the probe axis. From the combination of computed values of φ and ψ, one may then find C_0 and C_q, from which it is easy to compute the flow velocity V. This process may be automated by use of multi-variable curve-fitting algorithms. A typical operating range of five-hole probes is within a cone of 30° half-angle. For applications involving larger flow incidence, seven-hole probes have been developed, which reportedly perform well within half-cone angles of 75° [54]. Other authors [55–57] have examined the dependence of multi-hole probe response on Reynolds number, shear, wall proximity, turbulence, compressibility, and other factors.

Fast-response probes: In most applications, static-pressure tubes are connected to manometers or pressure transducers through tubing, in which case their frequency response would be relatively slow, typically of the order of 0.1 Hz. Nevertheless, pressure fluctuations have been measured with static tubes in which a fast-response pressure transducer was placed near the static orifices. A much-quoted early design [46] not only contained a microphone embedded inside the tube, but also had a nose tip that was mounted to the main probe body through four piezoelectric elements that measured the velocity component normal to the tube axis; analogue circuitry provided compensation to the static-pressure reading for turbulence effects. More recently, in-flow fluctuating pressure has been measured by the placement of miniature piezoresistive (semiconductor-type) pressure transducers inside pressure tubes and multi-hole probes [20, 22]. Four thin piezoelectric elements are arranged in a Wheatstone bridge configuration and mounted on a pressure-sensing membrane, and the sensor output is normally compensated for temperature sensitivity. A different concept is exploited in *bleed-type pressure tubes* [58, 59]. These tubes are connected to a constant-pressure air supply, and the pressure is regulated such as to produce laminar flow through the tube and the static-pressure orifices. The tubes contain a hot-film sensor, which measures flow velocity. The instantaneous pressure difference across the tube is computed through its linear relationship to flow velocity in the tube. The probe must be calibrated, and its output is compensated to improve its frequency response. All probes just described are subjected to errors that are due to vibrations, internal motions, and external turbulent fluctuations. If the flow velocity is measured simultaneously by other means, the instantaneous pressure reading may, in principle, be corrected for turbulence effects; the magnitude of such corrections remains an issue of controversy.

8.5 Dynamic response and testing of pressure-measuring systems

Dynamic response of liquid-in-glass manometers: Although liquid-in-glass manometers are not used to measure rapidly changing pressures, they do tend to exhibit oscillatory response, and so it is of interest to understand their dynamic characteristics and use them as a guide for selection or design of a manometer. The following is a summary of a simplified analysis [15], which has been roughly confirmed experimentally.

Consider a liquid-in-glass manometer, as sketched in Fig. 8.1(a), containing a liquid with a density ρ and connected on both sides to a gas. Let d be the diameter of the tube and L the total length of the tube part that contains the liquid. The unsteady flow of the liquid inside the tube could be quite complex and different from fully developed laminar or turbulent pipe flow. If, however, one assumes the flow to have a fully developed, steady, laminar profile, and further neglects secondary effects such as surface tension and gas inertia, it is possible to show that the motion of the liquid in the manometer has the dynamic response of a second-order system with undamped natural frequency

$$\omega_n = 1.2\sqrt{\frac{g}{L}}, \tag{8.30}$$

and damping ratio

$$\zeta = \frac{9.8\nu}{d^2}\sqrt{\frac{L}{g}}. \tag{8.31}$$

The consideration of turbulent flow leads to a non-linear system, which does not lend itself to analytical description. Further simplifying the problem by assuming that the liquid flow is oscillatory with a frequency ω and an amplitude Δz, but has the same wall shear stress as fully developed, stationary, turbulent pipe flow, allows one to determine a quasi-linear response with the following properties: undamped natural frequency

$$\omega_n = 1.4\sqrt{\frac{g}{L}}; \tag{8.32}$$

this is only 14% higher than the laminar flow value, which increases confidence in the use of either expression (8.30) or expression (8.32) for rough predictions; equivalent damping ratio

$$\zeta_e = \left(\frac{9.8\nu}{d^2}\sqrt{\frac{L}{g}}\right)\left(\frac{1}{1200}\frac{\omega\Delta z d}{\nu}\right)^{3/4}, \tag{8.33}$$

which is a function of the frequency (essentially equal to ω_n) and amplitude of oscillation; the characteristic Reynolds number $\omega\Delta zd/\nu$ of the oscillation should be greater than 1200 if the flow is to be turbulent over most of the cycle, and so the turbulent damping ratio should be greater than the laminar one; furthermore, this relationship shows that damping would decrease as the oscillations continue, which means that the amplitude of oscillation would diminish slower from one cycle to the next than that of a second-order system.

Dynamic response of transducer–tube systems: The dynamic response of a complex pressure-measuring system would generally be non-linear and difficult to analyse accurately. Nevertheless, it would be instructional to present a few simplified cases, in which transducer–tube combinations have been modelled as low-order linear dynamic systems. Let us consider a remotely mounted pressure transducer, such as shown in Fig. 8.6(a), connected to a small pressure tap through a tube having a diameter d and a length l. The transducer is assumed to contain a deformable elastic membrane or diaphragm, mounted within an internal pressure chamber or in contact with the fluid in an external cavity. Let V be the volume of the pressure chamber or external cavity. An important parameter is the change of fluid-filled volume of the pressure chamber or cavity as a result of the diaphragm deformation. This is expressed in terms of the *compliance* C of the transducer, defined as the change of volume per unit change of applied pressure. The transducer compliance is sometimes provided by the transducer manufacturer and may also be estimated theoretically or experimentally.

Effects that may influence the pressure-measuring system response include these:

- Volume changes that are due to deformation of the diaphragm; with the exception of extreme cases, volume changes that are due to expansion or contraction of the tube and the cavity may be neglected.
- Fluid compressibility.
- Restoring forces that are due to diaphragm elasticity; again, restoring forces that are due to tube and cavity wall elasticity may usually be neglected.
- Acceleration ('inertial forces') of the fluid and the diaphragm.
- Frictional forces between the fluid and the solid walls and between solid components.

Depending on the transducer–tube configuration and the flow conditions, some or all of the preceding factors would have to be considered [8, 15]. A few typical cases of dynamic analysis of transducer–tube combinations [15] are presented in the following.

The simplest dynamic model of such systems assumes that the fluid is incompressible and neglects inertia effects. This applies mainly to liquids and relatively slow pressure changes. The pressure sensed by the transducer, considered as the system output, is related to the input pressure through a first-order system relationship with a time constant

$$\tau = \frac{128\mu l C}{\pi d^4}. \tag{8.34}$$

This expression clearly shows the extreme sensitivity of the time constant on the tube diameter and explains why small-diameter tubes should be avoided. It also shows that the tube length should be maintained as short as possible.

An extension of the preceding analysis to include inertia effects, albeit also based on significant simplifications, results in a second-order system response with a natural frequency and a damping ratio respectively equal to

$$\omega_n = \frac{0.767d}{\sqrt{\rho l C}}, \quad \zeta = \frac{15.6\mu}{d^3}\sqrt{\frac{lC}{\rho}}. \tag{8.35}$$

When fluid compressibility effects are accounted for, additional sources of nonlinearity are introduced. With the assumption of small pressure changes about a reference value p_r, it is possible to predict a second-order system response with a natural frequency and a damping ratio respectively equal to

$$\omega_n = \frac{\sqrt{\frac{\gamma p_r}{\rho}}}{l\sqrt{\frac{1}{2} + \frac{V}{\pi d^2 l/4}}}, \quad \zeta = \frac{16\mu l}{d^2\sqrt{\gamma p_r \rho}}\sqrt{\frac{1}{2} + \frac{V}{\pi d^2 l/4}}, \tag{8.36}$$

where γ is the specific heat ratio of the gas.

Dynamic testing of pressure transducers: In the majority of engineering applications, the dynamic responses of pressure transducers and pressure-measuring systems can be approximated by a second-order system response, in which case it would be sufficient to determine the natural frequency and damping ratio of the system. These can be estimated most conveniently by the monitoring of the step response of the system exposed to a sudden change in pressure. A useful device for the generation of step-like changes in pressure is the *shock tube*, namely a tube containing a metallic or plastic diaphragm, which separates gas under sufficiently high pressure from low-pressure gas. At a certain time, the diaphragm is punctured and a shock wave propagates through the low-pressure side until it reaches the end of the tube where the pressure system under test is mounted. This method is good for testing systems with natural frequencies as high as 250 kHz, or even higher. Simpler pressure transducer testing devices, also consisting of two chambers containing gas under different pressures and separated by a diaphragm or a fast-opening valve are available commercially, and may also be fabricated relatively easily. Compared with shock tubes, such devices are suitable for much lower dynamic ranges, not exceeding a few kilohertz. For liquid systems, one may also utilize the water-hammer pressure wave generated in a pipe by a fast-closing valve.

When the type of response of the pressure-measuring system is unknown, it is advisable to determine its frequency response rather than its step response. To do so, it is necessary to generate a sinusoidal pressure variation with adjustable frequency and, if possible, amplitude. This can be achieved, within the frequency range of up to approximately 10 kHz, in a sealed chamber filled with a liquid or gas, by driving a diaphragm through a piston–connecting rod mechanism or an electrodynamic shaker. Another arrangement is to release the chamber pressure through an orifice that is periodically aligned with each of multiple holes on a rotating disk. One may also utilize a variety of arrangements by which either the transducer is oscillated in front of a steady fluid stream or the stream is periodically interrupted. Each of these methods is subjected to interference and must be applied with caution. In general, it is highly recommended to test the performance of such apparatus vs. a high-quality, high-frequency-response pressure transducer.

QUESTIONS AND PROBLEMS

1. Consider a vertical U-tube manometer, as shown in Fig. 8.1(a), consisting of a glass tube with an inner diameter of 0.3 mm. Fluid 1 is air, fluid 2 is mercury, and fluid 3 is water, all at 20 °C. Point A is open to a standard atmosphere, whereas point B is at an elevation $z_B = 988$ mm. If $z_c = 743$ mm and $z_D = 345$ mm, determine the absolute water pressure at point B. Compute all capillary corrections and apply them to your result.
2. Determine and plot the capillary rise or fall of manometric liquid in a glass tube with inner diameters in the range between 1 and 20 mm for the liquid combinations contained in Table 8.1. Discuss your observations.
3. Consider a vertical U-tube manometer, as shown in Fig. 8.1(a), consisting of a glass tube with an inner diameter of 0.3 mm and connected to air on both sides. The manometric fluid occupies a length of 950 mm. Estimate the undamped natural frequency, damping ratio and damped natural frequency of this manometer, if the manometric fluid is (a) water, (b) mercury, or (c) glycerol, all at 20 °C. If a pressure difference of 100 Pa is suddenly applied on the manometer, plot the elevation difference of the manometer free surfaces vs. time. Which fluid would you recommend and why?
4. A pressure transducer that has been calibrated statically is used to measure pressure fluctuations in a turbulent flow. Discuss briefly the concerns that you have and any means to remove such concerns.
5. Measurements in a high-speed flow of helium have provided a stagnation temperature of 500 K, a stagnation pressure of 671 kPa, and a static pressure of 400 kPa. Determine the flow velocity and the Mach number. Also estimate the error in velocity if Bernoulli's equation were used instead of more appropriate relationships.

6. Consider a fully developed, turbulent air flow at 20 °C in a smooth pipe with an inner diameter $D = 500$ mm. Wall static pressure is measured with two pressure taps, an upstream one with a diameter $d_1 = 2$ mm and a second one with a diameter $d_3 = 4$ mm, located 1000 mm downstream of the first one. The static pressure at the first tap is approximately 106.3 kPa, whereas the pressure difference between the two taps is 95 Pa. (a) Estimate the wall shear stress assuming that there is no tap-size error; (b) estimate the bulk velocity of the flow in the pipe assuming that there is no tap-size error (*Hint:* Find the velocity by iteration using the Moody diagram to determine the friction coefficient); (c) estimate corrections for the two tap readings; (d) if you find these corrections to be significant, correct the estimated wall shear stress and bulk velocity; iterate until you are satisfied with the accuracy of your results.
7. A Pitot tube with an inner diameter of 0.85 mm and an outer diameter of 1.22 mm is used to measure total pressure in the wake of a nearly two-dimensional object in air flow at a temperature of 20 °C. The difference Δp between the tube reading at different distances y of the tube's centre from the wake axis and the static pressure measured accurately by other means is presented in the following table.

y [mm]	Δp [Pa]	y [mm]	Δp [Pa]	y [mm]	Δp [Pa]
0	86.4	6	110.7	16	127.8
1	91.6	7	113.4	18	129.4
2	96.2	8	115.9	20	130.6
3	100.4	10	120.0	25	132.6
4	104.2	12	123.2	30	133.7
5	107.6	14	125.8	35	134.3

(a) Determine the velocity variation across the wake, neglecting the displacement effect; (b) estimate appropriate corrections for the displacement effect; and (c) plot the uncorrected and corrected velocity profiles and discuss possible differences between them.
8. Consider a circular Pitot tube with an inner diameter of 0.85 mm and an outer diameter of 1.22 mm inserted in horizontal flow of glycerol at an elevation 200 mm below the free surface, where you may assume that the pressure is hydrostatic. The temperature is 20 °C, and the atmospheric pressure is standard. Estimate the flow velocity as a function of the total pressure indicated by the tube over the speed range between 10^{-3} and 1 m/s. Plot this estimate in logarithmic axes and fit an appropriate empirical expression to your results to facilitate computing the flow velocity from pressure measurements.
9. A hypodermic needle with an external diameter of 0.5 mm and an internal diameter of 0.3 mm is used to measure the total pressure in an open-channel flow of a homogeneous mixture of 80% glycerol and 20% water at 20 °C. The channel depth is 100 mm. The tube readings (gauge pressure), measured with a micromanometer at different elevations y from the bottom, are presented in the following table. Assume that the static pressure in the channel is hydrostatic. The density and kinematic

viscosity of the mixture can be found from tables as 1208.5 kg/m^3 and 49.57 $\times$ 10^{-6} m^2/s.

y [mm]	p_0 [Pa]	y [mm]	p_0 [Pa]
0.5	1179.3	50	1854.7
2	1167.8	60	2140.4
6	1137.3	70	2001.3
10	1106.8	80	1735.2
20	1087.1	90	1309.3
30	1141.1	95	913.3
40	1409.4		

Estimate, applying appropriate corrections if required, the velocity variation across the channel. Discuss the level of confidence in these results and assign uncertainty limits.

REFERENCES

[1] T. G. Beckwith, R. D. Marangoni, and J. H. Lienhard V. *Mechanical Measurement* (5th Ed.). Addison-Wesley, Reading, MA, 1993.

[2] R. P. Benedict. *Fundamentals of Temperature, Pressure and Flow Measurements* (2nd Ed.). Wiley Interscience, New York, 1977.

[3] E. Ower and R. C. Pankhurst. *The Measurement of Air Flow*. Pergamon, Oxford, UK, 1977.

[4] P. W. Harland. *Pressure Gauge Handbook*. Marcel Dekker, New York, 1985.

[5] B. E. Richards (Editor). *Measurement of Unsteady Fluid Dynamic Phenomena*. Hemisphere, Washington, DC, 1977.

[6] R. J. Emrich (Editor). Fluid dynamics. In L. Marton and C. Marton (Editors-in-Chief), *Methods of Experimental Physics*, Vols. 18A and B. Academic, New York, 1981.

[7] R. J. Goldstein. *Fluid Mechanics Measurements* (2nd Ed.). Taylor & Francis, Washington, 1996.

[8] T. Arts, H. Boerrigter, J.-M. Buchlin, M. Carbonaro, G. Degrez, R. Dénos, D. Fletcher, D. Olivari, M. L. Riethmuller and R. A. van den Braembussche. *Measurement Techniques in Fluid Dynamics* (2nd Ed.). Von Karman Institute for Fluid Dynamics, Rhode-Saint Genèse, Belgium, 2001.

[9] G. N. Peggs (Editor). *High Pressure Measurement Techniques*. Applied Science Publishers, London, 1983.

[10] J. H. Henry. *Pressure Measurement in Vacuum Systems*. Chapman & Hall, London, 1964.

[11] G. K. Batchelor. *An Introduction to Fluid Dynamics*. Cambridge University Press, Cambridge, UK, 1970.

[12] R. L. Panton. *Incompressible Flow*. Wiley, New York, 1984.

[13] R. E. Sonntag, C. Borgnakke, and G. J. Van Wylen. *Fundamentals of Thermodynamics* (5th Ed.). Wiley, New York, 1998.

[14] R. W. Fox, A. T. McDonald, and P. J. Pritchard. *Introduction to Fluid Mechanics* (6th Ed.). Wiley, New York, 2004.

[15] E. O. Doebelin. *Measurement Systems Application and Design* (5th Ed.). McGraw-Hill, New York, 2004.

[16] H. F. Olsen. *Elements of Acoustical Engineering*. Van Nostrand, Princeton, NJ, 1957.
[17] W. P. Mason. *Physical Acoustics*. Academic, New York, 1964.
[18] S. Handel. *A Dictionary of Electronics*. Penguin Books, Harmondsworth, Middlesex, UK, 1962.
[19] M. Gad-el-Hak (Editor). *The MEMS Handbook*. CRC Press, Boca Raton, FL, 2002.
[20] R. W. Ainsworth, R. J. Miller, R. W. Moss, and S. J. Thorpe. Unsteady pressure measurement. *Meas. Sci. Technol.*, 11:1055–1076, 2000.
[21] L. Löfdahl and M. Gad-el-Hak. MEMS-based pressure and shear stress sensors for turbulent flows. *Meas. Sci. Technol.*, 10:665–686, 1999.
[22] P. Kupferschmied, P. Köppel, W. Gizzi, C. Roduner, and G. Gyarmathy. Time-resolved flow measurements with fast-response aerodynamic probes in turbomachines. *Meas. Sci. Technol.*, 11:1036–1054, 2000.
[23] M. Sayer and A. Mansingh. *Measurement, Instrumentation and Experiment Design in Physics and Engineering*. Prentice-Hall of India, New Delhi, 2000.
[24] C. M. Van Atta. *Vacuum Science and Engineering*. McGraw-Hill, New York, 1965.
[25] J. F. O'Harlon. *A User's Guide to Vacuum Technology*. Wiley-Interscience, New York, 1980.
[26] A. Roth. *Vacuum Technology*. North-Holland, Amsterdam, 1982.
[27] W. H. Howe. What's available for high pressure measurement and control. *Control. Eng.*, 2:53, 1955.
[28] R. E. Franklin and J. M. Wallace. Absolute measurements of static-hole error using flush transducers. *J. Fluid Mech.*, 42:33–48, 1970.
[29] B. J. McKeon and A. J. Smits. Static pressure correction in high Reynolds number fully developed turbulent pipe flow. *Meas. Sci. Technol.*, 13:1608–1614, 2002.
[30] R. Shaw. The influence of hole dimensions on static pressure measurements. *J. Fluid Mech.*, 7:550–564, 1960.
[31] J. L. Livesey, J. D. Jackson, and C. J. Southern. The static hole error problem. *Aircraft Eng.*, 34:43–47, 1962.
[32] R. E. Rayle. Influence of orifice geometry on static pressure measurements. Master's thesis, MIT, Cambridge, MA, 1949.
[33] R. E. Rayle. Influence of orifice geometry on static pressure measurements. ASME Paper 59-A-234, American Society of Mechanical Engineers, New York, 1959.
[34] W. J. Rainbird. Errors in measurements of mean static pressure of a moving fluid due to pressure holes. Tech. Rep. DME/NAE, *Quarterly Bulletin of the Division of Mechanical Engineering*, National Aeronautical Establishment, National Research Council of Canada, 1967.
[35] P. Bradshaw and D. G. Goodman. The effect of turbulence on static-pressure tubes. Tech. Rep. and Memo. No. 3527, Aeronautical Research Council, London, 1966.
[36] M. Mackay. Static pressure measurement with surface-mounted disk probes. *Exp. Fluids*, 9:105–107, 1990.
[37] A. J. Smits and T. T. Lim (Editors). *Flow Visualization Techniques and Examples*. Imperial College Press, London, 2000.
[38] C. Mercer (Editor). *Optical Metrology for Fluids, Combustion and Solids*. Kluwer Academic, Dordrecht, The Netherlands, 2003.
[39] T. Liu, T. Campbell, S. Burns, and J. Sullivan. Temperature and pressure sensitive luminescent paints in aerodynamics. *Appl. Mech. Rev.*, 50:227–246, 1997.
[40] B. M. McLachlan and J. H. Bell. Pressure-sensitive paints in aerodynamic testing. *Exp. Therm. Fluid Sci.*, 10:470–485, 1995.

[41] W. W. Willmarth. Unsteady force and pressure measurements. *Annu. Rev. Fluid Mech.*, 3:147–170, 1971.

[42] S. H. Chue. Pressure probes for fluid measurement. *Prog. Aerosp. Sci.*, 16:147–223, 1975.

[43] W. H. Rae and A. Pope. *Low-Speed Wind Tunnel Testing*. Wiley, New York, 1984.

[44] D. W. Bryer and R. C. Pankhurst. *Pressure-Probe Methods for Determining Wind Speed and Flow Direction*. National Physical Laboratory, London, 1971.

[45] S. Tavoularis. Techniques for turbulence measurement. In N. P. Cheremisinoff, editor, *Encyclopedia of Fluid Mechanics*, Vol. 1, Chap. 36, pp. 1207–1255. Gulf Publishing, Houston, TX, 1986.

[46] T. E. Siddon. On the response of pressure measuring instrumentation in unsteady flow. Tech. Rep. UTIAS Rep. 136, University of Toronto, Institute for Aerospace Studies, 1969.

[47] F. A. L. Winternitz. Simple shielded total-pressure probes. *Aircraft Eng.*, pp. 313–317, October 1958.

[48] A. D. Young and J. N. Maas. The behaviour of a pitot-tube in a transverse total-pressure gradient. Tech. Rep. and Memo. No. 1770, Aeronautical Research Council, London, 1936.

[49] S. Tavoularis and M. Szymczak. Displacement effects of square-ended pitot tubes in shear flows. *Exp. Fluids*, 7:33–37, 1989.

[50] F. A. Macmillan. Experiments on pitot tubes in shear flow. Tech. Rep. and Memo. No. 3028, Aeronautical Research Council, London, 1957.

[51] A. Quarmby and H. K. Das. Displacement effects on pitot tubes with rectangular mouths. *Aeronaut. Q.*, 20:129–139, 1969.

[52] P. M. Ligrani, B. A. Singer, and L. R. Baun. Miniature five-hole pressure probe for measurement of three mean velocity components in low-speed flows. *J. Phys. E*, 22:868–876, 1989.

[53] A. L. Treaster and H. E. Houtz. Fabricating and calibrating five-hole probes. In R. A. Bajura and M. L. Billet, editors, *Proceedings of the Fluid Measurements and Instrumentation Forum*, pp. 1–4. AIAA/ASME 4th Fluid Mechanics, Plasma Dynamics and Lasers Conference, Atlanta, 1986.

[54] K. N. Everett, A. A. Gerner, and D. A. Durston. Seven-hole cone probes for high angle flow measurement: Theory and calibration. *AIAA J.*, 21:992–998, 1983.

[55] A. L. Treaster and A. M. Yocum. The calibration and application of five-hole probes. *ISA Trans.*, 18(3):23–34, 1979.

[56] N. Sitaram, B. Lakshminarayana, and A. Ravindranath. Conventional probes for the relative flow measurement in a turbomachinery rotor blade passage. *J. Eng. Power*, 103:406–414, 1981.

[57] M. Samet and S. Einav. Directional pressure probe. *Rev. Sci. Instrum.*, 55:582–588, 1984.

[58] B. W. Spencer and B. G. Jones. A bleed-type pressure transducer for in-stream measurement of static pressure fluctuations. *Rev. Sci. Instrum.*, 42:450–454, 1971.

[59] B. G. Jones. A bleed-type pressure transducer for in-stream fluctuating static pressure sensing. *TSI Quarterly*, 7(2):5–11, April–June, 1981.

9 Measurement of flow rate

This chapter covers common instrumentation and techniques that are used to measure *bulk flow rates*, namely the amounts of fluids that pass through a certain cross section of a pipe, duct, channel, or other flow conduit per unit time. Bulk flow measurement is not concerned with local velocity variations across the cross section, nor with short-time (e.g., turbulent) fluctuations. Measured flow rates can be either *mass flow rates* $\dot{m}$, or, when dealing with liquids or low-speed gases, *volume flow rates* Q. Flow-rate measurement is an essential activity in a variety of industries and utility services, but it is also employed regularly in the fluid mechanics laboratory, notably in the important role of monitoring and controlling the experimental conditions. The operation of flow-rate measurement systems is based on diverse physical principles; with some exceptions, such systems require calibration or empirical corrections. The following presentation is mainly concerned with bulk flow measurement in 'simple' flows, which are single phase and either steady or very slowly varying. Some of these methods can be extended to flows of multi-phase fluids, slurries, and granular materials, but the reader is advised to consult specialized manufacturers and references when dealing with such media. More details on general methods for the measurement of flow rate and specific instruments can be found in several books [1–5], handbooks [6–9], and manufacturer's catalogues (e.g., Ref. 10).

9.1 Direct methods

The simplest flow-rate measurement methods are *direct*, which means that they measure a typical flow velocity or the amount of discharged fluid over a period of time. Such methods are more suitable for liquid than for gas flows. For example, one may obtain a rough measurement of the bulk velocity of flows in water tunnels and open channels by timing the motion of suspended or floating objects. For flows of non-volatile liquids in an open-loop configuration, one can measure the volume flow rate by timing the filling of a container by the discharge of the apparatus; similarly, one can measure the mass flow rate by weighing the discharged fluid. In such cases, one must take care that discharging of fluid has no appreciable effect on the operation of the system, as for example would be the case if removal of liquid from the loop resulted in lowering the head of a feeding tank or shifting the operating point of a pump. Direct flow-rate measurement methods, applicable to liquid and gas flows in both open- and closed-loop configurations, include the use of positive-displacement flow meters, discussed in the next section.

9.2 Positive-displacement flow meters

Positive-displacement (PD) flow meters are devices that isolate fixed volumes of the fluid flowing into their inlet in sealed compartments and then discharge them to the outlet. Neglecting leakage and other possible defficiencies, one can easily compute the volume flow rate from the size and number of compartments in the device and the measured rate of repetition of the cycle of their operation. In many cases, the same instruments can be configured to measure the total volume of fluid that passes through them over a time interval, and, for this reason, PD meters are commonly used to monitor the consumption of water, natural gas, and hydrocarbon fuels. PD meters may be operating passively, receiving power from the flowing fluid, or driven by an external source to create the fluid motion, in which case they are called *metering pumps*. There is a great variety of designs of PD meters, most of which can be classified as *rotary*, *reciprocating*, or *nutating*. The important parameters for their operation are the *leakage* and the *pressure loss* across them. Sealing relies on capillary action in narrow gaps between meshing parts; thus leakage depends on the speed of operation, the viscosity of the fluid, and the wear of the moving components and the housing. To minimize leakage, the components are manufactured under small tolerances and the clearances between meshing parts are kept low; this necessitates the use of clean fluids, and in most cases the meter is accompanied by a filter. PD meters are suitable for fluids with relatively wide ranges of viscosity. Increasing viscosity improves the sealing action but also increases pressure losses, which have to be maintained as low as possible to avoid significant loading of the fluid system. Temperature variation affects the operation of PD meters in two ways: by affecting the viscosity of the fluid, with implications for leakage and pressure loss, and by affecting the density of the fluid, which is of concern when one is converting volume flow rate to mass flow rate. Manufacturers would normally provide charts describing the operation characteristics of each model, such as the pressure loss for different flow rates and fluid viscosities as well as correction factors and uncertainties at different speeds. A few representative designs of PD meters are subsequently discussed [1, 3, 6, 9, 11, 12].

PD flow meters for liquids

Nutating disk meters: The main element of these meters is a disk that is rotating in a nutating (precessing, wobbling) fashion, while both its sides are partially in contact with a dual conical housing [Fig. 9.1(a)]. Fluid enters through the inlet port facing one side of the disk during half of the cycle, in isolation from the outlet port; it is then swept by the precessing motion of the disk to the outlet during the following half of the cycle, while being in isolation from the inlet port.

Reciprocating-piston flow meters: These meters contain a number of plungers or pistons, driven by a wobble plate and sweeping the volumes of corresponding

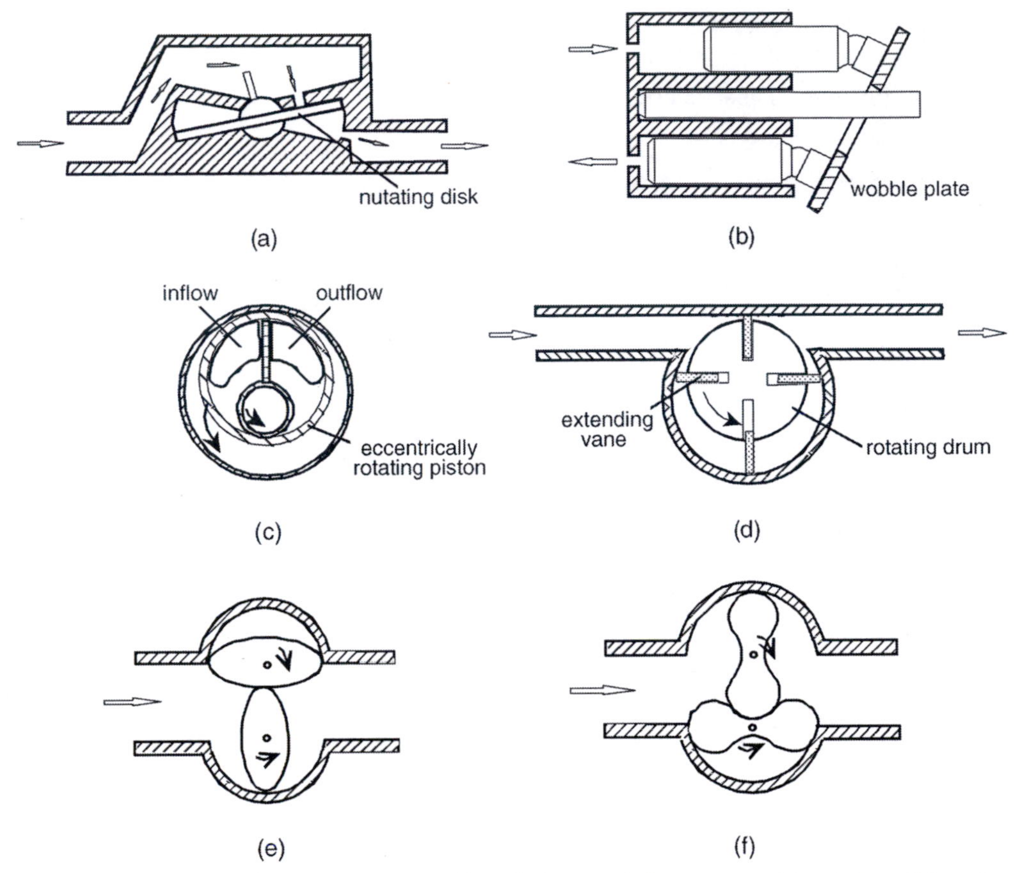

Figure 9.1. Sketches of representative positive displacement flow meters: (a) nutating disk meter, (b) reciprocating-piston meter, (c) rotary-piston meter, (d) rotary-vane meter, (e) oval gear meter, and (f) roots meter.

cylinders, while at the same time opening and closing input and output ports or valves [Fig. 9.1(b)].

Rotary-piston flow meters: These meters contain a cylindrical drum, which is mounted eccentrically inside a cylindrical housing and rotates with its outer surface in contact with the housing, while its inner surface maintains contact with an inner cylinder, coaxial with the housing [Fig. 9.1(c)].

Rotary-vane flow meters: Flat vanes are inserted into matching slots around the perimeter of a rotating cylindrical drum, located eccentrically within the housing. Centrifugal action or springs cause the vanes to slide out of the slots until they come into contact with the housing, thus isolating a volume of the flowing fluid and transporting it from the inlet towards the outlet [Fig. 9.1(d)].

Rotor meters: These meters contain rotating meshing elements of different shapes, including oval gears [Fig. 9.1(e)], circular gears, helical gears, and lobes. The *rotary-abutment meters* contain both specially shaped rotors and rotating vanes. In these devices, fluid is trapped in the space between the rotating elements or between an element and the housing and is pushed towards the outlet in isolation from the input.

PD flow meters for gases

Roots-type flow meters: This is a trademark name that describes a particular design of a lobe meter [Fig. 9.1(f)], developed for use with gases.

Diaphragm-type flow meters: These meters are commonly used in domestic gas lines. They contain bellows that fill up with gas during part of the cycle and discharge it to the outlet during a subsequent part; the gas flow from the inlet to the outlet is controlled by sliding valves, and the motion of the bellows is linked to a mechanism that counts the cycles.

Liquid-sealed drum-type flow meters: Aslo known as *wet gas meters*, these devices consist of a hollow drum rotating within a cylinder partly filled with a liquid, which provides the sealing action.

9.3 Venturi, nozzle, and orifice-plate flow meters

Also known as *restriction* or *obstruction flow meters*, these are devices that force the flow through a restriction, thus increasing its velocity and decreasing its pressure. The flow rate is estimated from a measured pressure difference and an empirical correction coefficient. To describe the idealized response of these devices, consider steady, uniform, inviscid, incompressible flow, in the absence of body forces, flowing within a circular tube with a diameter D and guided to a restriction with a diameter d. Then one may use continuity and Bernoulli's equation to relate the ideal volume flow rate Q_{id} to the pressure drop Δp between the two cross-sections as

$$Q_{id} = \frac{\pi D^2/4}{\sqrt{1-(d/D)^4}}\sqrt{\frac{2\Delta p}{\rho}}. \tag{9.1}$$

To account for deviations from the idealized behaviour, one may introduce an empirical discharge coefficient C_d to compute the actual volume flow rate as

$$Q = C_d \frac{\pi D^2/4}{\sqrt{1-(d/D)^4}}\sqrt{\frac{2\Delta p}{\rho}}. \tag{9.2}$$

The value of C_d depends mainly on the geometry of the apparatus and the Reynolds number, with $0 < C_d < 1$. For Reynolds numbers sufficiently large for the flow to be in the fully turbulent regime, C_d becomes insensitive to Reynolds number and depends only on the shape of the device. It is desirable to keep its value as large as possible in order to reduce the permanent pressure loss in the flow meter. On the other hand, large values of the discharge coefficient may be achieved only with carefully shaped and relatively long devices, which tend to be bulkier and more expensive than devices with lower C_d. Common low-loss restriction flow meters are the *Venturi tubes* [Figs. 9.2(a)

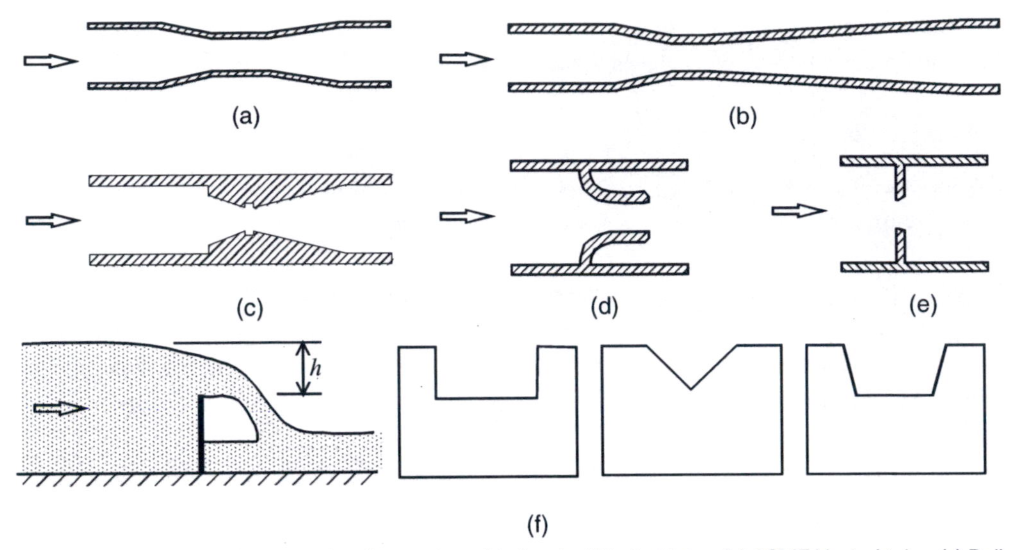

Figure 9.2. Sketches of obstruction flow meters: (a) classical Venturi tube, (b) ASME Venturi tube, (c) Dall tube, (d) flow nozzle, (e) orifice plate, and (f) weirs.

and 9.2(b)] and the *Dall tubes* [Fig. 9.2(c)], whereas relatively high-loss meters include *flow nozzles* [Fig. 9.2(d)] and *orifice plates* [Fig. 9.2(e)]. The designs of such devices are regulated by various standards [e.g., ISO and American Society of Mechanical Engineers (ASME) standards], so that they may be used interchangeably and without the need for individual calibration. A limitation of obstruction flow meters is their relatively narrow dynamic range, which is a result of the non-linearity in the Q–Δp relationship. Thus the sensitivity of these flow meters decreases rapidly as the flow rate drops below about 25% of the full-scale value.

9.4 Open-channel flow measurement

The volume or mass flow rate of liquids in open channels or partially filled pipes and ducts is often measured with the use of direct methods (see Section 9.1). Another common approach is the use of flow restrictions, including *weirs* and *Venturi flumes* [3,7,13–15].

Weirs consist of obstructions positioned across the channel and over which the liquid is forced to flow. There are several types of weirs, both sharp crested and broad crested, with rectangular, V-shaped, or trapezoidal openings [Fig. 9.2(f)]. The most common type used in fluid mechanics and hydraulics laboratories is the sharp-crested, V-notch weir. An approximate expression for the flow rate over such weirs is [15]

$$Q \approx 2.5 \tan \frac{\theta}{2} H^{5/2}, \tag{9.3}$$

where θ is the full angle of the notch (usually equal to 90°) and H is the easily measurable head of the liquid above the weir, namely the vertical distance between the free surface and the lowest point in the weir opening.

Venturi flumes are converging–diverging channel constrictions, analogous to Venturi tubes used in pipes flows. They generally have low pressure losses and are available in

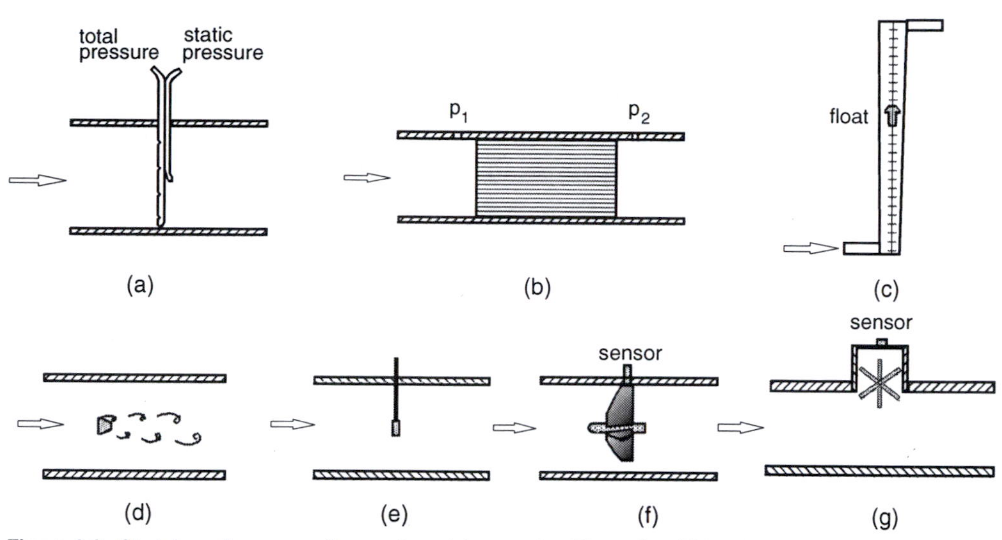

Figure 9.3. Sketches of common flow meters: (a) averaging Pitot tube, (b) laminar flow elements, (c) rotameter, (d) vortex-shedding flow meter, (e) drag flow meter, (f) turbine flow meter, and (g) paddlewheel flow meter.

several different designs, including flumes with rectangular, trapezoidal, and U-shaped cross sections; commonly used designs for sewage and irrigation flows are the *Parshall* and *Palmer–Bowlus flumes*. Their advantages over weirs is that they do not cause a water back-up and are less likely to be affected by deposited solids that may be transported by the water.

9.5 Averaging Pitot tubes

These instruments consist of a tube spanning the cross section of the pipe and having multiple frontal openings such that it measures a total pressure roughly averaged over the cross section [Fig. 9.3(a)]; they also have a second tube, facing backwards and monitoring the local static pressure. The volume flow rate is estimated from the pressure difference Δp as

$$Q = C_d \frac{\pi D^2}{4} \sqrt{\frac{2\Delta p}{\rho}}, \tag{9.4}$$

where C_d is an empirical correction coefficient accounting for deviations from the ideal response. Simplicity of operation and low cost are the main advantages of averaging Pitot tubes. Their limitations include the need for clean fluid and a narrow dynamic range, extending to only about 30% of full scale.

9.6 Laminar flow elements

Laminar flow elements are pipe sections or devices of a larger diameter that contain tube bundles or relatively long honeycombs [Fig. 9.3(b)]. The fluid is subdivided to pass through these elements, which are sufficiently narrow for the Reynolds number in each element to be lower than the transitional value of about 2300. Thus the flow in the elements is laminar and the pressure drop across the elements is related to the volume flow rate through the Hagen–Poiseuille expression, which is linear, in contrast with the quadratic expression for turbulent pipe flow. For fully developed laminar flow in a circular tube of length l and diameter D, this expression becomes

$$\Delta p = \frac{128\mu l}{\pi D} Q. \tag{9.5}$$

Because of their linear response, these flow meters are suitable for very low flow rates. Their disadvantages are large frictional pressure losses and bulkiness. They also tend to be clogged by impurities in the flow.

9.7 Rotameters

Rotameters, or, more generally, *variable-area flow meters*, are simple and versatile devices that can be used with a wide variety of liquids and gases over wide ranges of flow rates. They consist of a vertical tube, tapered such that its cross section increases linearly upwards, and a 'float', which is pushed upwards by the flowing fluid and stops at a position at which the drag, the boyancy, and the weight are in balance [Fig. 9.3(c)] [1, 3, 7]. The height of the float is proportional to the flow rate, which is displayed in appropriate units on a scale engraved on the tube. Variable-area flow meters are not very sensitive to fluid viscosity and can be corrected for density variations. They are popular because they require no external power, can be positioned near pipe bends, and present relatively low pressure losses. They are fairly accurate, except in the lower end of their scale, typically below 10% of full-scale reading. Different tube and float materials, sizes, and shapes are available for different applications. Tubes are commonly made of glass or transparent plastic, but stainless steel variations with magnetic sensing of the float position are also available for corrosive liquids or high temperatures and pressures.

9.8 Vortex-shedding flow meters

The main component of the *vortex-shedding flow meters* [1, 3, 9] is a bluff object immersed in the flowing fluid and spanning the pipe cross section [Fig. 9.3(d)]. Their operation is based on the periodic shedding of vortices (*von Kàrmàn vortex street*) from the edges of the object; this occurs at a frequency f (in cycles per second), which is related to the frontal width h of the object and the flow velocity V. In dimensionless form, the shedding frequency is called the *Strouhal number*

$$\mathrm{S} = \frac{hf}{V}. \tag{9.6}$$

For Reynolds numbers greater than a certain value (typically about 5000), the Strouhal number maintains an essentially constant value in the range 0.14–0.21, depending on the shape of the object and independent of V. The shedding frequency is detected by a variety of means, including piezoelectric pressure transducers, strain gauges, self-heated resistance elements, and ultrasonic beams.

9.9 Drag flow meters

Also referred to as *target flow meters*, *drag flow meters* [6, 10] [Fig. 9.3(e)] are based on the relationship between the drag force F_D on an immersed bluff object and the flow velocity. In general,

$$F_D = \frac{1}{2} C_D \rho A V^2, \tag{9.7}$$

where C_D is the drag coefficient and A is the frontal area of the object, namely the area of its projection on a plane normal to the flow velocity. C_D is essentially constant for an object with sharp corners immersed in turbulent flow at sufficiently large Reynolds numbers, typically greater than about 1000. Thus the volume flow rate through a pipe would be given as

$$Q = k\sqrt{F_D}, \tag{9.8}$$

where k is a constant. In practice, the *target,* which is a disk-like object, is inserted in the pipe and mounted on a support instrumented with strain gauges or linear variable-differential transformers (LVDTs), which measure the drag force through deflection. Such instruments are very sensitive and bidirectional and can be used at high pressures and with a variety of fluids. As the target is usually positioned in the centre of the pipe, they do not get easily clogged by suspended impurities.

9.10 Turbine flow meters

Turbine flow meters measure the volume flow rate of fluids in pipes as proportional to the angular velocity of an immersed vaned rotor [3, 7, 10]. A very common type utilizes an axial turbine with its axis aligned with the pipe centreline [Fig. 9.3(f)]. The passage of each rotating blade is sensed electromagnetically by an externally mounted sensor, and the flow rate is given by

$$Q = kn, \tag{9.9}$$

where n is the number of pulses per unit time provided by the sensor and k is a constant depending on the impeller design and size, the pipe diameter, and the number of blades. Turbine flow meters are subject to significant pressure losses and are prone to cavitation when used with high-speed low-pressure liquids. A low-cost version, called the *paddle-wheel flow meter* [Fig. 9.3(g)], utilizes a partially immersed rotor, with its axis normal to the flow direction. Besides flow rate, turbine flow meters may also provide the total fluid volume that passed through over a time interval. Common domestic water meters are of the turbine type.

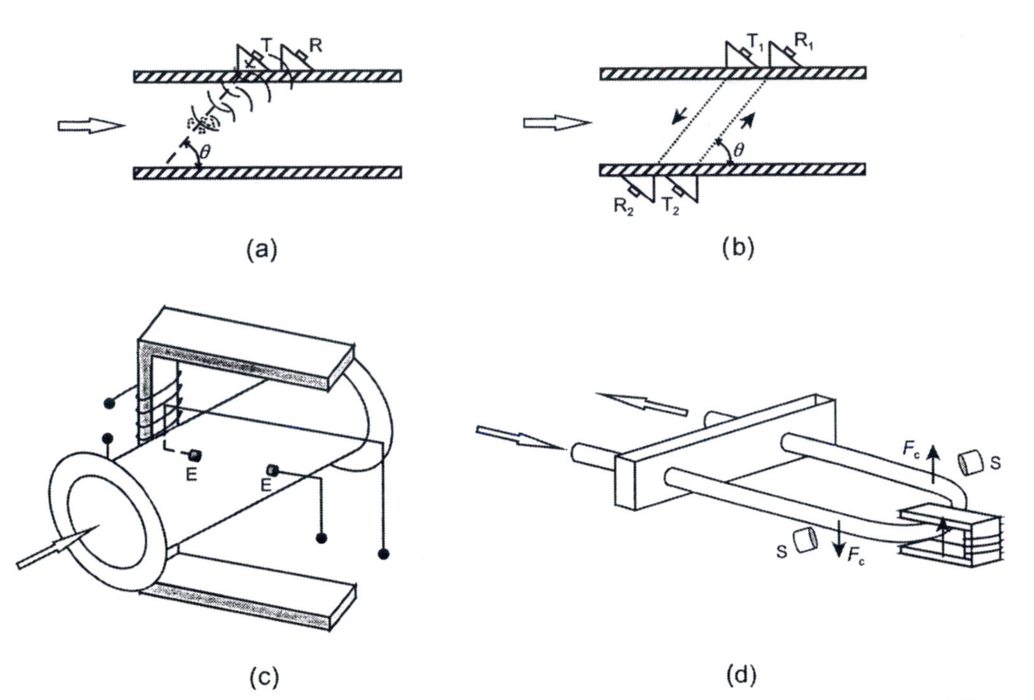

Figure 9.4. Sketches of representative flow meters: (a) Doppler untrasonic flow meter, (b) time-of-flight ultrasonic flow meter, electromagnetic flow meter, and (d) Coriolis flow meter.

9.11 Ultrasonic flow meters

Ultrasonic flow meters [1,3,9] utilize high-frequency (typically of the order of 10 MHz) pressure waves to compute the volume flow rate of liquids in pipes. There are two distinct types of such meters: the *Doppler flow meters* and the *time-of-flight flow meters*.

A representative Doppler flow meter consists of two piezoelectric crystals, a transmitter T, which transmits an ultrasonic wave through the pipe, and a receiver R, which receives the ultrasound reflected by solid particles or gas bubbles transported by the flowing fluid [Fig. 9.4(a)]. The frequency f_r of the reflected sound is shifted from the frequency f_t of the transmitted sound by an amount Δf, called the *Doppler shift*, which is proportional to the velocity V of the reflector, as

$$\Delta f = f_t - f_r = \frac{2 f_t \cos\theta}{c} V, \tag{9.10}$$

where c is the speed of sound. Such devices are calibrated to provide an output that is equal to the average velocity of the fluid in the pipe, assuming that the flow is fully developed. They are non-invasive and can be handheld or strapped to the outside of a pipe.

A representative time-of-flight flow meter consists of two externally mounted pairs of piezoelectric transducers. Each transmitter emits sound waves towards the corresponding receiver, one of which is located downstream of its mate and the other is upstream of it [Fig. 9.4(b)]. Each transmitter emits a sound pulse each time the corresponding receiver receives the previous one. Because sound waves are transported by the flowing fluid, sound propagates faster downstream than upstream and the frequencies of pulsation of the two pairs differ by an amount

$$\Delta f = \frac{2\cos\theta}{l} V, \tag{9.11}$$

where l is the distance between the transducers of each pair. This configuration makes the flow measurement independent of the speed of sound and thus flow temperature.

9.12 Electromagnetic flow meters

These instruments provide the volume flow rate of electrically conducting liquids in pipes. Their operation is based on *Faraday's law of electromagnetic induction*, which states that, when a conductor with length l moves with speed V in a direction normal to the direction of a magnetic field with magnetic flux density B, an electric potential E is generated across it as

$$E = BlV. \tag{9.12}$$

Practical electromagnetic flow meters [1, 3, 10] consist of an insulated pipe section of the same diameter D as the pipe of interest, surrounded by an alternating or pulsed magnetic field and having two surface electrodes embedded on the wall across a diameter normal to the magnetic field direction [Fig. 9.4(c)]. The voltage difference between these electrodes is related to the volume flow rate as

$$E = \frac{4kB}{\pi D} Q, \tag{9.13}$$

where k is a numerical coefficient. Electromagnetic flow meters have an accuracy that exceeds 0.5% and are not overly sensitive to the velocity profile. On the other hand, they are bulky, heavy, and relatively expensive.

9.13 Coriolis flow meters

Coriolis flow meters [1, 3, 7, 16] were developed relatively recently, but have become increasingly popular in a variety of industries because of their versatility and their capacity to measure true mass flow rate, essentially independent of fluid properties and flow conditions. There are several different geometrical designs, all based on the Coriolis force principle. Consider a fluid element with mass δm flowing with velocity V in a tube that rotates with angular velocity ω about an axis normal to its own axis; assume that the fluid element is at a radial distance r from the axis of rotation and, during time $\delta t = \delta r / V$, moves away from it to a radial distance $r + \delta r$. Then the angular momentum $(\delta m)\,\omega r^2$ of this fluid element would increase to $(\delta m)\,\omega\,(r + \delta r)^2 \approx (\delta m)\,\omega(r^2 + 2r\delta r)$. This increase of angular momentum is attributed to a torque rF_c, where $F_c = 2(\delta m)V\omega$ is called the *Coriolis force*. The direction of the Coriolis force is circumferential and opposite in sense to the direction of rotation, for outward motion. In vectorial notation, the Coriolis force is written as $\overrightarrow{F}_c = 2(\delta m)\overrightarrow{V} \times \overrightarrow{\omega}$. It is also written as $\overrightarrow{F}_c = (\delta m)\,\overrightarrow{a}_c$, where $\overrightarrow{a}_c$ is called the *Coriolis acceleration*. The flowing fluid receives this force from the tube walls; by reaction, the fluid applies a force upon the containing tube, which is equal in magnitude and direction to the Coriolis force, thus affecting the tube motion.

Practical flow meters do not rotate the tube but set it in vibration at its natural frequency by subjecting it to an alternating magnetic field. As representative of this class of instruments, Fig. 9.4(d) shows a sketch of the *U-tube Coriolis flow meter*. The fluid is passed through a bent tube, whose ends are clamped, while its tip is set to vibration. The instantaneous angular velocity and therefore the Coriolis force increase towards the tip. The two legs of the tube receive forces in opposite directions, and thus the tube is twisted in one sense during half of the cycle and in the opposite sense during the other half. The twist angle is measured by magnetic or optical position sensors that sense the time delay Δt between the passage of the two legs through a transverse plane. This time delay is related to the mass flow rate as

$$\Delta t = \frac{8r_t^2}{K_s}\dot{m}, \tag{9.14}$$

where r_t is the radius of the tube and K_s is a constant that, ideally, depends only on the tube material. Small deviations from this relationship may be caused by multi-phase effects and other variations in fluid properties. Even so, Coriolis flow meters are suitable for conventional as well as contaminated and non-Newtonian fluids.

9.14 Thermal mass flow meters

Thermal mass flow meters are used to measure the mass flow rate of gases. They are not used for liquid flows due to the much higher power required to heat a liquid than a gas. For relatively low-mass flow rates, the entire gas stream is passed through the meter, whereas, at higher flow rates, only part of the gas is heated by passing it through a bypass tube. There are two types of such instruments, the *heated-tube flow meters* and the *immersion-probe flow meters* [10].

In the heated-tube flow meters, the flowing gas is passed through a piece of tube that is heated electrically and is instrumented with two temperature sensors, commonly thermocouples or resistance temperature detectors (RTDs). The first sensor is located upstream of the heated section and the other one is downstream of it. The rate of heat transfer $\dot{H}$ to the fluid is

$$\dot{H} = \dot{m}C_P\Delta T, \tag{9.15}$$

where $\dot{m}$ is the mass flow rate of the gas, C_P is its specific heat under constant pressure, and ΔT is the temperature difference across the heated section. Thus the mass flow rate for a given gas can be measured from measurements of $\dot{H}$ and ΔT. Manufacturers supply instruments with an output that has been calibrated in air, nitrogen, or some other gas. When used with different gases, this output has to be corrected by multiplying it by the ratio of specific heats of the two gases.

Immersion probe flow meters consist of a probe with two RTDs connected in a Wheatstone bridge configuration. One RTD is used to measure the gas temperature, and the other is provided with a current so that it is heated to a temperature higher than the gas temperature by a fixed amount ΔT. The electric power required to heat the second sensor is related to the mass-weighted velocity ρV of the gas by a non-linear relationship, called *King's law*. Electronic circuitry is employed to linearize the output

Table 9.1. Flow meters

Flow meter type	Dirty fluid[1]	Dynamic range	Pressure loss[2]	Uncertainty[3]	Upstream pipe[4]	Viscosity effect[2]	Cost[2]
PD	N	10:1	H	±0.25 r	none	H	M
Venturi	Y	4:1	L	±1 fs	5–20	H	M
Nozzle	Y	4:1	M	±1–2 fs	10–30	H	M
Orifice plate	Y	4:1	M	±2–4 fs	10–30	H	L
Weir (V-notch)	Y	100:1	VL	±2–5 fs	none	VL	M
Parshall flume	Y	50:1	VL	±2–5 fs	none	VL	M
Pitot	N	3:1	VL	±3–5 fs	20–30	L	L
Rotameter	Y	10:1	M	±0.5 r	none	M	L
Vortex	Y	10:1	M	±1 r	10–20	M	H
Drag	Y	10:1	M	±1–10 fs	10–30	M	M
Turbine	N	20:1	H	±1 r	5–10	H	H,M
Doppler	Y	10:1	none	±5 fs	5–30	none	H
Time-of-flight	N	20:1	L	±1–5 fs	5–30	none	H
Electromagnetic	Y	40:1	none	±0.5 r	5	none	H
Coriolis	Y	10:1	L	±0.4 fs	none	none	H
Thermal	Y	10:1	L	±1 fs	none	none	H

[1] Y: yes, N: no
[2] H: high, M: medium, L: low, VL: very low
[3] Percentage of full scale (fs) or reading (r)
[4] In diameters

so that it is proportional to ρV for a given gas. To obtain the mass flow rate, one has to multiply this output by the pipe cross-sectional area. Corrections for use with different gases are also available.

9.15 Selection of flow meter

Considering the diversity of designs and properties of flow meters and the wide ranges of flow conditions encountered in a fluid mechanics laboratories, one should compare carefully the different options that are available before purchasing a flow meter. Although it is possible that several devices may be equally suitable for a given application, it is also quite certain that several others would be totally unsuitable. As an aid towards the selection of the optimal flow meter, Table 9.1 is provided; it is based on information supplied by different manufacturers and contained in the website http://www.geocities.com/ull_km1980/flowmeterselectionguide.html

QUESTIONS AND PROBLEMS

1. Describe a method that can be used to measure the flow rate of water in a steel pipe without requiring any modification of the piping system.
2. After searching through manufacturers catalogues and the Internet, select specific flow meter models that would be suitable for measuring the flow rate in the following systems. Estimate the required range by appropriate arguments. In cases for which

you cannot find appropriate products, describe briefly the design and specifications of a flow meter that would be suitable for the task.

a. Natural gas through a main pipe of 180-mm diameter.
b. Blood during surgery, through a tube of 8-mm diameter.
c. Helium through a tube of 2-mm diameter.
d. Paper pulp through a pipe with 500-mm diameter.
e. Water through a rectangular open channel with a height of 2 m and a width of 3 m.

REFERENCES

[1] R. C. Baker. *Flow Measurement Handbook*. Cambridge University Press, Cambridge, UK, 2000.
[2] R. P. Benedict. *Fundamentals of Temperature, Pressure and Flow Measurements* (2nd Ed.). Wiley Interscience, New York, 1977.
[3] R. A. Furness. *Fluid Flow Measurement*. Longman, London, 1989.
[4] R. J. Goldstein. *Fluid Mechanics Measurements* (2nd Ed.). Taylor & Francis, Washington, 1996.
[5] A. T. J. Hayward. *Flowmeters*. Macmillan, London, 1979.
[6] N. P. Cheremisinoff (Editor). *Encyclopedia of Fluid Mechanics*. Gulf Publishing, Houston, TX, 1986.
[7] J. G. Webster. *The Measurement, Instrumentation and Sensors Handbook*. CRC Press, Boca Raton, FL, 1999.
[8] J. Saleh (Editor). *Fluid Flow Handbook*. McGraw-Hill, New York, 2002.
[9] S. P. Parker (Editor-in-Chief). *Fluid Mechanics Source Book*. McGraw-Hill, New York, 1987.
[10] Omega Engineering. *The Flow and Level Handbook*. Omega Engineering, Stamford, CT, 2001.
[11] Anonymous. 1969 guide to process instruments elements. *Chem. Eng.*, 76(12):137–164, 1969.
[12] R. C. Baker and M. V. Morris. Positive-displacement meters for liquids. *Trans. Inst. Meas, Control*, 7:209–220, 1986.
[13] F. M. White. *Fluid Mechanics* (4th Ed.). McGraw-Hill, New York, 1999.
[14] P. Ackers, W. R. White, J. A. Perkins and A. J. M. Harrison. *Weirs and Flumes for Flow Measurement*. Wiley, New York, 1980.
[15] E. F. Brater and H. W. King. *Handbook of Hydraulics* (6th Ed.). Mc-Graw-Hill, New York, 1976.
[16] K. O. Plache. Coriolis/gyroscopic flowmeters. *Mech. Eng.*, 101(3):36–41, 1979.

10 Flow visualization techniques

Some of the most important discoveries in fluid mechanics were largely achieved by flow visualization, using pioneering versions of modern techniques. As milestones in the development of visual and optical methods, one could mention the very detailed observations and sketches of river and channel flows by Leonardo da Vinci (1452–1519), Antonie Van Leewenhoek's observation and timing of the motion of blood cells in order to measure blood flow velocity (1689), Ernst Mach's use of the reference beam interferometer and other optical methods to study gas dynamics phenomena (1878), Osborne Reynolds' experiments with dye in water, demonstrating the difference between laminar and turbulent pipe flows (1893), and Ludwig Prandtl's surface flow visualization of flows around cylinders and airfoils by using aluminium filings (1930s). The distinction between flow visualization and associated measurement methods of flow velocity, temperature, pressure, and composition is often unclear, although it is generally understood that flow visualization is of a qualitative nature and not primarily concerned with measurement. Current interest on flow visualization has been intense, and a great diversity of methods, each with many variants, is available to the modern researcher. In the present chapter, some of only the most popular methods are briefly outlined; the reader is referred to the voluminous literature for additional methods and more details [1–6].

10.1 Overview

Under certain conditions, some flow characteristics, such as fluid boundaries and even internal motion patterns, can be seen and recorded without any externally introduced disturbances or special instrumentation. Such flows may be called *naturally visible*. This may happen when the fluid contains surfaces with abrupt changes of refractive index, as in the case of liquid–gas flows, mixtures of immiscible liquids, liquids with a free surface, and regions of phase change, including condensation, boiling, cavitation, melting, and solidification.

Flow visualization techniques are available for nearly all sorts of flows and fluids, in the full range between stagnation and hypersonic speeds. Such techniques are commonly classified into two general classes:

Marker techniques, which infer patterns of fluid motion from the visible motion of foreign materials contained in the fluid; this class also includes surface marker techniques, identifying fluid velocity, temperature, and pressure patterns from visible changes of solid wall properties; marker techniques are available for the majority of liquid and gas flows, although not suitable for low-density gases.

Optical techniques, which utilize variations of the refractive index within the fluid or radiation emission from fluid atoms; most of these techniques are applicable to gas flows; variations of refractive index are associated with changes of density, and so such techniques are suitable for the study of compressible flows, natural convection, flames, combustion, and, more generally, chemical reactions; radiation emission techniques are applicable to very low-density gas flows.

As a rough selection guide, Table 10.1 lists some of the most common flow visualization techniques for liquid and gas flows and the typical ranges of flow speeds to which they apply.

10.2 Marker techniques

Tufts: *Tufts* are short pieces of yarn, string, or other flexible material, fastened at one end on an immersed surface or some thin support inside the flow [7]. They may be used in both gaseous and liquid flows. Their main utility is as flow direction indicators; they may also identify flow separation regions, flow instability, and transition to turbulence. Under certain conditions, one may be even able to extract some quantitative results from tuft motion patterns, although at a very limited resolution. The properties that affect the performance of tufts are their density, stiffness, and length. Obviously the dynamic response of tufts improves with diminishing values of all these properties.

Common materials for tufts are wool, various other natural and synthetic fibres, and even paper strips. For an enhanced visibility, one may dip the tufts in a fluorescent dye, illuminate them with a suitable monochromatic light, and then record the images by using an optical filter that removes the oncoming light while permitting the fluorescent light to pass. One disadvantage of conventional tufts is that they are easily set in flapping motion when exposed to unsteady or highly turbulent flows. Small rigid *cones*, whose apex is attached to the surface through a thread, and *hinged tufts*, consisting of hinged sections of stiff materials, are reputed to present considerably less flapping. On the other extreme, relatively long tufts, called *streamers*, have also been used to indicate streamline patterns near and away from a surface.

In addition to being attached on a surface, tufts may be introduced in midstream by various devices: a *tuft screen* is a mesh of thin wires stretched across a stream and with tufts epoxied on the mesh nodes; a *tuft wand* is a thin rod that may be easily traversed through a flow and having a tuft or a streamer attached to its tip; *pin tufts* are tufts attached to thin needles protruding from a surface.

Surface marking methods: These techniques provide the visualization of flow direction and flow patterns near solid surfaces in contact with the fluid, based on visible

Table 10.1. Common flow visualization techniques (M1 means that the Mach number equals 1)

Techniques for liquids	Speed range
Tufts: surface tufts, in-flow tufts, streamers, tuft screen	0.05–2 m/s
Surface marking:	
oil dots	0.5–4 m/s
oil film	0.1–25 m/s
electrolytic etching	0.01–0.1 m/s
Continuous dye injection (streakline marking):	0.5 mm/s–10 m/s
Particle tracing (pathline marking):	
suspended solid markers, droplets, bubbles	0.1 mm/s–30 m/s
floating solid markers	0.5 mm/s–5 m/s
Line marker generation (timeline marking):	
hydrogen bubbles	5 mm/s–10 m/s
thymol blue	<0.1 m/s
photochromic	<0.1 m/s
electrolytic precipitation	0.5 mm/s–0.1 m/s
Optical methods	
shadowgraph	n.a.
Schlieren	n.a.
interferometry	n.a.
Techniques for gases	**Speed range**
Tufts: surface tufts, in-flow tufts, streamers, tuft screen	0.1 m/s–M1
Surface marking:	
oil dots	20 m/s–M10
oil film	5 m/s–M6
sublimation	10 m/s–M2
soluble chemical film	0.01–4 m/s
temperature-sensitive paint	150 m/s–M6
Continuous smoke injection (streakline marking):	0.1 m/s–M1
Particle tracing (pathline marking):	
suspended solid markers, droplets, bubbles	1–20 m/s
Line marker generation (timeline marking):	
smoke wire	0.3 m/s–8 m/s
sparks	2 m/s–M8
Optical methods	
shadowgraph	70 m/s–M4
Schlieren	2 m/s–M3
interferometry	70 m/s–M10

effects of flow on applied surface coatings [3, 8]. They utilize relationships between near-wall flow and wall shear stress, pressure, temperature, and mass transfer rate. Such relationships are sometimes unclear, and so the interpretation of surface maps is often subject to ambiguities. On the other hand, several of these methods can be extended or refined to actually measure the corresponding wall properties.

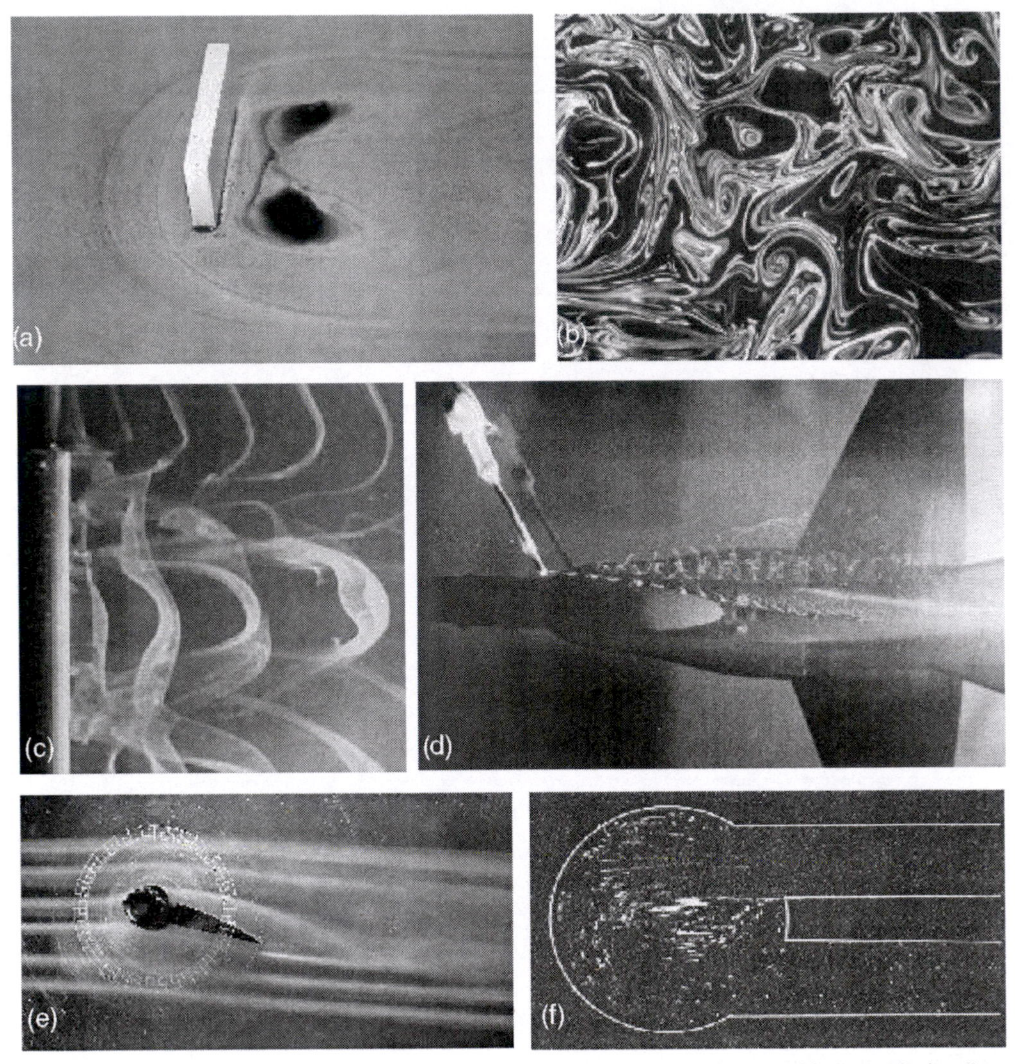

Figure 10.1. Flow visualization with the use of relatively simple marker techniques: (a) oil-streak visualization of flow adjacent to a plane surface with a bluff protrusion; the mixture consists of a suspension of 3 ml of graphite powder in 70 ml of penetrating oil (photograph by Sean Bailey); (b) laser-sheet visualization of fluorescent dye in a highly turbulent shear flow (photograph by Sebastien Marinau-Mes); (c) visualization of water flow past a cylinder with a step change in diameter by use of electrolytic precipitation of lead (photograph by Warren Dunn); (d) hydrogen-bubble visualization of the vortex produced by the leading-edge extension of a fighter aircraft model (photograph by Warren Dunn); (e) smoke-line (oil-mist) visualization of flow separation from an airfoil (photograph by Warren Dunn); and (f) visualization of pulsatile flow in a ventricular-assist-device model by use of laser-illuminated polysterene particles impregnated with a fluorescent dye (photograph by Jean-Baptiste Vergniaud).

A relatively simple method that is commonly used in aerodynamics is the *oil-streak* method, which is applicable to both gas and liquid flows. The surface of an immersed object is coated with a paint consisting of a pigment (e.g., TiO_2, china clay, lamp black, copier toner, or fluorescent chrysene) suspended in some mineral oil (e.g. kerosene, diesel oil, or light oil). When exposed to the flow, the pigment coagulates and is deposited on the surface in the form of small lumps, in the wake of which the oil collects, forming short streaklines. At the end of the test, these patterns can be observed and photographed to provide information about flow direction, flow separation, and reattachment and transition to turbulence [see, for example, Fig. 10.1(a)]. Although an analysis of the relationship between oil patterns and flow properties is available [9], the oil-streak method is generally used for qualitative purposes. It is also recognized that, in high-speed flows, the paint coating thickness may be comparable with the boundary-layer thickness, and thus the coating may distort the wall flow. A related method, called the *oil-film* method, utilizes interferometry to measure wall shear stress (Section 14.6). A variance of the oil-streak method is the *oil-dot* method, which consists of applying the paint in dots rather than coating the entire surface. An advantage of this method is that it may be applied to surfaces with complex shapes, such as aircraft models.

A variety of techniques utilize change of colour of an immersed surface that is due to *mass transfer* towards or from it. Such techniques are mainly used to identify regions of turbulence, which enhances the mixing process. In liquid flows, this technique consists of coating the surface with a substance that reacts with a reagent dissolved in the flowing stream. A related technique exploits *electrochemical* deposition or release of a substance from a surface serving as the cathode or anode of an electrolytic configuration. In gas flows, mass transfer from a surface is utilized in the form of *sublimation* of a solid coating (e.g., hexachloroethane or naphthalene) or *evaporation* of a volatile liquid coating. In some cases, mass transfer is sufficiently substantial to leave *relief patterns* on the surface. This is the case of the *ablation* technique, used in high-speed gas flows, and the *plaster of Paris* ($CaSO_4$) technique, used in water flows.

Surface marking methods include the use of special surface coatings whose appearance is sensitive to pressure, temperature, or both [3, 4, 10–12]. These methods are also used for the measurement of pressure and temperature and are discussed in more detail in Sections 8.3 and 12.4. *Pressure-sensitive paints* (*PSPs*) contain fluorescent molecules whose luminescence is inversely proportional to the pressure, as well as sensitive to temperature. Similar materials, but with a relatively high temperature sensitivity compared with their pressure sensitivity, are known as *temperature-sensitive paints* (*TSP*). A different type of coating contains metallic compounds that ungergo chemical changes accompanied by colour changes when exposed to a particular temperature over a certain time. Calibration provides the temperature variation as a function of colour for a given exposure time. During testing, the colour of the paint, which has been applied on thermally insulated walls, is monitored and recorded so that high-temperature regions may be identified. This technique is applicable only to high-speed flows, which generate sufficient temperature differences. Another type of temperature-sensitive coating consists of *liquid crystals*, which are either applied on the surface in the form of a paint or bonded to it in thin sheet form. The change of colour of the liquid crystal at a characteristic temperature can be used for both qualitative and quantitative purposes [4, 13]. Besides temperature mapping (Section 12.4), liquid-crystal coatings have also been used to measure wall shear stress (Section 14.6). Temperature-sensitive coatings and liquid crystals can be used to visualize regions of flow separation and transition to turbulence. For this purpose, the surface is heated and these regions are identified by colour change that is due to a change that is in convective cooling rate.

Dyes: One may create, in liquid flows, visible regions of a colour different from that of the main liquid by either injecting a dye solution into the stream or by producing a coloured region in the fluid by electrolytic or photolytic methods.

Dye injection: Dye solutions are usually injected into a liquid stream through wall orifices or through hypodermic tubes in midstream [14]. As much as possible, the dyed stream should be injected at the local speed of the flow (*isokinetic injection*). Another requirement for the injected solution is to match, if possible, the density of the fluid (*neutrally buoyant injection*). Examples of commonly used dyes are milk, ink, food colourings, and various chemical dyes, such as Congo red, methylene blue, and crystal violet. The choice of dye depends, among other factors, on its visibility to the particular light source that is available. To visualize complex motions and mixing, it is preferable to use two or more dyes of different colour. A special class of chemical dyes is the *fluorescent dye*, notably Rhodamine 6G, Rhodamine B, and fluorescein disodium; when illuminated by a laser or other monochromatic light source, and with the incident light removed by an optical filter, they remain visible through their fluorescence, while light reflections from immersed objects or the walls and windows of the apparatus become invisible [Fig. 10.1(b)]. The *fluorescent dye-layer* method has been used in towing tanks to visualize near-wall and separated flow phenomena; in this technique, dye layers of different colours are applied on horizontal planes before the towing of a model starts [15]. *Reacting dyes* are particularly effective for visualizing mixing streams; they consist of a pair of acidic and basic solutions, whose reaction produces a salt solution of a distinct colour. A disadvantage of dye injection is that, if continued over an extended period of time, it would saturate the liquid in the apparatus and the dye visibility would deteriorate. Certain dyes (e.g., milk) would also foul if they remain in the apparatus, and some chemical dyes may be toxic or corrosive and must be avoided. Dye injection is most effective in laminar flows, with speeds up to a few centimetres per second. Because the molecular diffusivity of most dyes is relatively large, they would diffuse rapidly in turbulent flows and become ineffective beyond a small distance from their injection point. In such cases, it may be worthwhile to select a dye with a relatively small diffusivity. Special *shear-thickening dyes* have been developed, which produce streaks influenced mainly by large-scale, low-shear motions in the fluid and are insensitive to small-scale motions [16].

Electrochemical methods: In this type of technique, a coloured region of the fluid is produced by a chemical reaction triggered by an electric current. The *thymol blue technique* [17] utilizes a solution of thymol blue (pH indicator) dye in distilled water. In an acidic environment (pH < 8.0), this dye is orange-yellow but it changes colour to blue when the pH increases beyond 9.6. The water in the apparatus also contains an acid (HCl) and an electrolyte (NaOH), at proportions such that the pH is just below 8.0. A thin platinum wire (typically with a diameter of about 10 μm) is stretched across the flow and connected to the negative pole of a dc power supply, while the positive pole is connected to another electrode in contact with the water. When a current pulse is supplied to the wire, it acts as a cathode, releasing hydrogen ions in the solution, some of which combine to form molecular hydrogen. This leaves an excess of negative OH ions in the vicinity of the wire, which makes the solution alkaline and causes the dye to turn blue within a thin cylinder surrounding the wire. Thus this method creates a timeline, which is transported downstream with the flow. Because the marked fluid is locally neutrally buoyant, the thymol blue method has been used effectively in stratified and rotating flows, mainly related to atmospheric and oceanic flow phenomena. A similar method of producing timelines is the *tellurium wire* technique [18]. In this approach, a cathode, consisting of a thin tellurium wire, is supplied with current pulses. Tellurium ions are released into the flow and form a black colloidal suspension. The fluid is alkaline water ($9 < pH < 10$), also containing some H_2O_2, which releases O_2 for stabilizing the suspension. A related approach is *electrolytic precipitation* [19]. Part of an immersed surface is coated with lead, tin, solder, or copper and connected as an anode. Following a current pulse, a white crystalline salt is released in the form of a cloud of microscopic spherical particles. This method is particularly effective for the visualization of separating boundary layers and vortex formation behind bluff objects [Fig. 10.1(c)].

Photolytic methods: Finally, coloured fluid can be produced locally by *photocatalysis* [20, 21]. The working fluid contains a *photochromic* indicator (e.g., pyridines or spyrans), which is normally colourless. Typically, when a molecule of such substances absorbs a photon in the ultraviolet range, it temporarily changes its absorption properties, becoming capable of absorbing photons in the red-green range of the spectrum. When subsequently illuminated by white light, regions containing such molecules become visible, taking a dark blue appearance. Such substances can be dissolved in alcohol, kerosene, and other solvents, but not in water, which limits the usability of this method to relatively small, contained facilities. The reaction is reversible, as the substance returns to its original state following a typical time of a few seconds. This property makes this method suitable for closed-circuit experiments, as it does not contaminate the fluid. A common application of the method utilizes a pulsed laser beam, which produces a timeline that can be followed in time to yield velocity profiles. It is also possible to mark small parcels of the fluid by focussing a pulsed light beam.

Hydrogen bubbles: When current flows between two electrodes separated by water, it will cause its *electrolysis*, releasing hydrogen gas at the cathode and oxygen gas at the anode, according to the chemical reaction

$$H_2O \;\rightarrow\; H_2 + \frac{1}{2}O_2. \tag{10.1}$$

The gas will be attached to the electrodes until buoyancy or drag from a flowing fluid detaches it in the form of bubbles. Although both gases can serve as flow markers, hydrogen is preferable, first because it is produced at twice the volume rate than that of oxygen, and second because oxygen would oxidize many conducting materials. Most commonly, a thin metallic wire, made of platinum, nickel, or stainless steel with typical diameters in the range 10–100 μm, is used as a cathode, whereas a much thicker carbon rod is used as the anode. When a stream flows past a thin cathode, hydrogen bubbles would be released, continuously forming a sheet. The bubble diameter d_h would decrease with increasing flow speed and decreasing cathode current, but, for rough estimates, one may take it as equal to one half the wire diameter. The rate of mass production of hydrogen by a current I is, according to *Faraday's law of electrolysis*,

$$\dot{m} = kI, \tag{10.2}$$

where $k = 1.04 \times 10^{-8}$ kgs^{-1}A^{-1}. The volume flow rate would be

$$Q = \frac{\dot{m}}{\rho} = \frac{\dot{m} R_h T}{p_h}, \tag{10.3}$$

where $R_h = 4124$ J/(kg K) is the hydrogen gas constant, T is the absolute temperature, and p_h is the pressure inside the bubble. The latter can be estimated from the local water pressure p_w (e.g., the hydrostatic pressure) as

$$p_h \approx p_w + \frac{4\sigma_h}{d_h}, \tag{10.4}$$

where the surface tension σ_h of hydrogen–water combination may be taken as about 0.072 N/m, equal to that of air–water combination. Then, one may calculate the number of bubbles N_h released per unit time t and per unit length l of the wire as [22]

$$\frac{N_h}{tl} = 6 \frac{R_h T}{(p_w + 4\sigma_h/d_h)\, d_h^3} \frac{kI}{l}, \tag{10.5}$$

For a good visibility of the bubble sheet, one requires a sufficient bubble density and thus sufficient current. Every other factor being equal, it is clear that the required current would increase proportionately to the flow speed, but bubble visibility can improve dramatically by optimizing the illumination setting (see Section 5.3). Typical current values used in practice are between 0.02 and 1 A. The electrode voltage that must be applied to produce the desired current is equal to the current times the water-path electric resistance. The latter depends on the distance between electrodes, the quality of surface of the electrodes, the water temperature, and the presence of electrolytes. Untreated tap water usually contains a sufficient amount of salts to ensure good conductivity. To further reduce the water-path resistance, it is customary to add a small amount of sodium sulfate (Na_2SO_4). Required voltages are in the range between a few volts and several hundred volts, and one must be concerned about possible electric hazards.

The use of a *straight wire* supplied with a dc current would produce a continuous sheet, which may be suitable for many flow visualization purposes [Fig. 10.1(d)], but cannot differentiate between different velocities on the visualized plane. Supply of periodic short-current pulses would essentially mark timelines, which may be followed in time to devise flow patterns. Another variation of the method is to release bubbles only at selected locations along the wire. One may consider insulating the wire in parts while leaving short uninsulated sections in between, but this approach will likely result in large bubbles forming at the boundaries and disturbing the process. A more successful approach is to use a *kinked wire*, produced, for example, by squeezing the wire between two fine gears; in this case, bubbles are generated on the entire wire length but released only at the apexes, thus introducing streaklines with substantial bubble density. Another approach is to use a *ladder-type probe*, with the bubbles produced along the 'steps' of the ladder, which are aligned with the flow. Combination of current pulsation and pointwise bubble release produces a grid of small bubble markers, which can serve as combined *timelines–streaklines* mapping the three-dimensional flow velocity variation.

The hydrogen-bubble technique has been used extensively in water flows and, to a lesser degree, in flows of aqueous solutions and mixtures. In particular, this technique has been successfully applied to the study of low-Reynolds-number phenomena by use of water–glycerol mixtures, with glycerol concentrations exceeding 90% [23]. Although primarily used for qualitative purposes, hydrogen-bubble images can be analysed to yield flow velocity fields and even some turbulence statistics [24–26]. Postprocessing of simultaneous records of two perpendicular views allows the reconstruction of three-dimensional motions and velocity vectors [27]. To validate the accuracy of such calculations, one needs to estimate the relative velocity between a typical bubble and the surrounding fluid and, particularly, the rise speed of the bubbles that is due to buoyancy (see Section 5.6).

Smoke, mists, and fogs: In flow visualization, *smoke* is a general term, which includes suspensions in air or other gases of solid products of combustion as well as liquid droplets (*mists*) and visible vapours and tracer gases. The suspended particles are usually small enough (typically less than 1 μm) to follow the fluid motion. Smoke visualization has been largely used in wind-tunnel studies, but it can be effectively used in any apparatus that uses gas as a medium. A disadvantage of this method is accumulation of smoke in the apparatus and the need for occasional cleaning of the windows and other exposed surfaces.

The most common methods for generating smoke are the following:

Burning: Common combustible materials, such as tobacco, incense, wood chips, and paper are suitable for many applications and a variety of *smoke generator* designs are available [28]. For improved performance; one may control the supply of oxygen in order to produce a denser smoke and pass the smoke through a filter to remove the larger particles before injection into the flow. Appropriate exhaust and ventilation are required if considerable amounts of smoke are used.

Vapourization and condensation: This is the most popular approach, usually involving mineral oils (e.g., kerosene). Smoke generators of this type are available at a relatively low cost. Oil flowing in a tube is heated by an electric heater and evaporates; then it mixes with a cool air stream, in which it condenses, forming a mist of tiny droplets, in the submicrometre range (*oil mist*). Large supplies of smoke must be cooled before injection into the flow in order to avoid buoyancy or damage to the test models or facility. For low-speed, small-scale air flows, one may use steam fog, dry ice (solid CO_2), or liquid nitrogen, which are non-polluting. Commercial fog generators are available to generate thick *fogs* of non-hazardous, water-based liquids in small or large quantities. A related method used exclusively in supersonic flows is the *vapour-screen method* [29]. For this method, moist air is supplied to the supersonic nozzle of the wind tunnel, and, as it expands, it condenses into a fog. As the air flows around a model, the concentration of droplets becomes non-uniform and thus regions of separation, wakes, etc., become visible.

Aerosol generation: A popular material is titanium tetrachloride ($TiCl_4$). In this method, the aerosol, created when compressed air is passed through liquid $TiCl_4$, is passed through water, which reacts with it, producing titanium oxide (TiO) in the form of a dense white mist. The use of this material in uncontrolled environments should be avoided as it is toxic. Other organic or inorganic materials have also been used in a similar fashion.

Injection of smoke in bulk into a stream may be useful in identifying flow boundaries; however, the connection between visible smoke patterns and flow characteristics is quite uncertain. Smoke can be injected from an orifice at a wall or through a thin tube, inside a stream, in the form of a streakline, called a *smoke line* [Fig. 10.1(e)]. Injection from a rake of tubes in steady flow, marking a set of streamlines, is a convenient method in visualizing air flows past models in wind tunnels. A technique that introduces a very limited amount of smoke in a facility is the *smoke wire method* [30,31]. For this method, a thin wire is stretched vertically across the wind tunnel and a small quantity of oil from a reservoir is let to drip along it. Because of surface tension, the oil forms small droplets attached to the wire at even distances from each other. A pulse of electric current to the wire heats it and evaporates the oil, forming a set of streaklines. A camera, triggered in synchronization with the electric pulse, is used to record the image.

Solid markers, bubbles, and droplets: Besides hydrogen bubbles, dyes, and smoke, a great variety of materials has been used as markers–tracers in flow visualization studies [3] [Fig. 10.1(f)]. One requirement for their selection is that they should follow the fluid motions, at least to the degree that is resolved by the observation–recording system. In this respect, the requirement is not as severe as for velocity measurement, and markers that follow only the large-scale motions of the fluid may be acceptable. Ideally, the markers should be neutrally buoyant, a condition relatively easily achievable in water and other liquid flows, but not in gas flows. One can visualize liquid free-surface motion by sprinkling it with various powders, such as talcum, aluminium flakes, lycopodium, and even paper shreds. Unless there is vigorous mixing, the solid particles would be suspended by surface tension and their motion would mark pathlines. The only neutrally buoyant markers used in gases are *helium-filled soap bubbles*, which may be produced by bubble generators to typical sizes of 1 mm [32]. The dynamic response of markers improves dramatically with decreasing size, at the expense of visibility. Therefore the optimal size of a marker is a compromise between dynamic response and visibility. The visibility of markers also depends on their reflectivity, which is proportional to the difference between their refractive index and that of the fluid. Thus preferred materials would include aluminium, glass, and droplets of liquids with a large refractive index. For enhanced visibility, solid markers may be coated with luminescent or fluorescent paints or even have a fluorescent substance incorporated in their chemical composition. In addition to the preceding requirements, markers that are hazardous to health or corrosive to the facility should be avoided.

The most common shape of flow markers is spherical, or approximately spherical, as a result of generation methods involving surface tension (e.g., polystyrene and glass beads, liquid droplets). Non-spherical shapes have also been used in many applications. These include irregular-shaped solids produced by the crushing or filing of a material, flakes, and fibres. The latter two types maintain relatively small inertia while being highly visible when viewed from certain angles. Flakes and fibres are aligned with surfaces of maximum shear, so they can also be used to indicate the direction of shear. A water-based suspension of microscopic crystalline platelets, known as *kalliroscope fluid*, is particularly effective for the viewing of complex vortical flows [33–35]. A limitation of this method, and other methods involving dense suspensions of markers, is that it renders the fluid opaque beyond a few centimetres away from the viewing window.

10.3 Optical techniques

In the following subsections, the simplified principles of operation of some classical optical flow visualization techniques are presented. More details and descriptions of advanced methods can be found in recent specialized texts [36, 37].

Light deflection in variable refractive-index media: When a light ray enters a medium that has a varying refractive index, it will follow a path that will minimize the time it takes to pass through this medium, according to Fermat's principle (see Section 5.2). Use of variational calculus leads to a set of differential equations for the coordinates of this path, whose solution can provide the path of the ray as a function of the refractive-index field [3, 38]. The simplified analysis subsequently presented illustrates the procedure, but avoids the use of advanced mathematical tools.

As illustrated in Fig. 10.2, consider that two closely spaced light rays travelling through medium 1 with uniform refractive index n_0 enter medium 2 with a variable refractive index $n(x, y)$, only slightly different from n_0 and such that $\partial n/\partial y > 0$. The extent of medium 2 is in the range $x_1 < x < x_2$, following which the rays exit into medium 3, which also has a uniform refractive index n_0. The distance between the two rays in medium 1 is δy, changing in medium 2 to δr, where r is the local radius of curvature of the ray. The symbols s and s' indicate the lengths along the two rays. The speed of light along the two rays would be $v(x, y)$ and $v(x, y) + \delta v(x, y)$; keep in mind that $n = c/v$, where c is the speed of light in vacuum. We can assume that the deflection of the rays is very small, so that $\delta s \approx \delta x$ and $\delta r \approx -\delta y$. Now consider a wave front A inside medium 2; during a small time interval δt, this wave front will move to position B, rotating by an angle $\delta\varphi$, expressed as

$$\delta\varphi = \frac{\delta s}{r} = \frac{v\delta t}{r} = \frac{\delta s'}{r + \delta r} = \frac{(v + \delta v)\,\delta t}{r + \delta r} = \frac{\delta v \delta t}{\delta r}, \tag{10.6}$$

from which we get

$$\frac{1}{r} = \frac{1}{v}\frac{\delta v}{\delta r} \approx -\frac{1}{v}\frac{\partial v}{\partial y} = -\frac{1}{c/n}\frac{\partial\,(c/n)}{\partial y} = \frac{1}{n}\frac{\partial n}{\partial y}. \tag{10.7}$$

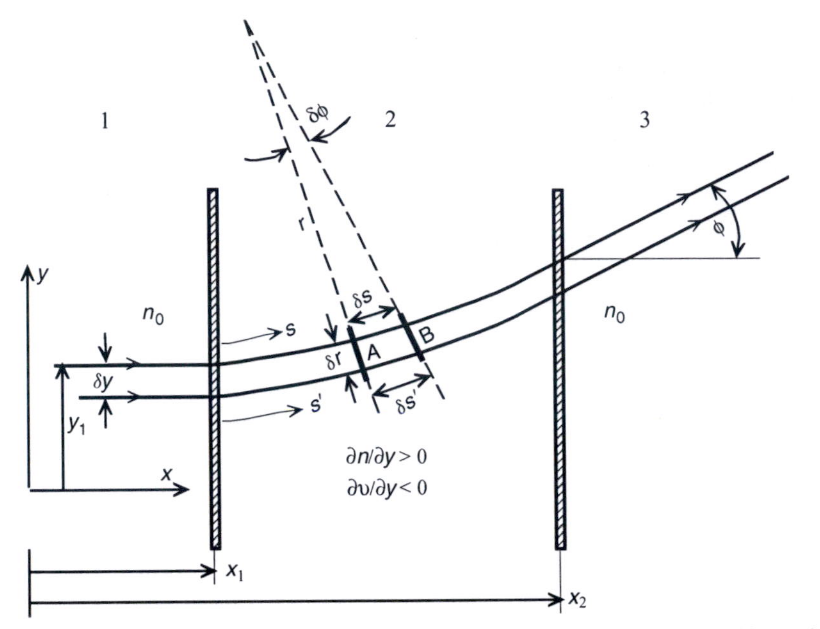

Figure 10.2. Sketch illustrating the passage of light rays through a medium with a variable refractive index; proportions have been distorted for clarity.

On the other hand, the radius of curvature of the path is given by

$$\frac{1}{r} = \frac{\mathrm{d}^2 y/\mathrm{d}x^2}{\left[1 + (\mathrm{d}y/\mathrm{d}x)^2\right]^{3/2}} \approx \frac{\mathrm{d}^2 y}{\mathrm{d}x^2}, \tag{10.8}$$

which provides a differential equation for the ray path as

$$\frac{\mathrm{d}^2 y}{\mathrm{d}x^2} \approx \frac{1}{n}\frac{\partial n}{\partial y}. \tag{10.9}$$

This equation, together with appropriate boundary conditions, can be solved for the path $y(x)$, $x_1 < x < x_2$. The angle $\delta\varphi$ also represents the rotation of the ray during time δt. The total deflection angle of the ray in medium 2 would approximately be

$$\varphi \approx \int_{x_1}^{x_2} \frac{\mathrm{d}x}{r} \approx \int_{x_1}^{x_2} \frac{1}{n}\frac{\partial n}{\partial y}\mathrm{d}x \approx \frac{1}{n_0}\int_{x_1}^{x_2} \frac{\partial n}{\partial y}\mathrm{d}x. \tag{10.10}$$

If n is a function of only y, one may get the further simplified expression

$$\varphi \approx \frac{1}{n_0}\frac{\mathrm{d}n}{dy}(x_2 - x_1). \tag{10.11}$$

If, on the other hand, n depends on all three coordinates, x, y, z, one has to consider angular deflections on both lateral directions as

$$\varphi_y \approx \frac{1}{n_0}\int_{x_1}^{x_2} \frac{\partial n}{\partial y}\mathrm{d}x, \quad \varphi_z \approx \frac{1}{n_0}\int_{x_1}^{x_2} \frac{\partial n}{\partial z}\mathrm{d}x. \tag{10.12}$$

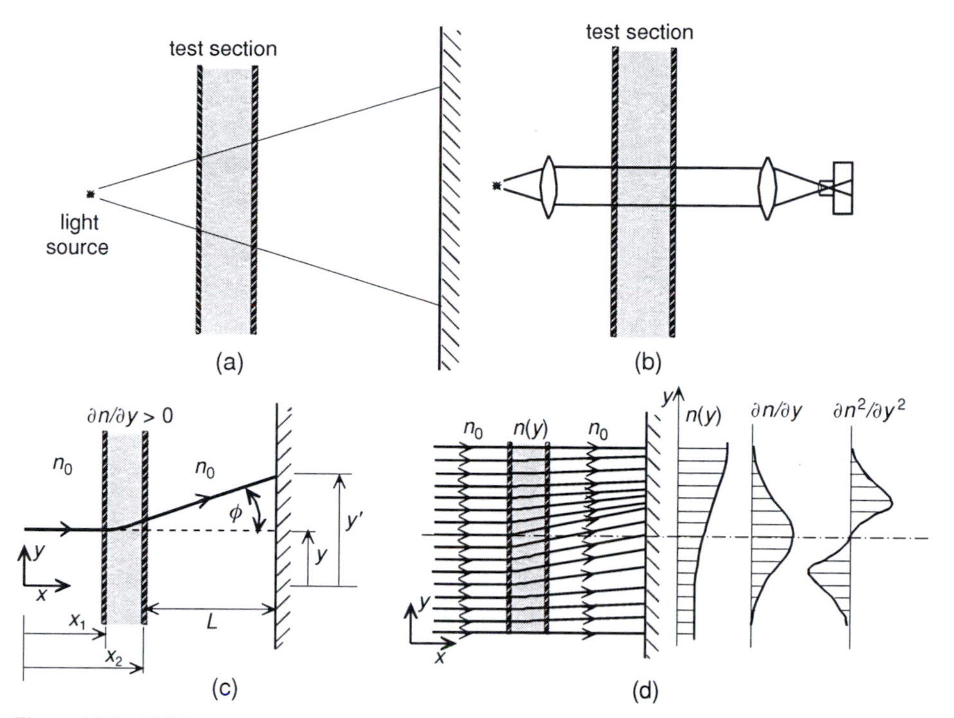

Figure 10.3. (a) Divergent-light shadowgraph, (b) collimated-light shadowgraph, (c) light beam deflection, and (d) light beam deflection through a mixing layer.

In gases for which the Gladstone–Dale formula $n = 1 + K\rho$ applies, one may rewrite approximation (10.10) as

$$\varphi \approx K \int_{x_1}^{x_2} \frac{\partial \rho}{\partial y} \mathrm{d}x \tag{10.13}$$

and similarly rewrite approximations (10.11) and (10.12) in terms of density ρ.

The shadowgraph method: This is a strictly qualitative flow visualization method, possible to implement by relatively simple means. Like other optical methods, it is based on the variation of refractive index within a medium, and thus it is applicable to variable-density flows. The shadowgraph operation may be explained by mainly geometrical optics concepts. A simple shadowgraph consists of a concentrated light source, approximating a point source, and a projection plane [Fig. 10.3(a)]. The diverging light from the source passes through the fluid and gets refracted by local refractive-index variations, so that its projection appears to have variations of light intensity with darker ('shadows') and brighter regions. Because it would be difficult to record the projected image, because of its large size, a more usable shadowgraph utilizes one lens to collimate the light and a second lens to collect the light after it crosses the region of interest and focus it into the lens of a camera [Fig. 10.3(b)]. To improve the contrast of shadowgrams, high-intensity point sources, such as sparks, would be preferable.

To understand the operation of the shadowgraph, consider the propagation of a light beam through a medium with $\partial n/\partial y > 0$, as shown in Fig. 10.3(c). The beam will be deflected by a total angle φ and reach the observation plane at a position y', rather than the position y corresponding to a uniform refractive index along the entire beam path. If the optical disturbances are small, φ is given by approximation (10.10) and one may calculate the beam deflection as

$$\Delta y = y' - y = L \tan\varphi \approx L\varphi \approx \frac{L}{n_0}\int_{x_1}^{x_2} \frac{\partial n}{\partial y}\mathrm{d}x. \qquad (10.14)$$

If the refractive index is independent of the spanwise direction z, the power of the undisturbed beam per unit span would be $I_0\mathrm{d}y$. Assuming negligible light absorption, the deflected beam power per unit span would be $I\mathrm{d}y' = I_0\mathrm{d}y$. Then the disturbance of the light power on the observation plane would be

$$\frac{\Delta I}{I_0} = \frac{I_0 - I}{I_0} \approx \frac{I_0 - I}{I} = \frac{\mathrm{d}y'}{\mathrm{d}y} - 1 \approx L\frac{\partial\varphi}{\partial y} \approx \frac{L}{n_0}\int_{x_1}^{x_2} \frac{\partial^2 n}{\partial y^2}\mathrm{d}x. \qquad (10.15)$$

This shows that the shadowgram contrast would be sensitive to the second derivative of the refractive index. When the refractive index is independent of x, Eq. (10.15) can be simplified to

$$\frac{\Delta I}{I_0} \approx \frac{L\,(x_2 - x_1)}{n_0}\frac{\partial^2 n}{\partial y^2}, \qquad (10.16)$$

which is a differential equation for n. If the light intensity were measured with a light densitometer or a photodetector, one would, in principle, be able to solve this equation to determine the variation of n and thus the density variation. This is not done, however, because this process would require a double integration, whose accuracy would be inadequate considering the relatively poor spatial resolution of shadowgrams. A more general expression for the light power disturbance caused by three-dimensional refractive-index fields is

$$\frac{\Delta I}{I_0} \approx \frac{L}{n_0}\int_{x_1}^{x_2} \left(\frac{\partial^2 n}{\partial y^2} + \frac{\partial^2 n}{\partial z^2}\right)\mathrm{d}x. \qquad (10.17)$$

As an illustration of shadowgraph application, consider a two-dimensional mixing layer of two fluid streams with different temperatures (or two different fluids), with the mean flow direction parallel to the z axis, as shown in Fig. 10.3(d). The same figure shows the variation of the refractive index and its first two derivatives. It may be seen that the largest deviations from the average light intensity on the observation plane would occur at locations at which $\left|\partial^2 n/\partial y^2\right|$ is maximum, whereas, at locations at which $\left|\partial^2 n/\partial y^2\right|$ is small, the light intensity would be nearly equal to the undistorted level.

Shadowgrams are a valuable tool for the qualitative assessment of high-speed gas flows and flows involving heating and cooling of fluids or mixing of different fluids. They are particularly successful in visualizing shock waves, and expansion and compression waves, and high-speed or heated wakes and jets [see example in Fig. 10.4(a)]. In interpreting shadowgrams of turbulent flows, one has to keep in mind that, because they are sensitive to second derivatives of the density, they tend to illustrate small-scale

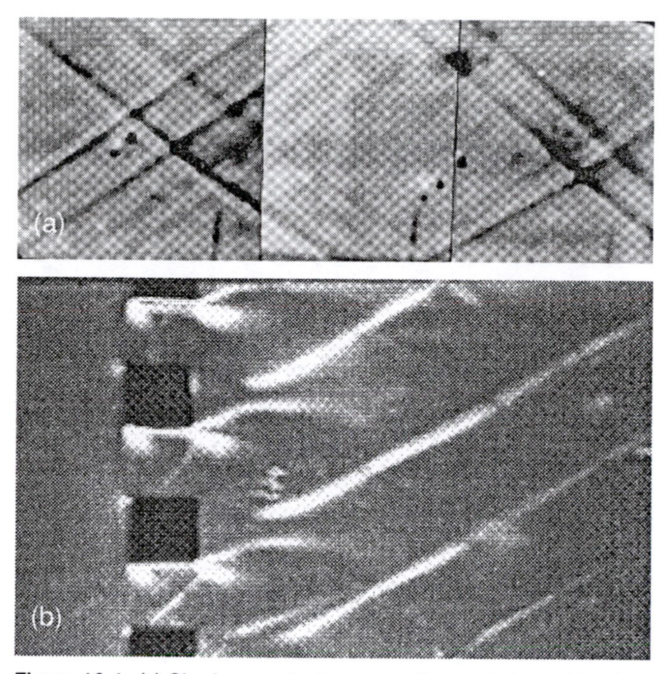

Figure 10.4. (a) Shadowgraph showing a diamond-shaped family of compression waves generated by a perforated plate; (b) Schlieren image showing oblique shock waves generated by a square-rod grid; in both cases, the obstructions were used to produce both supersonic flow and grid turbulence (photographs by Phil Zwart).

phenomena with much higher contrast than large-scale ones. The grainy appearance of shadowgrams is not necessarily an indication of isotropic turbulence structure but rather is an artefact of the method.

The Schlieren method: The Schlieren method is the most popular optical flow visualization method because of its good contrast and the relative ease by which it can be implemented. Its name is the plural of the German word *Schliere*, which indicates an optical inhomogeneity in glass. As indicative of its many variations, the operation principle of the classical *Töpler method* is presented. The optical system consists of a concentrated light source, a converging lens to collimate the light, and a collecting lens, called the *Schlieren head*, which collects the light after it passes the region of interest. If a light ray passes through a fluid in which the refractive index varies, as indicated in Fig. 10.5(a), it will be deflected by an angle φ, as discussed earlier in this section. As a result of this deflection, the ray will cross the focal plane of the collecting lens at a distance $\Delta y \approx f_S \varphi$ from the crossing point of the undistorted ray, where f_S is the focal distance of this lens. Now consider that the light source has a rectangular shape, which, in the absence of any distortion, would be focussed onto a rectangle with height h [shown by a dashed-line rectangle in Fig. 10.5(b)] on the focal plane of the Schlieren head. The distorted image of the source on the focal plane [shown by a solid-line rectangle in Fig. 10.5(b)] would be displaced by a distance Δy with respect to the undistorted image. The Schlieren method consists of using a sharp object, such as a knife edge or

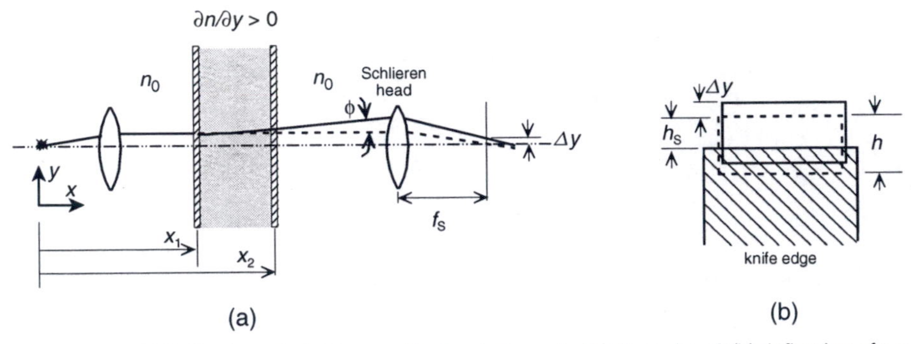

Figure 10.5. (a) Deflection of a light ray on the focal plane of a Schlieren head; (b) deflection of a rectangular light source image on the focal plane of a Schlieren head.

a razor blade, to partly block the focussed image of the source, such that only a rectangle with height $h_S < h$, corresponding to light intensity I_0, passes through. Because of light deflection, the distorted image will be blocked by a different proportion than the undistorted one. In the case of $\partial n/\partial y > 0$, as shown in Fig. 10.5, $\varphi > 0$ and the distorted source image would be deflected away from the edge, so that its light intensity I would be greater than I_0; in the case of $\partial n/\partial y < 0$, $\varphi > 0$ and the distorted image would be blocked more by the edge than the undistorted image, such that $I < I_0$. The relative difference in light intensity passing through the focal plane would be

$$\frac{\Delta I}{I_0} \approx \frac{\Delta y}{h_S} \approx \frac{f_S \varphi}{h_S} \approx \frac{f_S}{h_S} \frac{1}{n_0} \int_{x_1}^{x_2} \frac{\partial n}{\partial y} \mathrm{d}x. \tag{10.18}$$

If n were independent of x, this expression would be simplified to

$$\frac{\Delta I}{I_0} \approx \frac{f_S (x_2 - x_1)}{h_S n_0} \frac{\mathrm{d}n}{\mathrm{d}y} \tag{10.19}$$

The preceding expressions indicate that the Schlieren image intensity variation would be proportional to the refractive-index gradient in the direction normal to the knife edge. Compared with shadowgraph images, Schlieren images are generally sharper and more likely to be usable for quantitative estimates of density variation, as they would require a single, rather than double, integration. Approximation (10.18) shows that, for a given lens and test-section width, the sensitivity of the method is inversely proportional to h_S, which means that, for improved contrast, one would have to block a larger proportion of the focussed light source image. This, however, has a limit, because as h_S is decreased, there is an increasing risk that the distorted image would be either entirely unblocked or entirely blocked by the knife edge, which would produce uniform illumination. An example of a Schlieren image is shown in Fig. 10.4(b).

Besides the single knife edge just described, a variety of other partial obstructions have been used to obtain Schlieren images, including double-edge and circular ones. A variation of the basic method is the *colour Schlieren technique*, in which the knife edge is replaced with an optical filter consisting of parallel strips of different colours. A deflected light beam passing through the filter would appear to be of different colour than if it were undistorted, thus revealing refractive-index gradients. An advantage of the colour Schlieren technique over the knife-edge technique is that it can be applied to cases with relatively large light beam deflections.

Interferometry: The various available interferometric methods utilize the phase shift of light waves propagating through media of variable refractive index. Consider a light beam propagating through such a medium, as illustrated in Fig. 10.2. If the medium had a uniform refractive index n_0, the beam would cross the medium during a time

$$t_0 = \int_{x_1}^{x_2} \frac{1}{v_0} \mathrm{d}x = \frac{1}{c} \int_{x_1}^{x_2} n_0 \mathrm{d}x, \tag{10.20}$$

where v_0 is the (constant) light speed in the medium and c is the speed of light in vacuum. In the distorted medium, the corresponding light speed would be v and the corresponding time would be

$$t = \int_{x_1}^{x_2} \frac{1}{v} \mathrm{d}x = \frac{1}{c} \int_{x_1}^{x_2} n \mathrm{d}x. \tag{10.21}$$

Two light waves entering in phase into two media with equal thicknesses and slightly different refractive indices equal to n_0 and n, respectively, would exit these media with a phase difference

$$\Delta\theta \approx \frac{2\pi v_0}{\lambda}(t - t_0) \approx \frac{2\pi}{\lambda n_0} \int_{x_1}^{x_2} (n - n_0)\,\mathrm{d}x. \tag{10.22}$$

If n is independent of x, the phase shift would be simplified to

$$\Delta\theta \approx \frac{2\pi\,(x_2 - x_1)}{\lambda n_0}(n - n_0). \tag{10.23}$$

Thus the phase shift would be directly proportional to refractive-index variations, rather than variations of its derivatives, as in the shadowgraph and Schlieren methods.

A direct application of the preceding analysis is illustrated by the *Mach–Zehnder interferometry* method, which is the classical interferometric method in gas dynamics.

This method, illustrated in Fig. 10.6, consists of splitting a collimated light beam into two beams, one of which passes through the test section containing the flow of interest, while the second one (*reference beam*) passes through a pair of transparent plates, identical to the walls of the test section. The two beams are made to *interfere* with each other on the observation plane or the recording film, where the disturbed beam would arrive with a phase lag given by approximation (10.22), compared with the reference beam. The interfering planar light waves would combine to form a wave whose amplitude would be proportional to

$$\sin(2\pi vt) + \sin(2\pi vt - \Delta\theta) = 2\cos\left(\frac{\Delta\theta}{2}\right)\sin\left(2\pi vt - \frac{\Delta\theta}{2}\right). \tag{10.24}$$

The frequency response of a film, photodetector, or the human eye would be insufficient to resolve the light frequency, and so the captured power of the light on the recording plane would be proportional to $\cos^2(\Delta\theta/2)$. Therefore the recorded light intensity would vary depending on the phase shift in the test section. When

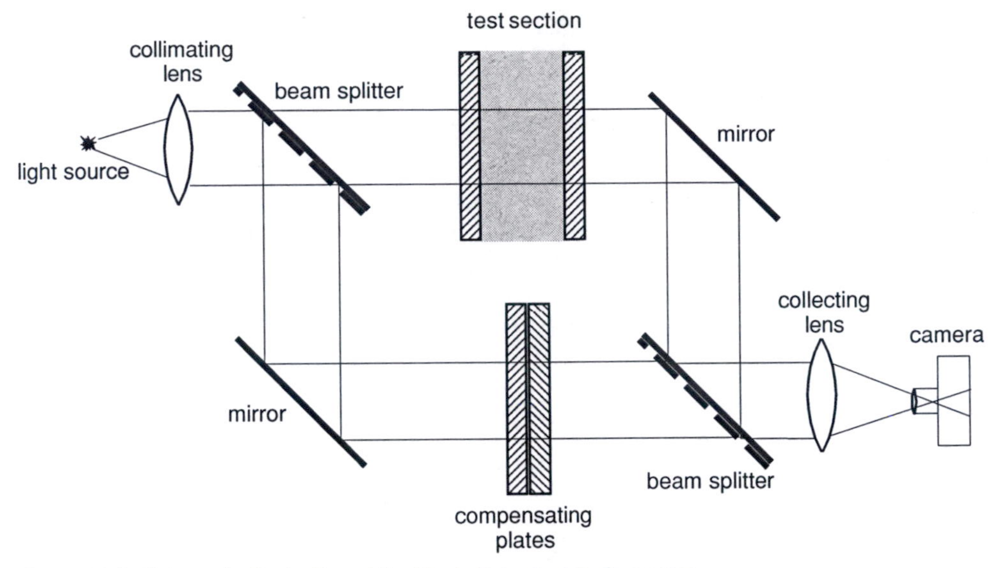

Figure 10.6. Schematic illustration of the Mach–Zehnder interferometer.

$\Delta\theta = (2i + 1)\pi, i = 0, \pm 1, \pm 2, \ldots$, the light intensity would vanish, whereas, when $\Delta\theta = 2i\pi, i = 0, \pm 1, \pm 2, \ldots$, the light intensity would be maximum. Thus, if the test section contains a fluid with varying refractive index, the interferogram would appear as a set of bright and dark *fringes*. Each fringe would be a contour of constant refractive index and the difference between the refractive indices of two adjacent dark (or bright) fringes would be

$$\Delta n = \frac{\lambda n_0}{x_2 - x_1}. \tag{10.25}$$

Therefore the interferogram may also be used for quantitative purposes, because the fringes would be contours of constant n, namely of constant density. If, for example, the medium is a gas for which the Gladstone–Dale formula (see Section 5.2) applies, the density difference between two fringes of the same kind would be

$$\Delta\rho \approx \frac{\lambda}{(x_2 - x_1) K}, \tag{10.26}$$

where K is the Gladstone–Dale constant. If the medium is a perfect gas under constant pressure, then the fringes would also be isotherms (i.e., contours of constant temperature). To find the absolute temperature on a fringe, one must specify the temperature T_r on some reference fringe. Such reference temperatures can be determined by theoretical means or by independent measurement, for example by use of a thermocouple. If N is the number of fringes between the location of interest and the reference fringe, the temperature can be found as

$$T = \left[\frac{N\lambda R}{(x_2 - x_1) K p} + \frac{1}{T_r}\right]^{-1}, \tag{10.27}$$

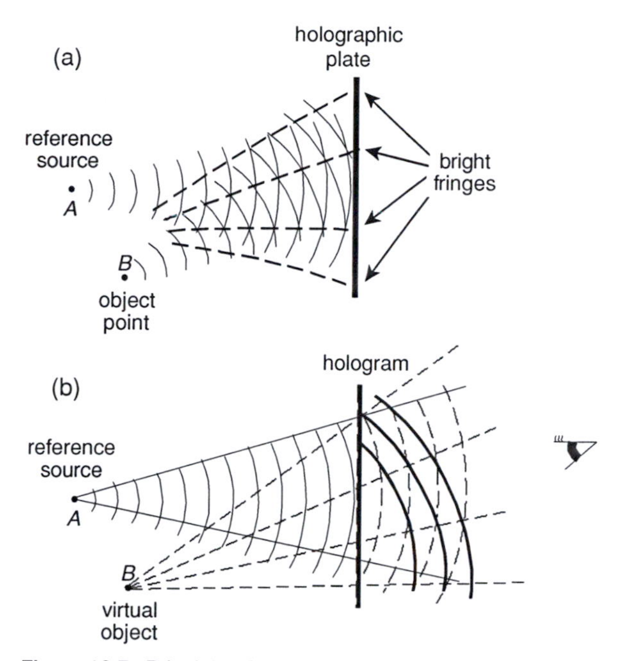

Figure 10.7. Principle of holography: (a) recording of image and (b) reconstruction of virtual object.

where R is the gas constant and p is the pressure. If refractive-index variation is due to composition changes, the fringes would represent isoconcentration contours. One can find the value of concentration on each contour by knowing a reference value and the concentration–refractive-index relationship for the mixture, assuming, of course, that the mixture is isothermal.

In the preceding setup, also known as *infinite-fringe interferometry*, there would be no fringes if the refractive index along the beam path were uniform. One can achieve improved resolution by introducing a variable phase shift to one of the beams by passing it through a prism. Superposition of the undistorted and distorted interferograms would produce a *moiré pattern*, which may be analysed to resolve the refractive-index variation; this technique is called *finite-fringe interferometry*. Mach–Zehnder interferometry requires careful alignment and high-precision optical components and test-section windows. For these reasons, it has been largely superseded by holographic interferometry, which is discussed in the following subsections.

Holography: Holography is an optical method capable of recording and reconstructing three-dimensional images. It was invented in 1948 by T. Gabor [39], who was awarded a Nobel Prize for this discovery in 1971. The basic principle of holography is illustrated in Fig. 10.7. Consider a monochromatic point source A, producing a coherent spherical light wave, termed the *reference wave*. Now consider an object, for simplicity also assumed to be a point B, which is illuminated by the point source and, through reflection and scattering, also produces a coherent spherical light wave, called the *object wave*. As the reference and object waves propagate through the medium, they interfere with each other, producing standing waves whose amplitude depends on the phase relationship.

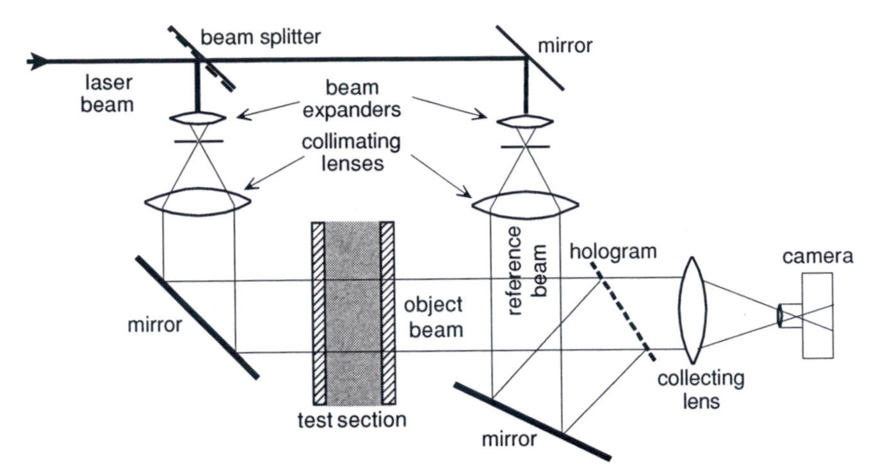

Figure 10.8. Schematic illustration of a typical holographic interferometry setup.

Where the two waves are in phase, which will be along hyperbolic surfaces, maximum light intensity will be produced. If a photographic plate is exposed to this light, it will record a pattern of interference fringes. This is the *recording* step of holography. The developed photographic plate is called the *hologram*. During *reconstruction*, the same source is used to illuminate the hologram. The hologram will then act as a diffractive grating (see Section 5.3), not only allowing the reference wave to pass through, but also reconstructing the phase and amplitude characteristics of the object wave. Thus an observer looking backwards through the hologram will see a *virtual object*, identical to the original one. This process applies not only to point objects, but also to any three-dimensional object, whose reconstruction allows its observation from different directions, like the original object. Although the concept of holography precedes the invention of lasers, its extensive application had to await the development of lasers, because of its dependence on a monochromatic, coherent light source. Holography is used in many applications and with many different variations [40–42], but our present interest is limited to configurations used in flow visualization and measurement of fluid temperature, composition, and velocity.

Holographic interferometry: A representative optical arrangement for holographic interferometry is shown in Fig. 10.8. A single, monochromatic laser beam is split, expanded, and collimated into an object beam and a reference beam, both of which are directed onto the photographic plate. The object wave crosses the test section and is phase shifted by the variable refractive-index fluid, while the reference wave bypasses the test section and is undistorted. When the hologram is illuminated by the reference beam, the image of the test section appears. Because the reconstruction of the object reverses all optical errors, there is no need for high-quality optical components, window materials, or alignment. Each hologram may contain not only a single object wave but several, recorded at different instances. During reconstruction, all objects would appear and their images would interfere with each other. This fact has been the basis of the *double-exposure* interferometric technique, used widely in heat and mass transfer.

Consider, for example, that one is interested in the temperature distribution in a heated flow. The photographic plate is exposed to two consecutive illumination steps: The first one, called a *comparison wave*, is produced by the unheated flow, and the second one, called a *measuring wave*, is produced by the heated flow. Reconstruction produces an interference pattern because of the heat-related changes in refractive index in the test section. This technique eliminates the effects of light distortion that are due to curvature of the test-section walls or other irregularities, but has the disadvantage that it can be used to produce only a single interferogram at a time, and so it is not suitable for transient flow phenomena. The *real-time* interferometric technique permits the continuous observation of the interference pattern. Like the double-exposure technique, it starts with exposing the photographic plate to a comparison wave, but then the plate is developed to produce a hologram. The hologram is repositioned precisely in its original position, and, when exposed to the measuring wave, it produces a time-continuous interference pattern, which can be recorded cinematographically or videoscopically.

Streaming birefringence [1, 3]: This is a method of flow visualization of certain liquids, which become *birefringent* (see Section 5.2) under the application of shear stresses. Such liquids include polymers and colloidal solutions of elongated crystals. When at rest, these liquids are optically isotropic, but, in shear flows, the long axes of the molecules or particles in the liquid tend to get oriented in directions normal to the velocity gradient, thus introducing birefringence. A light beam passing through such a medium will be separated into two plane-polarized components with polarization planes normal to each other. Both components propagate in the same direction, but with different phase speeds, so that, on exit from the medium, they have a phase shift with respect to each other. In other words, the refractive indices of the medium for the two components are different and superposition of the two exiting beams will introduce a fringe pattern, which may be correlated with flow characteristics. Methods introduced to detect the phase shift include the use of a polariscope (an optical instrument commonly used in photoelasticity experiments), Mach–Zehnder interferometry, and a scattering-based technique. Milling yellow dye solutions have been used widely as a birefringent liquid, because of their high sensitivity.

The Wollaston-prism technique [3, 37]: This is a hybrid method with elements of both the Schlieren method and interferometry. In a Schlieren optical setup, the knife edge is replaced with a *Wollaston prism*, which is a birefringent device (see Section 5.2) that splits the light beam through it into two linearly polarized components, propagating through the fluid with different phase speeds and at a very small angle to each other. When superimposed, the two components produce an interference pattern, which, similar to the Schlieren method, is sensitive to the refractive-index gradient, rather than to the refractive-index value. Both finite-fringe and infinite-fringe patterns can be obtained, depending on whether the prism is located on the collected beam focal plane or not.

Tomography [36]: The term *tomography* encompasses a host of sensing techniques and mathematical algorithms, which reconstruct three-dimensional images and fields of various properties of an object by processing multiple one-dimensional scans of the object. They are used extensively in medicine, process engineering, and many other areas of science and technology. A great variety of sensing techniques have been employed, based on absorption, refraction, scattering, and emission of various forms of radiation and sonic waves. Among the many commonly used tomographic methods are *positron emission tomography* (*PET*), *nuclear magnetic resonance* (*NMR*), *magnetic resonance imaging* (*MRI*), *computerized axial tomography* (*CAT*), *electric capacitance tomography* (*ECT*), *electric resistance tomography* (*ERT*), and *inductance electromagnetic tomography* (*EMT*) methods. Typically, a sensor records the cumulative effect of property change along a linear path across the object, a number of such records along intersecting paths on the same plane are obtained simultaneously or sequentially, and the multiple records are processed by a mathematical algorithm capable of reconstructing the local values of the desired property over a planar section of the object. The three-dimensional field of the property can be also reconstructed by processing multiple planar sections. In fluid mechanics research, tomographic methods are used for three-dimensional flow visualization and for the measurement of velocity, temperature, and concentration. Although the resolution and the accuracy of such methods are generally inferior to those of other measurement methods, they are valuable in revealing three-dimensional patterns, especially of transient and unsteady phenomena. Optical tomographic methods for three-dimensional flow visualization utilize refractive-index-based sensing, including interferometric and Schlieren techniques.

10.4 Radiation emission techniques

These are flow visualization and measurement techniques suitable for low-density gases, in which they have superior sensitivity to other methods. Their operation is based on the emission of visible radiation by gas molecules following their excitation by an electric field. Radiation emission techniques have been used regularly for flow visualization of shock waves and other flow patterns in supersonic and hypersonic wind tunnels or shock tubes, as well as for the measurement of temperature, pressure, density, and composition variations. Such techniques include the following ones [1,3]:

Spark tracing: This is suitable for absolute gas pressures between about 100 kPa and 1 Pa. Depending on the shape of the electrodes, the spark could form along a line, a plane, or other surface. The basic configuration consists of two electrodes pointed at each other across the flow. A high-voltage (in the tens of kilovolts), short-duration (0.1–1 ms) electric pulse applied across the electrodes causes a spark to form along the shortest path between them, ionizing the air along its path. Subsequent electric pulses generate sparks that travel along the original ionized path, which in the meantime has been convected downstream, thus marking timelines in the gas. The electrodes are usually needles or thin wires, between 8 and 130 mm apart. A thin wire stretched across the electrodes helps to create a straight initial timeline. The ionization persists for about 1 ms; therefore the frequency of electric pulses must be at least 1 kHz for a single timeline and higher for multiple timelines. This method is suitable for gas speeds between 1 m/s and several hundreds of metres per second. As with any other timeline method, spark tracing cannot resolve the motion of individual fluid particles and therefore cannot identify the flow direction. To remedy this, a method has been suggested in which the flow is seeded with AlN particles, which burn along the spark's path, marking short bright pathlines.

Glow discharge method: This is suitable for pressures between 10 Pa and 0.01 Pa. A strong electric field applied to the flow accelerates free electrons and ions that are contained in the gas. These particles collide randomly with neutral molecules, which are excited and emit radiation, which is proportional to the gas density. Thus density fields can be mapped from measured variations of light intensity or changes in colour.

Electron beam method: This method is the only means of visualizing gases in the extremely low-pressure range below 0.01 Pa. The flow is bombarded with high-speed electrons with the use of an *electron gun*, which excites the gas molecules to emit radiation, whose intensity is proportional to the local density.

10.5 Enhancement of flow visualization records

Computers and software are indispensable components of most flow visualization methods. They can be used to improve the quality of flow images and to assist the observer in the presentation and interpretation of visual patterns [1]. In this section, we are mainly concerned with qualitative flow visualization techniques or those with a relatively low resolution. The use of computers and related tools to extract quantitative information from images, as for example in PIV, is considered in the appropriate sections in later chapters. Similar techniques can be used to enhance and process not just optical images, but also two- and three-dimensional image-like plots, reconstructed from transducer measurements (e.g., multiple hot-wire signals) or from results of analytical calculations and numerical simulations.

Before being processed by a computer, flow images recorded on film or analogue video tape must be digitized, so that the information can be stored on the computer's memory or a digital storage device (e.g., CD-ROM, floppy diskette or digital magnetic tape). For the digitization, one may use a scanner, a CCD or CMOS camera, or a digitizing board. Depending on the resources available and the desired resolution, each grey-tone image is replaced with an array of $M \times N$ pixels, each containing the corresponding light intensity, approximated by one of 2^n (from $2^1 = 2$ to $2^8 = 256$) levels; colour images are replaced with three superimposed arrays, each containing the intensity of one of the three primary colours, red, green, or blue (see Section 7.3).

A digitized image can be filtered, the same way discrete-time series are, in order to remove certain features or accentuate others. Low-pass filtering can be achieved by the replacement of the original intensity of each pixel by an average over neighbouring pixels. Boxcar (i.e., unweighted) averaging or more sophisticated algorithms, in which variable weights are used, may be used for this purpose. Low-pass filtering reduces optical noise as well as softens the appearance of small-scale features of the image, while preserving the large-scale features. High-pass filtering can be achieved by subtraction of the low-pass filtered intensities from the original ones; this sharpens the appearance of the image, emphasizing boundaries between flow patterns and small-scale features; it also removes large-scale, systematic variations of intensity that are due to variable-brightness illumination.

At times it is desirable to enhance the contrast of a digitized image by replacing the intensity at each pixel with one of two (black and white) or more (black, white, and grey) levels. One achieves this by setting appropriate intensity thresholds and replacing values of the original intensity within a certain range with the limiting value. This method can be used to replace grey images with pseudocolour images by the selection of a colour for each range between two thresholds.

A common task of flow visualization is to identify the boundaries between flow regions with different features, a process called *edge detection*. The most precise definition of a boundary is achieved when it has a thickness of one pixel, which is rarely, if ever, the case in unprocessed digitized images. Optimized thresholding and low-pass filtering can be used for this purpose. The same methods can be used to identify lines, such as timelines, in the flow (*line reduction*). Other methods that emphasize changes of intensity are *gradient methods*, namely the computation and display of the intensity gradient and *differencing methods*, by which *bas-relief* illusions are produced when the image is inverted to create a negative, and the negative is shifted slightly in space and is superimposed on the positive.

Image reconstruction includes methods that combine information from different images in an attempt to recover features that are not contained in any single image. When periodic or quasi-periodic phenomena occur in a flow, as for example in vortex shedding from bluff objects, and phase information is recoverable from each image, one may *phase average* several images to enhance the appearance of common patterns, which in individual frames may be heavily distorted by random effects. One may also superimpose overlapping images of different flow regions to produce images of larger regions. If simultaneous images of the same flow patterns are taken from different orientations, they could be statistically correlated so that three-dimensional images may be reconstructed, which could be subsequently viewed from any orientation. Similarly, three-dimensional images can be reconstructed by the correlation of simultaneous images on parallel planes (tomography, see earlier section) or, in a convected frame, by the correlation of images on a fixed plane but at different time instances.

QUESTIONS AND PROBLEMS

1. Consider applying the hydrogen-bubble method to visualize the vertical profile of horizontal flow in a water channel with a depth of 0.80 m and a temperature of 20 °C. The bubble-producing wire has a diameter of 0.1 mm and has been kinked to form a sawtooth-like shape, in which each element is an equilateral triangle with a spacing of 4 mm along the wire axis. Assume that the wire is supplied with a dc of 0.4 A. Compute the number of bubbles that will be released per unit time from each apex of the wire at different depths from the free surface. Assuming that the bubbles

rise very slowly, estimate the sizes of bubbles released at different depths when they reach the free surface. Furthermore, making any simplifying assumptions that you find appropriate, estimate the vertical velocities of bubbles released at different depths as they rise through the water.

2. Discuss the simplest and least expensive flow visualization techniques that would be suitable for the following experiments. Identify all required materials and instrumentation.
   a. Locating regions of flow separation from a racing car model tested in (i) a wind tunnel and (ii) a water tunnel.
   b. Locating the positions of shock waves generated by a Pitot tube in a supersonic wind tunnel.
3. Describe briefly the applicability of the following flow visualization methods:
   a. The hydrogen-bubble method in a channel filled with silicon oil.
   b. The photocatalysis method in a water tunnel.
   c. The tufts method as a means of roughly estimating turbulence intensity in a wind tunnel.
4. A visualization study of low-Reynolds-number laminar flow of a mixture of 80% glycerol and 20% water at 20 °C uses glass beads of 20-μm diameter as markers. The density and kinematic viscosity of the mixture can be found from tables as 1208.5 kg/m^3 and 49.57×10^{-6} m^2/s, respectively. While the study is in progress, the supply of glass beads is exhausted and it becomes necessary to use different flow markers, which, however, should have at least the same level of precision in marking fluid pathlines.
   a. Compute the maximum size of (i) aluminium oxide particles, (ii) polystyrene beads, and (iii) hydrogen bubbles that could be used for this purpose. State clearly and evaluate all assumptions that you make.
   b. Assuming that a maximum deviation of 5% from the true flow velocity can be tolerated, determine the minimum horizontal and vertical flow velocities of flows that can be visualized adequately by the preceding particles.
   c. If the maximum horizontal flow velocity is 20 times larger than the value computed in part b, determine the maximum exposure time of the camera lens that would permit a 'clear' recording of individual images of each of the preceding particles, considering that such recording requires that the particle should not move by more than 25% of its diameter during recording.
5. A laser beam containing light with two wavelengths, $\lambda_1 = 509.7$ nm and $\lambda_1 = 912.5$ nm, passes through the square test section of a wind tunnel containing vertically stratified air flow, whose density variation can be approximated by the function

$$\rho = \rho_0 \left(1 - 0.2\frac{y}{h}\right),$$

where $\rho_0 = 1.2$ kg/m^3, y is the distance from the bottom wall, and $h = 0.300$ mm is the test-section height. The beam is observed on a plane wall located 2.5 m away from the test section. The Gladstone–Dale constants for air for the two wavelengths are $K_1 = 0.2274$ m^3/kg and $K_2 = 0.2239$ m^3/kg.

a. Explain what will happen to the laser beam as it passes through the test section.
b. Determine the distance between the positions of the two beam components on the observation plane. Would this distance depend on the vertical position of entry to the test section? Then, determine the phase shift between the two light components on the observation plane.

6. Consider horizontal, stratified, laminar air flow in a wind tunnel having glass walls. The tunnel's width is 0.5 m and its height is $h = 1$ m. The air density, measured with a sampling probe and fitted with a polynomial equation as a function of the vertical distance y from the bottom of the tunnel, is approximately

$$\rho = \rho_0 \left\{1 - 0.05 \left[3 \left(y/h\right)^3 + 4(y/h)^2 + 2y/h\right]\right\},$$

where $\rho_0 = 1.25\,\text{kg/m}^3$. Illumination is provided by a 'wide' laser beam with a wavelength of 589 nm. Determine and plot the variation of light intensity that would be obtained with (a) a shadowgraph, (b) a Töppler Schlieren system, and (c) a Mach–Zehnder interferometer. Present the plots in dimensionless form and make simplifying assumptions, when required. Now assume that the density variation was unknown. Would you be able to estimate it, based on the visual patterns provided by each of these three methods? Discuss any possible difficulties.

REFERENCES

[1] W.-J. Yang (Editor). *Handbook of Flow Visualization*. Taylor & Francis, Washington, DC, 1989.
[2] Japan Society of Mechanical Engineers. *Visualized Flow*. Pergamon, Oxford, UK, 1988.
[3] W. Merzkirch. *Flow Visualization* (2nd Ed.). Academic, New York, 1987.
[4] A. J. Smits and T. T. Lim (Editors). *Flow Visualization Techniques and Examples*. Imperial College Press, London, 2000.
[5] M. Gad-el-Hak. Visualization techniques for unsteady flows: An overview. *J. Fluids Eng.*, 110:231–243, 1988.
[6] M. Gad-el-Hak. Splendor of fluids in motion. *Prog. Aerosp. Sci.*, 29:81–123, 1992.
[7] J. P. Crowder. Tufts. In W.-J. Yang, editor, *Handbook of Flow Visualization*, pp. 125–175. Taylor & Francis, Washington, DC, 1989.
[8] R. Reznicek. Surface tracing methods. In W.-J. Yang, editor, *Handbook of Flow Visualization*, pp. 91–103. Taylor & Francis, Washington, DC, 1989.
[9] L. C. Squire. The motion of a thin oil sheet under the boundary layer of a body. Tech. Rep. AGARDograph, AGARD-AG-70, 7–28, Advisory Group for Aerospace Research and Development, Neuilly-sur-Seire, France, 1962.
[10] T. Liu, T. Campbell, S. Burns, and J. Sullivan. Temperature and pressure sensitive luminescent paints in aerodynamics. *Appl. Mech. Rev.*, 50:227–246, 1997.
[11] B. M. McLachlan and J. H. Bell. Pressure-sensitive paints in aerodynamic testing. *Exp. Therm. Fluid Sci.*, 10:470–485, 1995.
[12] C. Mercer (Editor). *Optical Metrology for Fluids, Combustion and Solids*. Kluwer Academic, Dordrecht, the Netherlands, 2003.
[13] N. Kasagi, R. J. Moffat, and M. Hirata. Liquid crystals. In W.-J. Yang, editor, *Handbook of Flow Visualization*, pp. 105–116. Taylor & Francis, Washington, DC, 1989.
[14] B. R. Clayton and B. S. Massey. Flow visualization in water: A review of techniques. *J. Sci. Instrum.*, 44:2–11, 1967.

[15] M. Gad-el-Hak. The use of the dye-layer technique for unsteady flow visualization. *J. Fluids Eng.*, 108:34–38, 1986.

[16] J. W. Hoyt and R. H. J. Sellin. A turbulent-flow dye-streak technique. *Exp. Fluids*, 20:38–41, 1995.

[17] D. J. Baker. A technique for the precise measurement of small fluid velocities. *J. Fluid Mech.*, 26:573–575, 1966.

[18] R. Eichhorn. Flow visualization and velocity measurement in natural convection with the tellurium dye method. *J. Heat Transfer*, 83:379–381, 1961.

[19] S. Taneda, H. Honji, and M. Tatsuno. The electrolytic precipitation method of flow visualization. In A. Asanuma, editor, *Flow Visualization*, pp. 209–214. Hemisphere, Washington, DC, 1979.

[20] A. T. Popovich and R. L. Hummel. A new method for non-disturbing turbulent flow measurements very close to a wall. *Chem. Eng. Sci.*, 22:21–25, 1967.

[21] M. Ojha, R. L. Hummel, R. S. C. Cobbold, and K. W. Johnston. Development and evaluation of a high resolution photochromic dye method for pulsatile flow studies. *J. Phys. E*, 21:998–1004, 1988.

[22] D. H. Thompson. Flow visualization using the hydrogen bubble technique. Tech. Rep. Aerodynamics Note 338, Aeronautical Research Laboratories, Australian Defence Scientific Service, Melbourne, Australia, 1973.

[23] F. W. Roos and W. W. Willmarth. Hydrogen bubble flow visualization at low Reynolds numbers. *AIAA J.*, 7:1635–1637, 1969.

[24] F. A. Schraub, S. J. Kline, J. Henry, P. W. Runstadler, and A. Littell. Use of hydrogen bubbles for quantitative determination of time-dependent velocity fields in low-speed water flows. *J. Basic Eng.*, 87:429–444, 1965.

[25] W. Davis and R. W. Fox. An evaluation of the hydrogen bubble technique for the quantitative determination of fluid velocities within clear tubes. *J. Basic Eng.*, 89:771–781, 1967.

[26] S. D. Bruneau and W. R. Pauley. Measuring unsteady velocity profiles and integral parameters using digital image processing of hydrogen bubble timelines. *J. Fluids Eng.*, 117:331–340, 1995.

[27] C. R. Smith and R. D. Paxson. A technique for evaluation of three-dimensional behavior in turbulent boundary layers using computer augmented hydrogen bubble-wire flow visualization. *Exp. Fluids*, 1:43–49, 1983.

[28] T. J. Mueller. Flow visualization by direct injection. In R. J. Goldstein, editor, *Fluid Mechanics Measurements* (2nd Ed.), pp. 367–450. Hemisphere, Washington, DC, 1996.

[29] I. McGregor. The vapour-screen method of flow visualization. *J. Fluid Mech.*, 11:481–511, 1961.

[30] T. Corke, D. Koga, R. Drubka, and H. Nagib. A new technique for introducing controlled sheets of smoke streaklines in wind tunnels. In *Proceedings of International Congress on Instrumentation in Aerospace Simulation Facilities*, IEEE Publication 77 CH1251-81AES, pp. 74–80, 1977.

[31] H. Yamada. Use of smoke wire technique in measuring velocity profiles of oscillating laminar air flows. In T. Anasuma, editor, *Flow Visualization*, pp. 265–270. Hemisphere, Washington, DC, 1979.

[32] J. C. Kent and A. R. Eaton. Stereophotography of neutral He-filled bubbles for 3-d fluid motion studies in an engine cylinder. *Appl. Opt.*, 21:904–912, 1982.

[33] P. Matisse and M. Gorman. Neutrally buoyant anisotropic particles for flow visualization. *Phys. Fluids*, 27:759–760, 1984.

[34] O. Savas. On flow visualization using reflective flakes. *J. Fluid Mech.*, 152:235–248, 1985.

[35] S. T. Thoroddsen and J. M. Bauer. Qualitative flow visualization using colored lights and reflective flakes. *Phys. Fluids*, 11:1702–1704, 1999.

[36] F. Mayinger and O. Feldmann (Editors). *Optical Measurements* (2nd Ed.). Springer, Berlin, 2001.

[37] G. S. Settles. *Schlieren and Shadowgraph Techniques*. Springer, Berlin, 2001.

[38] F. J. Weyl. Analysis of optical methods. In R. W. Ladenburg, editor, *Physical Measurements in Gas Dynamics and Combustion*, pp. 3–25. Princeton University Press, Princeton, NJ, 1954.

[39] D. Gabor. A new microscopic principle. *Nature (London)*, 161:777–778, 1948.

[40] E. R. Robertson and J. M. Harvey (Editors). *The Engineering Uses of Holography*. Cambridge University Press, Cambridge, UK, 1970.

[41] H. J. Caulfield and Sun Lu. *The Applications of Holography*. Wiley-Interscience, New York, 1970.

[42] C. B. Collier, C. D. Burckhardt, and L. H. Lin. *Optical Holography*. Academic, New York, 1971.

11 Measurement of local flow velocity

The measurement of local flow velocity is a main objective of experimental fluid mechanics. It may be accomplished by direct measurement of the speed of flow markers or by a variety of other methods, which utilize the relationship between flow velocity and other properties of the flow. In general, one may classify most flow velocity measurement methods into one of the following categories.

- *Pressure difference* methods. These utilize analytical relationships between the local velocity and the static and total pressures. The *Pitot-static tube* is the most common representative of instruments that measure local flow velocity from a pressure difference. These methods were discussed in Section 8.4.
- *Thermal* methods. These compute flow velocity from its relationship to the convective heat transfer from heated elements. Common instruments of this type are the *hot-wire* and *hot-film anemometers*.
- *Frequency-shift* methods. These are based on the Doppler phenomenon, namely the shifting of the frequency of waves scattered by moving particles. The main instruments in this category are the *laser Doppler velocimeter* and the *ultrasonic Doppler velocimeter*, utilizing light and sound waves, respectively.
- *Marker tracing* methods. These trace the motion of suitable flow markers, optically or by other means. Common optical methods include *chronophotography* and PIV, whereas *pulsed-wire anemometry* is a method that traces heat emitted from local sources.
- *Mechanical* methods. These take advantage of the forces and moments that a moving stream applies on immersed objects. Examples of instruments of this type are the *vane*, *cup*, and *propeller anemometers*.

11.1 Thermal anemometry

Thermal anemometers [1–6] measure the local flow velocity through its relationship to the convective cooling of electrically heated metallic sensors. These include *hot-wires*, usable in clean air or other gas flows, and *hot films*, usable in both gas and liquid flows. A related method, *pulsed-wire anemometry* is, in a strict sense, a thermal tracer method,

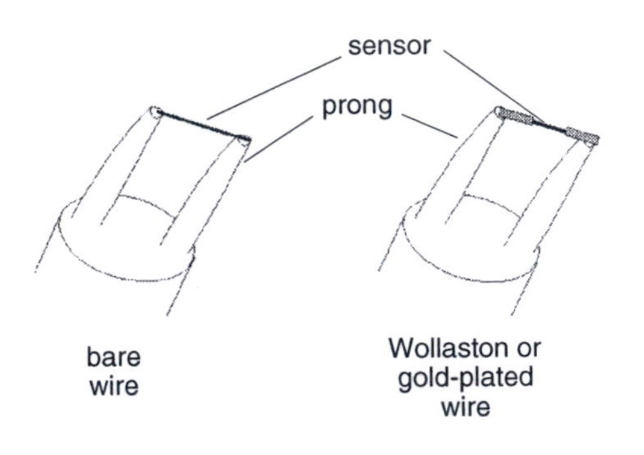

Figure 11.1. Sketches of typical hot-wire probes.

but it is included in this section as it utilizes instrumentation similar to that used by the other thermal anemometers.

Until the development of laser Doppler velocimetry (LDV) in the 1970s, thermal anemometry was essentially the only method having space, time, and amplitude resolutions that are sufficiently high for the measurement of turbulent characteristics down to the smallest scales of dynamic interest. Although the use of hot-wires as velocity sensors was anticipated by some work in the late 19th century, the seminal paper that clarified the relationship between heat transfer from a heated cylinder and flow speed was written by King in 1914 [7]. The resulting equation, known as *King's law*, is still commonly used. Early hot-wires were relatively thick and operated in the constant-current mode, which severely limited their frequency response. Turbulence measurement was made possible only following the use of frequency compensation circuits [8]. The next major development in thermal anemometry was the design of a workable constant-temperature electronic circuit [9, 10].

Sensor materials and mounting procedures: Although platinum, nickel, and various alloys have been used as hot-wire sensors, the materials that are almost exclusively used today are platinum alloys, mainly with 20% iridium or 10% rhodium, and tungsten. Pure platinum has excellent thermal properties but relatively low mechanical strength, which would result in frequent breakage caused by vibrations or aerodynamic loading. The platinum alloys have improved mechanical strength, although reduced thermal resistivity. Tungsten has sufficient mechanical strength, high thermal resistivity, and a high melting temperature, but it oxidizes at about 350 °C, so that it may not be used at extremely high overheats or in high-temperature flows. The length of sensing elements is usually 0.8 to 1.5 mm, and their diameter is commonly 5 μm; thicker sensors (7.5 μm) are used for higher strength in high-speed flows, whereas thinner sensors (2.5 μm) are used for the measurement of the turbulence fine structure.

A hot-wire is mounted at its two ends on thin metallic *prongs*, usually tapered and having diameters of less than 1 mm. Tungsten wire is platinum plated and spot-welded on the prongs, by discharging a capacitor through a needle-type electrode (see Fig. 11.1). Platinum alloys can be spot-welded as well. Spot-welded wires have the disadvantage of being active over their entire length and thus permitting temperature non-uniformity that is due to heat conduction to the prongs; they also suffer from increased aerodynamic interference of the prongs. These problems may be overcome by use of a wire that is active in only its middle region, whereas the two end regions are much thicker and thus

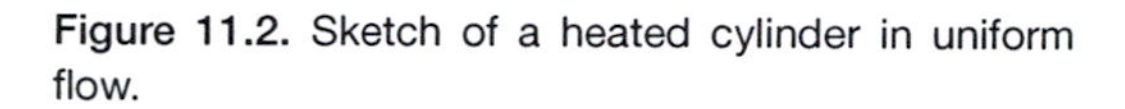

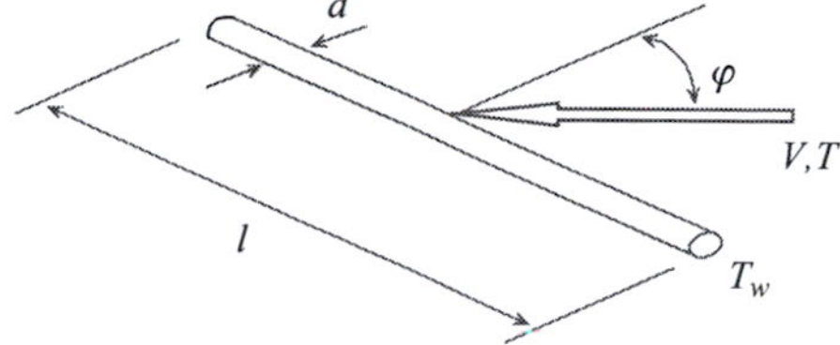

Figure 11.2. Sketch of a heated cylinder in uniform flow.

not subjected to appreciable Joule heating. One may achieve this by gold plating the ends of the wire in a special bath or by etching the middle portion of a *Wollaston* wire (consisting of a Pt or Pt-alloy core and a Ag coating, about 25 μm thick) in a thin jet of nitric acid at about 30% concentration. Then the gold-plated or silver-coated ends of the wire are soft-soldered on the prongs (see Fig. 11.1). Sensor material of all kinds as well as probe repair services are available commercially, although a laboratory that uses hot-wires routinely should be equipped with probe-repairing apparatus.

Hot-film sensors are manufactured by the depositing of a thin (about 0.1-μm thickness) film of platinum or nickel on an insulating substrate, usually made of quartz or mica; then a layer of quartz, of about 1-μm thickness, is added on the outside to protect and insulate the active film. Hot-film probes usually have cylindrical sensors, mounted on prongs like hot-wires but having a much larger diameter (typically 50 μm); probes with other shapes, including conical, wedge-type, and spherical shapes, are also available.

Heat transfer characteristics: First, consider a cylindrical heater with a diameter d, a length l, and at a uniform temperature T_w, immersed in a uniform, *steady* flow of a gas with a far-field temperature $T < T_w$ and a free-stream velocity V, inclined with respect to a plane normal to the cylinder axis by an angle φ (see Fig. 11.2). The (temperature) *overheat ratio* is defined as $a_T = (T_w - T)/T$. Heat transfer from the cylinder to the gas will, in general, take place in all possible modes: convection (natural, forced or mixed), conduction to supports at the end of the cylinder, and radiation. For the usual hot-wire anemometry (HWA) temperature range of a few hundred degrees centigrade, radiation losses are less than 0.1% of the convective losses and may be disregarded (exceptions include the study of combustion with the use of hot-wires heated to very high temperatures and measurements in very low-density flows, in which convection is low). The gas temperature in the vicinity of the cylinder will have values between T and T_w, and the gas flow properties will vary accordingly. It is a customary convention to define a *film temperature* as $T_f = \frac{1}{2}(T + T_w)$, and evaluate the gas thermal conductivity k, density ρ, kinematic viscosity ν, and thermal diffusivity γ at the film temperature. If $\dot{q}$ is the heat transfer rate from the wire, one may define the dimensionless heat transfer rate, called the *Nusselt number*, as

$$\mathrm{Nu} = \frac{\dot{q}}{\pi l k (T_w - T)}. \tag{11.1}$$

In the general case, the heat transfer will depend on several factors, which, in dimensionless form, may be expressed as

$$\mathrm{Nu} = \mathrm{Nu}(\mathrm{Re}, \mathrm{Pr}, \mathrm{Gr}, \mathrm{M}, \mathrm{Kn}, a_T, l/d, \varphi), \tag{11.2}$$

where $\mathrm{Re} = Vd/\nu$ is the Reynolds number, $\mathrm{Pr} = \nu/\gamma$ is the Prandtl number, $\mathrm{Gr} = g\alpha(T_w - T)d^3/\nu^2$ is the Grashof number (α is the coefficient of thermal expansion), $\mathrm{M} = V/c$ is the Mach number (c is the speed of sound), and $\mathrm{Kn} = \lambda/d$ is the Knudsen number (λ is the molecular mean free path; $\mathrm{Kn} = \sqrt{\frac{1}{2}\pi c_p/c_v}\,\mathrm{M}/\mathrm{Re}$). Restricting the analysis to air flows, one may disregard the dependence on Pr, which will be essentially constant. Most hot-wire applications satisfy the criterion $\mathrm{Gr} < \mathrm{Re}^{1/3}$ for the effects of natural convection to be negligible; this assumption is expected to become inaccurate for $\mathrm{Re} < 0.2$. Further considering only low-speed flows that may be treated as incompressible ($V < 100$ m/s; $\mathrm{M} < 0.3$), the dependence on M may also be neglected. Finally, hot-wires usually operate in the 'continuum regime' ($\mathrm{Kn} < 0.01$), or very close to the lower boundary of the 'slip flow regime' ($0.01 < \mathrm{Kn} < 1$), which also essentially eliminates their dependence on Kn.

Under such conditions, Eq. (11.2) is reduced to

$$\mathrm{Nu} = \mathrm{Nu}(\mathrm{Re}, a_T, l/d, \varphi). \tag{11.3}$$

Further considering cylinders with a large aspect ratio l/d (e.g., >100), so that end conduction effects may also be neglected, and with their axes normal to the flow ($\varphi = 0$), one may derive the considerably simplified relationship

$$\mathrm{Nu} = \mathrm{Nu}(\mathrm{Re}, a_T), \tag{11.4}$$

which permits relatively easy experimental determination. Following King's analysis, one may anticipate a relationship of the type

$$\mathrm{Nu} = (A + B\,\mathrm{Re}^n)(1 + \frac{1}{2}a_T)^m, \tag{11.5}$$

(notice that $1 + \frac{1}{2}a_T = T_f/T$) and, indeed, the well-known empirical response equation by Collis and Williams [11] gives, for the vortex-shedding regime,

$$\mathrm{Nu} = (0.24 + 0.56\,\mathrm{Re}^{0.45})(1 + \frac{1}{2}a_T)^{0.17}, \quad \text{for } 44 < \mathrm{Re} < 140, \tag{11.6}$$

and, for the attached-flow regime,

$$\mathrm{Nu} = 0.48\,\mathrm{Re}^{0.51}(1 + \frac{1}{2}a_T)^{0.17}, \quad \text{for } 0.02 < \mathrm{Re} < 44\,. \tag{11.7}$$

These response equations appear to hold universally for all hot-wires satisfying the assumptions previously made. In practice, however, the uncertainty in the determination of geometrical and material properties and the interference of many other factors make it necessary for each sensor to require individual calibration and its response to be expressed in terms of practical parameters, rather than dimensionless groups. In particular, for a given sensor and fixed overheat ratio and fluid properties, one needs to know the relationship between the voltage output E of the hot-wire operating circuitry (including the bridge, amplifier, filter, signal conditioner, and digitizer) and the flow velocity V, with a possible parameter being the local fluid temperature. Although polynomial and other expressions have been suggested, the most commonly used relationship remains a modified form of King's law, namely

$$\frac{E^2}{T_w - T} = A + BV^n, \tag{11.8}$$

with the constants A, B, and n determined by optimal fitting of this expression to fresh calibration results. The usual range of the exponent n is between 0.40 and 0.55 (0.50 was the original value proposed by King). Of course, the sensor average temperature cannot be measured directly, but it can be estimated from its relationship to the wire resistance R_w, directly measurable by the operating bridge. For metallic wires in the usual operating range, one may accurately employ the linear relationship

$$R_w = R_r \left[1 + \alpha_r (T_w - T_r)\right], \tag{11.9}$$

where α_r is the thermal resistivity coefficient at the reference flow temperature T_r. The value of α_r may be provided by the sensor supplier, but is best determined by temperature-resistance calibration in a heated calibration jet.

The response of hot-film sensors, both cylindrical and non-cylindrical, is more complex than that of hot-wires, especially because of their low aspect ratio and heat conduction to the substrate. However, such effects appear to be largely absorbable into the empirical coefficients in the modified King's law, which is thus applicable to all thermal sensors.

Now consider unsteady flow, in which the flow speed varies with time, but, for simplicity, let the flow temperature be constant. Let i be the electric current through the sensor. The power input to the wire is the Joule heating power $i^2 R_w$. Neglecting heat conduction and radiation, the heat losses from the wire consists only of the convective heat flux $\dot{q}$, which depends on the flow velocity and the sensor temperature, according to the relationships presented earlier. Then the energy equation applied to the sensor gives

$$mc\frac{\mathrm{d}T_w}{\mathrm{d}t} = i^2 R_w - \dot{q}(V, T_w), \tag{11.10}$$

where m is the mass of the sensor and c is its specific heat. Clearly this equation cannot be solved because it contains three unknowns: T_w, i, and V (R_w is directly related to T_w). To render this equation solvable, one must keep either the current or the sensor temperature constant, which may be achieved with the use of suitable electric circuits. The corresponding methods are, respectively, known as *constant-current anemometry* and *constant-temperature anemometry*. In the constant-temperature mode, the unsteady term in Eq. (11.10) vanishes and the unsteady response equation becomes identical to the steady response equation, so that a static calibration is sufficient. In the constant-current mode, the sensor temperature is permitted to vary and is therefore subjected to 'thermal inertia' effects, so that the anemometer will have a limited frequency response. For this reason, it is the constant-temperature mode that is used almost universally. The constant-current mode is used rarely, mostly in high-speed or high-temperature flows.

Constant-current anemometry: An approximately constant current i may be supplied to a sensor with resistance R_w by simply connecting it, in series with a large ballast

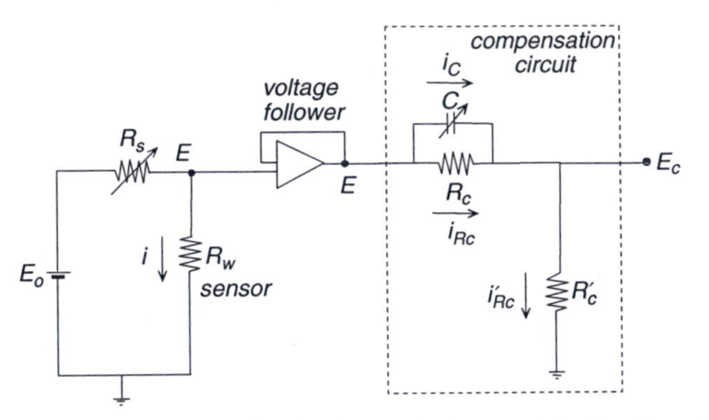

Figure 11.3. Sketch of a simple constant-current circuit with a simple frequency compensation circuit; the voltage follower has been added to prevent loading of the sensor by the compensation circuit.

resistor $R_s \gg R_w$, to a constant-voltage source E_0 such that $i = E_0/(R_s + R_w) \simeq E_0/R_s \simeq const$ (Fig. 11.3). The voltage output will be $E = i R_w$.

The unsteady energy equation presented in the previous section is highly non-linear. When linearized in the vicinity of an operating point, namely at a particular flow speed V_{op} and sensor temperature $T_{w_{\text{op}}}$, it leads to the following first-order differential equation:

$$\tau_w \frac{\mathrm{d}T_w}{\mathrm{d}t} + \left(T_w - T_{w_{\text{op}}}\right) = K_T(V - V_{\text{op}}). \tag{11.11}$$

This equation is characterized by a time constant τ_w, which is proportional to the overheat ratio, and a static sensitivity, K_T. Because the voltage E is proportional to R_W, which, in turn, is linearly related to T_w, the linearized E–V relationship will also be governed by a first-order differential equation, with the same time constant, as

$$\tau_w \frac{\mathrm{d}E}{\mathrm{d}t} + \left(E - E_{\text{op}}\right) = K(V - V_{\text{op}}). \tag{11.12}$$

A change in the flow velocity (assuming a constant flow temperature, for simplicity), will affect the convective cooling of the sensor, whose temperature T_w will change according to a first-order system response. Typical values of τ_w, measured by superimposing a low-amplitude square-wave-type voltage on E_0, are of the order of 1 ms for thin hot-wires and 10 ms for slim cylindrical hot-films. Such values are far too large for the wire to resolve turbulent motions. In a flow with variable speed or temperature, the overheat ratio a_T, and thus τ_w, will vary as well. For $i = \text{const}$, increased overheating at low speeds will result in burnout, whereas, at high speeds, reduced overheating will result in low static sensitivity K of the sensor.

To improve the frequency response of a constant-current anemometer, one may use a *frequency compensation circuit*, like the simple R–C circuit shown in Fig. 11.3. To analyse this circuit, notice that $i_c + i_{R_c} = i_{R_c'}$, which leads to the linear differential equation

$$R_c C \frac{\mathrm{d}(E - E_c)}{\mathrm{d}t} + (E - E_c) = E_c \frac{R_c}{R_c'} \tag{11.13}$$

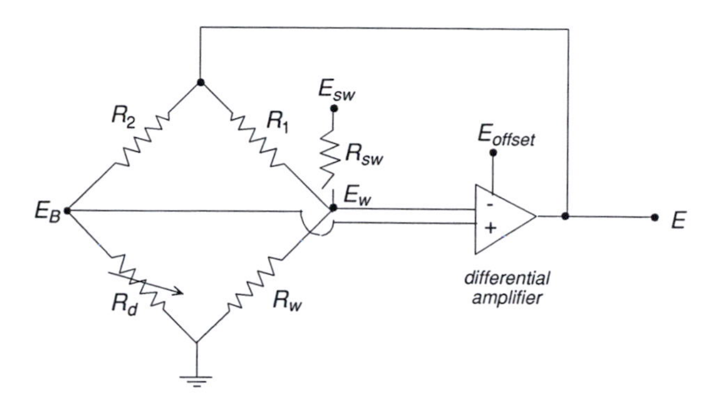

Figure 11.4. Sketch of a constant-temperature circuit.

To derive the response equation of the entire compensated anemometer system, one may combine differential equation (11.13) with the previous one, Eq. (11.12). When the compensation circuit components are selected such that $R_c C = \tau_w$ and $R_c' \ll R_c$, the response equation simplifies to

$$\tau_w \frac{R_c'}{R_c} \frac{\mathrm{d}E_c}{\mathrm{d}t} + (E_c - E_{\mathrm{op}} \frac{R_c'}{R_c}) = K \frac{R_c'}{R_c} (V - V_{\mathrm{op}}). \tag{11.14}$$

This equation shows that the compensated system is, approximately, a first-order system. Its time constant is $\tau_w R_c'/R_c \ll \tau_w$, and thus it has a far better frequency response than the uncompensated system. However, because τ_w depends on the flow velocity, in order to achieve the preceding response, one must readjust the R–C circuit (e.g., by adjusting a variable capacitor) at each speed of operation, which makes the use of this anemometer in shear flows quite cumbersome.

Constant-temperature anemometry: In this mode, the current through the sensor is continuously adjusted through an electronic feedback system and in response to changes in convecting cooling, such that the sensor's resistance and consequently its temperature, remain essentially constant. As a result, the unsteady heat transfer equation is identical to its steady form and the dynamic response of the anemometer is the same as its static response within a wide frequency range. By proper selection and adjustment of the electronic feedback system, it is possible to extend the frequency response of the unit beyond the usual ranges of most turbulent and transient flows.

For an explanation of the system's operation, consider the typical constant temperature anemometer sketched in Fig. 11.4. The sensor R_w comprises one leg of a Wheatstone bridge, the opposite leg of which consists of an adjustable decade resistor array R_d. To supply most of the available power to the sensor, and not to the passive leg, the *bridge ratio* R_2/R_1 is fixed at a relatively large value, 10 or 20, although other values in the range 1–50 have also been used for specific purposes (a lower bridge ratio results in a higher frequency response). The two midpoints of the bridge are connected to the inputs of a high-gain, low-noise differential amplifier, whose output is fed back to the top of the bridge. If $R_2/R_d = R_1/R_w$, then $E_B - E_w = 0$, and the amplifier output will be zero. However, if R_d is increased to a value R_d', the resulting bridge imbalance will generate an input imbalance of the amplifier. Then the amplifier will produce an output voltage, which will create some current through both legs of the bridge. The additional current through the hot-wire will create additional Joule heating, which will tend to increase its temperature and thus its resistance, until the resistance increases sufficiently to balance the bridge once more. A problem with this method is that, with a balanced bridge, the amplifier's output voltage would be zero and no current would flow through the system at the operating condition. To avoid this situation, the amplifier's input is offset by a voltage E_{offset}, so that some current flows through the bridge at all times. This produces an imbalance of the bridge, such that the sensor's resistance appears to be $R_w\,(1+\delta)$, namely higher than its actual value by a factor $\delta \ll 1$, called the *imbalance parameter*.

The dynamic response of the constant-temperature anemometer is non-linear. Under certain assumptions, one may formulate a third-order linear model, which has been studied to some extent. However, the operation of this system is usually analysed and optimized within the context of a second-order model, which is approximately valid under common operating conditions. According to this model, the natural frequency ω_0 of the system is expressed as

$$\omega_0 = \sqrt{2a_R K_0/(\tau_a \tau_w)}, \tag{11.15}$$

where a_R is the resistance overheat ratio (see definition below), K_0 is the system gain parameter, τ_a is the amplifier time constant, and τ_w is the sensor time constant. For a modern amplifier with a high gain–bandwidth product and a typical hot-wire sensor, this natural frequency is quite high, exceeding 100 kHz, and thus adequate for practically any flow measurement. The damping ratio ζ of the system can be expressed as

$$\zeta = \frac{1}{2}(1 + K_0\delta)\sqrt{\tau_w/(2K_0\tau_a a_R)}. \tag{11.16}$$

To avoid oscillations of the output, which may cause burnout of the sensor, while maintaining the widest possible frequency response, it is recommended to use a critically damped ($\zeta = 1$) or a very slightly underdamped system. Overdamping introduces non-linearities, which further complicate the dynamic response and introduce measurement errors. For a given sensor, overheat ratio, and amplifier, an adjustment of the damping ratio may only be accomplished by adjustment of the imbalance parameter δ, which is possible with the use of additional controls associated with the amplifier's offset.

The operating procedure of this anemometer involves the following steps:

Measurement of the 'cold' resistance R: The sensor, connected to the bridge as discussed, is inserted into the flow stream, so that it takes the flow temperature T, at which its resistance is R. One measures this resistance by disconnecting the feedback ('open-loop' configuration), feeding the bridge with some external small voltage, and adjusting the decade resistor R_d until the bridge is balanced.

Application of an overheat: The value of the decade resistor is increased to a value $R_d' = R_d(1 + a_R) = R_d R_w/R$, where the *resistance overheat ratio* $a_R = (R_w - R)/R$ is related to but distinct from the temperature overheat ratio $a_R = (T_w - T)/T$. Typical values for a_R are 0.5 to 0.9 for hot-wires and 0.1 to 0.2

for hot-films; higher values may result to damage to the sensor (e.g., oxidization of tungsten sensors at 350 °C), whereas lower values will produce low-velocity sensitivity.

Adjustment of the anemometer's dynamic response: The feedback is connected again ('closed loop' configuration). From now on, the circuitry will strive to maintain the sensor's resistance at the value R_w, irrespective of any changes in the flow velocity. The adjustment of the damping ratio is accomplished with the square-wave test. For this test, the sensor is inserted into a stream with a speed comparable with the average speed of the experiment, but free of velocity fluctuations. A square-wave-type voltage E_{sw}, fed to the sensor through a large resistor R_{sw}, produces a weak current, with an amplitude equal to about 20% of the sensor's current and superimposed on it. The voltage output of the system is observed on an oscilloscope, and some control is adjusted finely until the output oscillations cease (critically damped system) or nearly cease (slightly underdamped system).

Calibration of the anemometer: The sensor is inserted in a low-turbulence calibration facility, usually a jet or a wind tunnel, and its output is recorded for different values of the flow velocity; the flow temperature is also monitored. A response equation, usually the modified King's law, is fitted to the calibration data.

Operation of the anemometer: Finally, the calibrated probe is carefully inserted into the flow of interest for the collection of the experimental results. Because the frequency response of the constant-temperature anemometer is high, the voltage output will contain a significant amount of electronic noise, which must be reduced by low-pass filtering of the output signal before it is processed.

Various effects and error sources:

Velocity orientation effect: In the previous analysis of the cylindrical heated sensor response, it has been assumed that the flow velocity has a direction normal to the sensor axis. For a general velocity orientation, one may define an *effective cooling velocity* V_{eff}, as the magnitude of the (hypothetical) velocity normal to the sensor that would produce the same convective heat transfer as the actual velocity vector. First, consider a sensor long enough for the effects of prong interference to be negligible and cooled by a flow with a velocity vector with magnitude V and direction forming an angle φ with the normal to the sensor axis (Fig. 11.2). If one assumes that cooling of the cylinder is due to the normal component only, then,

$$V_{eff} = V \cos \varphi. \tag{11.17}$$

However, in reality, there would be some cooling even if the flow velocity were tangential to the sensor. To account for tangential cooling, the preceding relationship can be modified as

$$V_{eff} = V\sqrt{\cos^2 \varphi + k^2 \sin^2 \varphi}, \tag{11.18}$$

where the empirical coefficient k^2 has to be determined by calibration of each particular probe, exposed to a flow with a constant magnitude and changing direction. Typical values of k^2 for hot-wire probes are between 0.05 and 0.20. Because of the small value of k^2, and considering also the small value of $\sin\varphi$ at small angles, tangential cooling effects are negligible for small flow–probe axis misalignment.

Prong interference effects: Interference of the prongs and the probe body produce additional complications of the heat transfer characteristics. For example, for a sensor mounted to two prongs with identical lengths and its axis normal to the axis of the probe body, a stream in the binormal direction (namely a direction normal to both the sensor axis and the probe body axis) will produce a higher rate of cooling than a stream with the same velocity magnitude but in the normal direction (namely normal to the sensor and parallel to the probe axis). Again, the only practical way to account for this effect is to perform a direct calibration of each particular probe by varying its orientation within a three-dimensional envelope. The usual relationship that is fitted to such calibration data is

$$V_{\text{eff}} = \sqrt{V_N^2 + k^2 V_T^2 + h^2 V_B^2}, \tag{11.19}$$

where V_N, V_T, and V_B are, respectively, the normal, tangential, and binormal velocity components, as already defined. Typical values of h^2 for hot-wire probes are between 1.1 and 1.2. To minimize prong and probe body interference effects, it is recommended to utilize long and thin (preferably with a diameter less than $50d$) prongs, spaced as widely as possible, and the thinnest possible probe body. Tapered prongs and probe bodies are also recommended. Prong and probe interference, as well as axial conduction effects (see subsequent subsection) are lower for Wollaston-type or gold-plated sensors than for bare sensors. When orientation effects are significant, one must take particular care to maintain the same relative probe axis–velocity orientation during calibration and during measurement.

Heat conduction effects: The previous analysis of the thermal anemometer response was based on the assumption that the contribution of heat conduction to the total heat transfer was negligible. For cylindrical sensors, this requires a very large aspect ratio l/d, ideally $l/d \to \infty$. The effect of end conduction has been studied analytically by various authors. An often-quoted analytical solution, derived on the assumption of axial conduction alone, provides the variation of the temperature $T_w(z)$ along the sensor as a function of the distance z from the middle of the sensor, as

$$\frac{T_w(z) - T}{T_{w\infty} - T} = 1 - \frac{\cosh(z/l_c)}{\cosh(l/2l_c)}, \tag{11.20}$$

where T is the flow temperature and $T_{w\infty}$ is the (uniform) temperature of a sensor with the same material and diameter d, but infinite length, subject to the same electric current as the finite sensor. The *cold length* l_c is defined as

$$l_c = d\sqrt{\frac{1}{4}(k_w/k)(1 + a_R)/\text{Nu}} \tag{11.21}$$

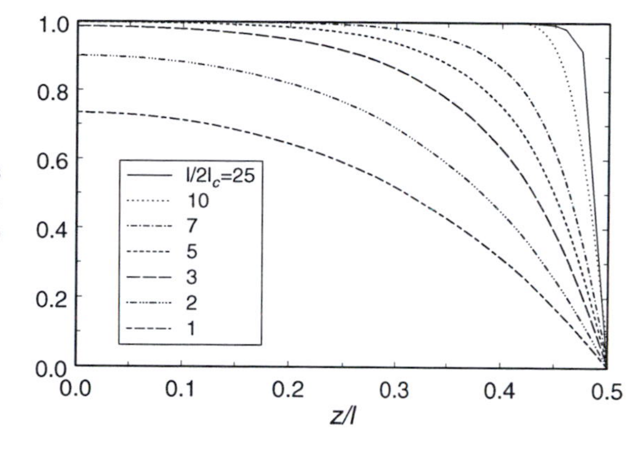

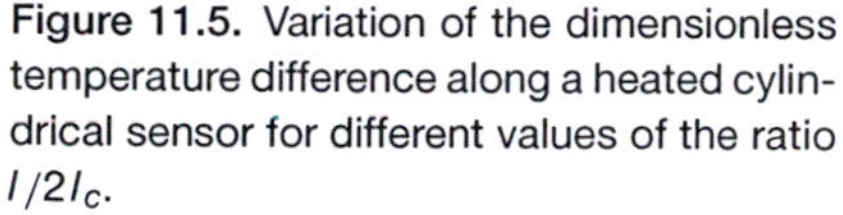
Figure 11.5. Variation of the dimensionless temperature difference along a heated cylindrical sensor for different values of the ratio $l/2l_c$.

and thus depends on the sensor operating conditions (k_w is the thermal conductivity of the sensor and k is the thermal conductivity of the fluid). The dimensionless temperature difference along a heated cylindrical sensor has been plotted in Fig. 11.5. It may be seen that a finite sensor would have an identifiable section with uniform temperature only when $l/2l_c \gtrsim 7$. For shorter sensors, the temperature along the sensor will vary continuously between the flow temperature value, at the two ends, and a maximum temperature, in the middle. Equation (11.21) indicates that the cold length decreases with decreasing thermal conductivity of the wire; therefore, from the viewpoint of minimizing heat conduction effects, it would be preferable to use a sensor material with as low heat conductivity as possible, while also maintaining a good mechanical strength and a relatively high electric conductivity. Among the three most popular hot-wire materials, namely tungsten, platinum, and platinum-10% rhodium, it is the latter that has the lowest k_w and that should be preferred when it is necessary to use short sensors. As an example, consider a typical tungsten sensor with $d = 5\ \mu$m and $l = 1.25$ mm, which gives $l/d = 250$. For typical low-speed operating conditions, $l_c \simeq 30d$, which results in $l/2l_c \simeq 4$. This means that such a sensor would have no uniform temperature region at all. When the wire operates in the constant-temperature mode, the feedback process will force its resistance to a fixed value R_w that corresponds to some average temperature T_{wa}, intermediate between T and the maximum temperature in the middle of the sensor. For the preceding example of a typical sensor and operating conditions, the maximum temperature in the middle of the sensor would be about 30% higher than the average value and the heat conduction losses would be about 15% of the total losses and therefore not negligible. One could reduce conduction losses by increasing the aspect ratio l/d, either by increasing l or by decreasing d. Both approaches would result in lower mechanical strength of the sensor, whereas an increase in l would degrade the sensor's spatial resolution. Obviously the optimal l/d ratio for a particular flow would be achieved as a compromise among several conflicting requirements. Hot-film sensors would be subject to lower axial conduction heat losses than hot-wires with the same length, because of their much smaller thickness. On the other hand, the radial conduction losses to the substrate are usually very significant and, for this reason, heat transfer from hot-films is more complex than heat transfer from hot-wires. Although it has been recognized that heat conduction effects are not negligible in thermal anemometry, the common approach (with the exception of a few very sophisticated studies) has been to disregard the explicit influence of heat conduction in analytical expressions and to absorb its effects empirically into the coefficients of the calibration equation (e.g., the modified King's law, particularly the exponent n). A recent study has further demonstrated that end conduction effects are expected to decrease significantly as the Reynolds number, based on sensor diameter, increases [12]. This is of considerable benefit in turbulence research, as the need for higher spatial resolution of the sensor becomes more pressing at higher Reynolds numbers.

Compressibility effects: As the flow speed increases, roughly for $M > 0.6$, the velocity and temperature fields around the sensor become quite complex, and simple relationships, such as King's law, are not sufficient to describe the hot-wire response. Compressible air flows may be characterized by three independent variables, for example, the flow velocity V, the density ρ, and the total (or stagnation) temperature T_0, while the pressure is related to the other properties by the perfect-gas law. In consequence, the hot-wire output is sensitive to variations of all three parameters. For simplicity, one may linearize the hot-wire response equation to express the output voltage fluctuations as the sum of contributions of fluctuations in V, ρ, and T_0, each multiplied by the respective corresponding sensitivities, S_V, S_ρ, or S_{T_0}. Then, in general, it would be impossible to measure velocity fluctuations from a single hot-wire output. In clearly supersonic flows ($M > 1.2$), $S_V = S_\rho$, and one may combine the density and velocity contributions into the *mass flow rate* ρV. Then, for a flow with a constant total temperature, one may use an empirical extension of King's law, in the form

$$E^2 = A + B\,(\rho V)^n \tag{11.22}$$

(with $n \simeq 0.55$) to measure the product ρV. Measurement of velocity or density separately is possible only under special conditions, notably when pressure fluctuations can be neglected. In high subsonic ($0.7 < M < 0.9$) and transonic ($0.9 < M < 1.1$) flows, all three sensitivities are different from each other, and an independent calibration to variations of each of these three parameters becomes necessary [13]. In low-density flows, one must also account for the effects of slip, as expressed by the value of the Knudsen number.

Temperature-variation effects: The previously presented thermal anemometer response equations, for example the modified King's law, contain explicitly the flow temperature T and are therefore adequate for determining its influence on the sensor output and for applying appropriate corrections, when necessary. For example, corrections for temperature differences would be required if a sensor were calibrated in a flow with a temperature T_1, but used in a flow with a temperature T_2. Such corrections are relatively straightforward when the flow temperature is uniform and contains no fluctuations. When, however, the flow temperature varies from position to position or contains turbulent fluctuations, corrections may be applied only if the local or instantaneous temperature is measured simultaneously with the velocity. This may be accomplished by the positioning of a temperature transducer [e.g., a thermocouple or a thermistor for slowly varying temperatures or a 'cold-wire' for turbulent temperature fields) and use of its output to substitute for the correct temperature in the modified King's law. Compared with the hot-wire sensitivity to velocity fluctuations, its sensitivity to temperature fluctuations decreases as the overheat ratio increases. Therefore one may reduce or eliminate the need for temperature corrections by operating the sensor at the highest possible overheat. At extremely low overheats, corresponding to a current through the sensor of less than 0.3 mA, the thermal anemometer becomes totally insensitive to velocity variations and becomes, instead, a *resistance thermometer*, namely a temperature transducer. Under such conditions, the sensor is known as a *cold-wire*.

Composition effects: The composition of the flow also affects the convective heat transfer from a thermal anemometer inasmuch as it affects the heat conductivity of the surrounding fluid. The sensitivity to composition, relative to the velocity sensitivity, depends on the sensor size, shape, material, and overheat ratio. In principle, it is possible to measure both velocity and concentration of a species (although not simultaneously) in a binary gas mixture by the running of the same, properly calibrated, sensor at two significantly different overheats. A more straightforward, although also more cumbersome, procedure is to use two closely positioned sensors with vastly different sensitivities to composition, and then combine their outputs to recover both the velocity and the composition. Three sensors may be used, at least in principle, to measure velocity, temperature, and composition at once.

Reverse flow and high-turbulence effects: A limitation of hot-wire–hot-film sensors is that their output voltage cannot resolve velocity orientation. For example, an infinitely long cylindrical sensor will produce the same output voltage for any orientation of the velocity vector on the normal plane. Thus forward flow cannot be distinguished from reverse flow of the same magnitude. Furthermore, even in the case of no flow, a thermal anemometer will produce a non-zero output, attributed to cooling by natural convection associated with the rising warm fluid near the sensor. In a highly turbulent flow (for turbulence intensities higher than 25%), reverse flow will occur statistically some of the time. This will result in an overestimation of the mean velocity and the velocity variance. Although some correction procedures for such effects have been developed, the use of thermal anemometers in reversing, recirculating, or highly turbulent flows is generally discouraged.

Multi-sensor probes: A single cylindrical sensor, approximately normal to the average local flow velocity will measure the normal velocity component. If the same sensor were inclined with respect to the mean velocity, it would provide an output corresponding to the effective cooling velocity, which is insufficient to resolve either the magnitude or the direction of the actual flow velocity. An approach, known as *slanted wire anemometry*, is to mount a single cylindrical sensor at an angle (e.g., at 45°) to the probe body axis and to take a number of measurements, one set after the other, by keeping the sensor

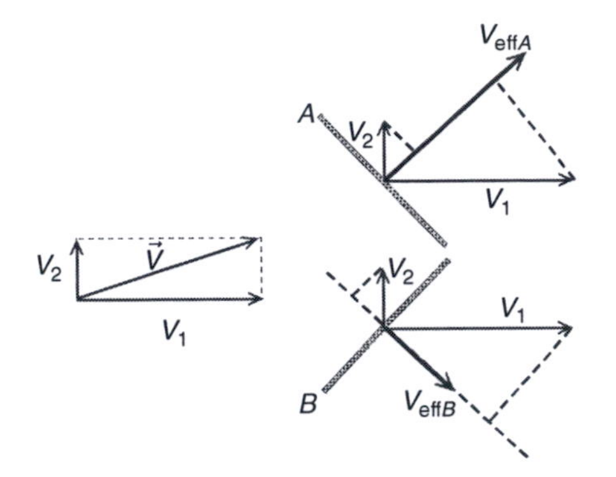

Figure 11.6. Sketch illustrating the effective cooling velocities of the two sensors of a cross-wire probe (for clarity, one sensor has been shifted vertically).

midpoint fixed and rolling the probe body so that the sensor orientation varies. Each set of readings is analysed to provide equations for those velocity components that affect the corresponding effective cooling velocity and the resulting system of coupled equations is solved to determine all components of the mean velocity as well as some or all of the turbulent stresses. This method is cumbersome and cannot provide simultaneous values. It is mainly used in near-wall studies, where multi-sensor probes would have inadequate spatial resolution.

A very common design, capable of resolving simultaneously two velocity components, is the *cross-wire* (X-wire) *anemometer* (Fig. 11.6), and its variance, the *V-wire anemometer*. This consists of two sensors, each mounted on a separate pair of prongs and operated by separate circuits, positioned normal to each other and inclined by nominally $\pm 45^\circ$ with respect to the common probe body axis. The probe is calibrated in a stream parallel to the probe axis to establish relationships between each voltage output and the corresponding effective cooling velocity. When the probe is put in a stream and aligned with the mean flow direction such that the average binormal velocity is nearly zero, the effective cooling velocities, neglecting tangential cooling and assuming identical sensors inclined at exactly $\pm 45^\circ$ with respect to the flow direction, would be

$$V_{\text{eff}A} = \frac{\sqrt{2}}{2}(V_1 + V_2), \tag{11.23}$$

$$V_{\text{eff}B} = \frac{\sqrt{2}}{2}(V_1 - V_2), \tag{11.24}$$

from which one may calculate the two velocity components as

$$V_1 = \frac{\sqrt{2}}{2}(V_{\text{eff}A} + V_{\text{eff}B}), \tag{11.25}$$

$$V_2 = \frac{\sqrt{2}}{2}(V_{\text{eff}A} - V_{\text{eff}B}). \tag{11.26}$$

More accurate values of the effective cooling velocities can be found by angular calibration of the probe, which consists of rotating the probe on a plane parallel to the two sensors and measuring both the tangential cooling coefficients in Eq. (11.18) and the exact angles of inclination of the two sensors. A variety of more accurate calibration methods for cross-wires are also available [1].

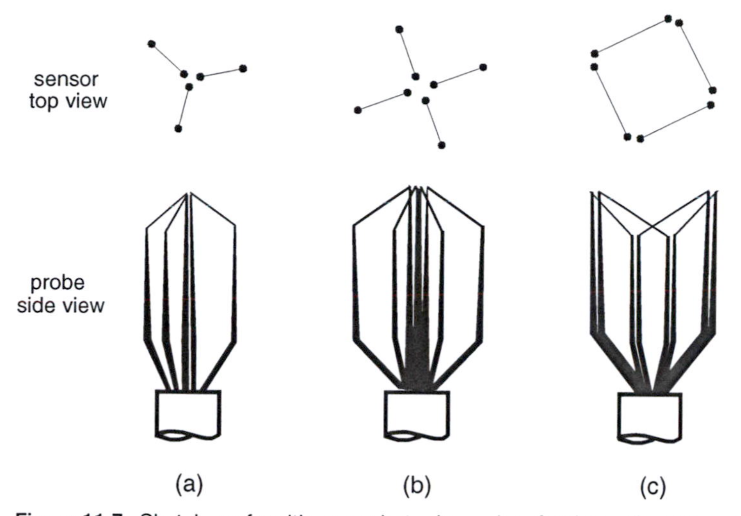

Figure 11.7. Sketches of multi-sensor hot-wire probes for three-dimensional velocity measurement: (a) a three-sensor probe and (b) and (c) two four-sensor probes; the probe shown in (c) may be also used for streamwise vorticity measurement.

Various geometric configurations of three-sensor [14–16] or four-sensor [17–19] probes have also been used to measure simultaneously all three velocity components (see Fig. 11.7). These probes require a three-dimensional calibration and a sophisticated signal analysis. Other multi-sensor probe designs have been used to measure velocity derivatives as well as vorticity components [20, 21]. The construction, calibration, and use of multi-sensor probes are subject to a number of subtle complications and should be the task of skillful experimenters.

Pulsed-wire anemometry [1, 22–24]: The *pulsed-wire anemometer* has been specifically developed for velocity measurements in high-turbulence intensity and reversing flows, in which the hot-wire anemometer cannot be used. The typical *pulsed-wire anemometer* probe consists of a central wire and two sensing wires, placed on either side of the central wire and normal to it (Fig. 11.8). It is sensitive to the velocity component normal to all wires, capable of measuring both its magnitude and sense of direction. For its operation, the central wire is supplied with a pulsed current, such that it produces a wake with a periodically changing temperature, and the sensing wires, which operate as resistance thermometers (cold wires; see Section 12.2), sense the passing of the heated wake of the central wire. One measures the flow velocity by timing the interval between the application of a current pulse and the sensing of temperature rise by either of the resistance thermometers. As the two sensor signals are measured independently, it is easy to determine the instantaneous flow direction. Designs for both in-flow and near-wall velocity measurements, as well as a variant capable of measuring wall shear stress, have been proposed in the literature. Pulsed-wire anemometers have not found wide application because their spatial and temporal resolutions are markedly inferior to those of conventional hot-wire probes. Their main advantage of measurement in reversing flows has also been largely undermined by optical methods, including LDV and PIV.

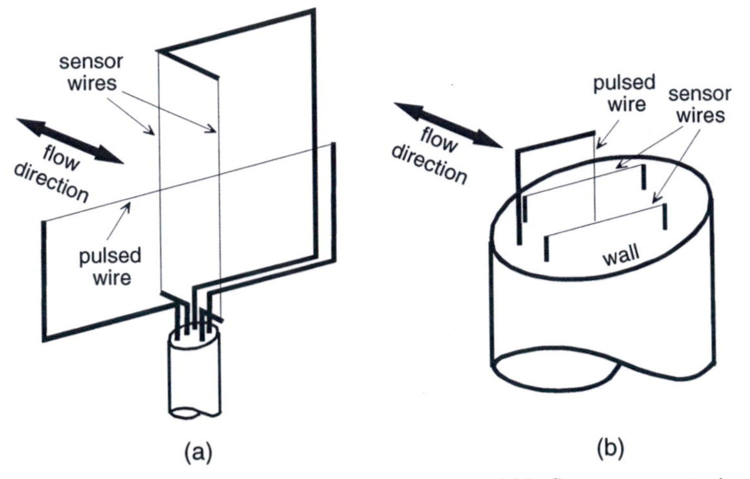

Figure 11.8. Two pulsed-wire anemometer probes: (a) in-flow measurement probe and (b) near-wall measurement probe.

11.2 Laser Doppler velocimetry

LDV, alternatively referred to as *laser Doppler anemometry* (*LDA*), was conceptualized in 1964 [25] and witnessed rapid and intensive development to become one of the most popular local velocity measurement methods in the fluid mechanics laboratory. Although a variety of LDV configurations and a multitude of related components and accessories have been developed, in the present review only the basic principles of LDV operation and some of the most widely used components and techniques are discussed [26–30].

Doppler shift of light frequency: Consider a monochromatic, coherent, linearly polarized, and collimated laser beam with a wavelength λ and frequency ν, being transmitted through a fluid containing suspended discrete particles with sizes that are not very small compared with λ and travel with speed $\vec{V}$ (Fig. 11.9). When one of these particles intersects the beam, it will adsorb some of the light, which can be viewed as plane electromagnetic waves, and reemit it, following Mie-scattering laws. At some distance from the particle, the emitted radiation can be viewed as spherical electromagnetic waves.

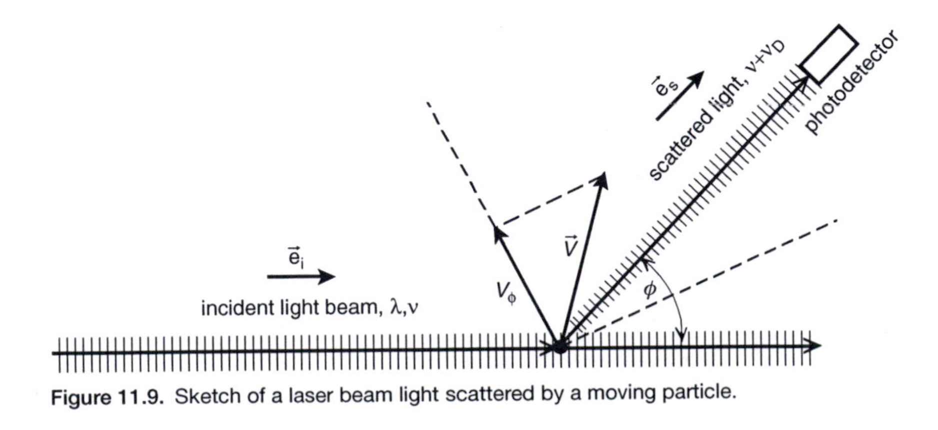

Figure 11.9. Sketch of a laser beam light scattered by a moving particle.

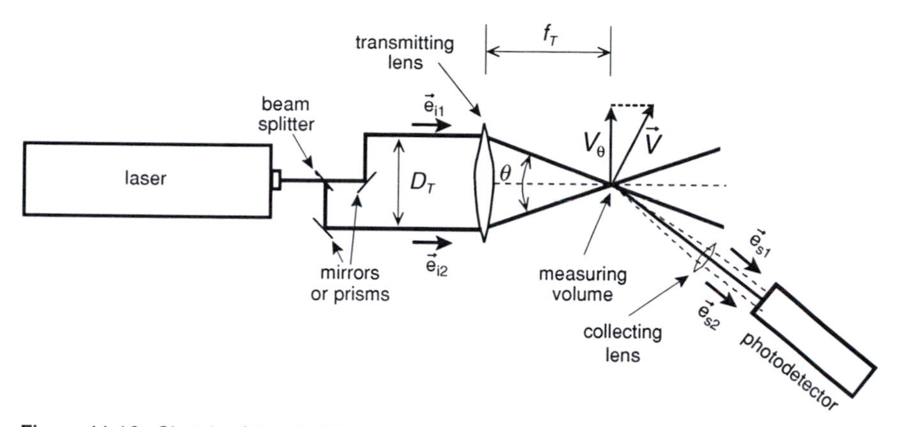

Figure 11.10. Sketch of the dual-beam laser Doppler configuration.

The frequency $\nu + \nu_D$ of these waves as seen by a small photodetector will be different from the incident frequency by an amount ν_D, called the *Doppler frequency*. This phenomenon is known as *Doppler* (or *Doppler–Fizeau*) *phenomenon*. The Doppler frequency depends on only the speed of the particle, the light wavelength, and the scattering angle ϕ, as

$$\nu_D = \frac{\vec{V} \cdot \left(\vec{e_s} - \vec{e_i}\right)}{\lambda} = \frac{2\sin(\varphi/2)}{\lambda} V_\phi, \tag{11.27}$$

where $\vec{e_i}$ and $\vec{e_s}$ are, respectively, the unit vectors parallel to the incident and the scattered beams and V_ϕ is the projection of the particle velocity vector on the direction normal to the bisector of the angle ϕ. Thus, given the parameters λ and ϕ, one may, in principle, determine the velocity V_ϕ from a measurement of the Doppler frequency ν_D, without any need for calibration or other considerations. In practice, because $\nu_D \ll \nu$, it is essentially impossible to measure it accurately as the difference of the incident- and the scattered-light frequencies and some modification of this fundamental configuration is required (a possible exception is the case of extremely high-speed flows). Among the few available workable configurations, the one used by most researchers and the usual one in commercial LDV systems is the *dual-beam* configuration. Other configurations that have been employed are the *reference beam* and *dual-scatter* configurations.

Dual-beam LDV: The basic single-component, forward-scatter, dual-beam LDV is illustrated schematically in Fig. 11.10. The monochromatic, coherent, linearly polarized beam of a laser is split by a *beam splitter* into two parallel beams, which are focussed onto a small control volume by the transmitting lens, forming an angle θ with each other. A small particle crossing the intersection volume with velocity $\vec{V}$ scatters the light of both beams, each Doppler shifted according to Eq. (11.27). Both scattered beams are collected by a small-aperture collecting lens and superimposed on a photodetector. Assume first that the intensity of each scattered beam collected by the photodetector varies sinusoidally as $A_i \sin 2\pi\,[(\nu + \nu_{Di})\,t]$, $i = 1, 2$. Optical mixing of these beams on the photodetector (*heterodyning process*) produces an output voltage E that is proportional to the square of the combined light intensity, as

$$\begin{aligned} E &\sim \{A_1 \sin 2\pi\,[(\nu + \nu_{D1})\,t] + A_2 \sin 2\pi\,[(\nu + \nu_{D2})\,t]\}^2 \\ &= A_1^2 \sin^2 2\pi\,[(\nu + \nu_{D1})\,t] + A_2^2 \sin^2 2\pi\,[(\nu + \nu_{D2})\,t] \\ &\quad + A_1 A_2 \{\cos 2\pi\,[(\nu_{D1} - \nu_{D2})\,t] - \cos 2\pi\,[(2\nu + \nu_{D1} + \nu_{D2})\,t]\}\,. \end{aligned} \tag{11.28}$$

Now, considering that the frequency response of the photodetector is insufficient to resolve any fluctuations with frequency of the order of magnitude of ν (typically, of the order of 10^{14} Hz), but sufficient to resolve fluctuations with the much lower *Doppler frequency difference*

$$\nu'_D = \nu_{D1} - \nu_{D2}, \tag{11.29}$$

it is easy to see that the photodetector output (*Doppler signal*) would be

$$E \sim a + b \cos 2\pi\,\nu'_D t. \tag{11.30}$$

The coefficients a and b would depend on the position of the particle within the measuring volume and thus on time, but their variation would be much slower than the period of Doppler oscillations. The Doppler frequency difference can be easily found as

$$\begin{aligned} \nu'_D &= \frac{\vec{V} \cdot (\vec{e_{s1}} - \vec{e_{i1}})}{\lambda} - \frac{\vec{V} \cdot (\vec{e_{s2}} - \vec{e_{i2}})}{\lambda} \\ &= \frac{\vec{V} \cdot (\vec{e_{i2}} - \vec{e_{i1}})}{\lambda} = \frac{2 \sin(\theta/2)}{\lambda} V_\theta, \end{aligned} \tag{11.31}$$

where V_θ is the projection of the particle velocity vector on the direction normal to the bisector of the angle θ between the two incident beams. Unlike Eq. (11.27), Eq. (11.31) is independent of the observation angle. The photocurrent generated by the photodetector will be fluctuating with the Doppler frequency ν'_D, which can be measured by various means, thus allowing the direct computation of V_θ as

$$V_\theta = \frac{\lambda}{2 \sin(\theta/2)} \nu'_D. \tag{11.32}$$

A great advantage of Eq. (11.32) is that it is linear and contains no undetermined constants, thus eliminating the need for calibration and errors that are due to instrumentation drifting. The proportionality coefficient $\lambda/[2 \sin(\theta/2)]$ is called the *calibration factor*, although no calibration is actually necessary.

The measuring volume: The laser beam coming out of the laser has a Gaussian distribution of intensity (see Section 5.3). The *laser beam diameter* d_e (namely the diameter of the circle within which the light intensity is higher than $1/e^2 \simeq 13.5\%$ of the maximum intensity) first diminishes, it reaches a minimum, called the *waist diameter*, and then it increases. The transmitting lens focusses the beam to an even smaller cross section, essentially on its focal plane. Optimal performance, resulting in the smallest possible beam cross section, will be achieved if the beam is focussed on its waist. The *focussed*

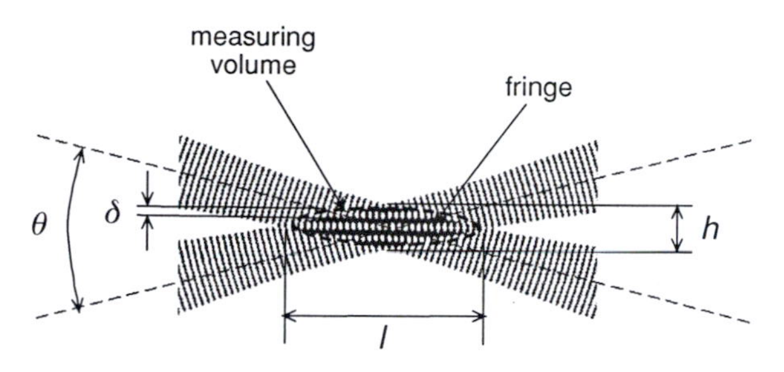

Figure 11.11. Measuring volume and fringe pattern in the dual-beam configuration.

beam diameter will be

$$d_{fe} \simeq \frac{4 f_T \lambda}{\pi d_e} \tag{11.33}$$

where f_T is the *focal distance* of the lens. When properly positioned, the two focussed beams in the dual-beam configuration will intersect on their waists to form a volume within which the total local light intensity would be equal to the sum of the two local beam intensities. The *measuring volume*, defined as the space within which the light intensity is higher than $1/e^2$ times its maximum value, has an ellipsoidal shape with the following dimensions (Fig. 11.11):

$$\text{width:} \quad d_{fe}; \tag{11.34}$$

$$\text{height:} \quad h = \frac{d_{fe}}{\cos(\theta/2)}; \tag{11.35}$$

$$\text{length:} \quad l = \frac{d_{fe}}{\sin(\theta/2)}; \tag{11.36}$$

$$\text{volume:} \quad \frac{\pi d_{fe}^3}{6\cos(\theta/2)\sin(\theta/2)}. \tag{11.37}$$

The preceding relationships show that, for intersection angles $\theta < \pi/2$, the measuring volume length would be its longest dimension. In fact, as θ is reduced to permit measurements away from the lens, the length of the volume increases dramatically, compared with the beam diameter, thus restricting the spatial resolution of the system.

Fringe model: Considering the two beams in the dual-beam configuration as travelling electromagnetic waves with the same frequency and phase, but with directions of propagation that are inclined with respect to each other, one may deduce that their interference within the measuring volume will produce a number of interference surfaces (standing waves). If the beams intersect at their waists, the two travelling waves will be essentially plane and the interference surfaces will be parallel planes, oriented as shown in Fig. 11.11. These surfaces are called *fringes* and appear as bright slabs, flanked by dark slabs. The *fringe spacing* is equal to the calibration factor,

$$\delta = \frac{\lambda}{2\sin(\theta/2)}, \tag{11.38}$$

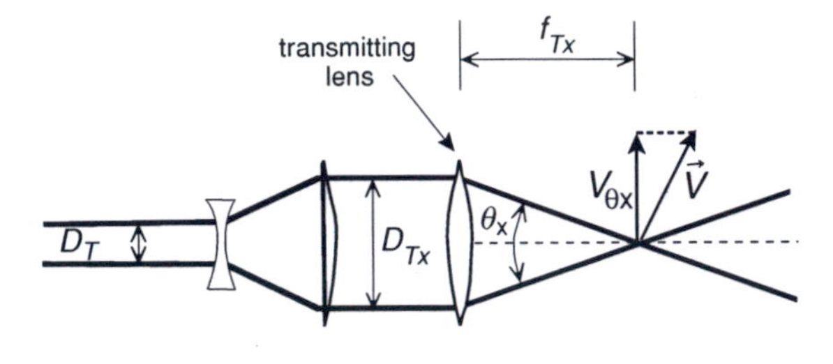

Figure 11.12. Sketch of a beam expander configuration.

and the *fringe number* within the measuring volume is

$$N = \frac{4}{\pi} \frac{D_T}{d_e}, \tag{11.39}$$

where $D_T = 2 f_T \sin(\theta/2)$ is the beam separation before the beams enter the transmitting lens. Although known to lead to certain inconsistencies, the fringe model may be used to explain the dual-beam LDV operation. For example, the Doppler signal fluctuations may be viewed as the result of fringe crossings by the particle. The frequency of fringe crossings is V_θ/δ, which is identical to the Doppler frequency difference, given by Eq. (11.31).

Beam expansion: A *beam expander* is a commonly used LDV accessory, whose function is to reduce the size of the measuring volume. This improves the spatial resolution of velocity measurement while also improving the amplitude resolution as a result of increased light power density within the measuring volume. The beam expander consists of a diverging lens and a converging lens, in addition to the transmitting lens, which has a larger aperture than that in the basic LDV system (Fig. 11.12). It is characterized by the *beam expansion ratio* $E_x > 1$. This increases the beam diameter from the original value d_e to the expanded-beam value $d_{ex} = E_x d_e$ and the beam separation from the original value D_T to the expanded-beam value $D_{Tx} = E_x D_T$. The expanded-beam intersection angle θ_x can be found as

$$\theta_x = 2 \sin^{-1} \left[\frac{E_x f_T}{f_{Tx}} \sin(\theta/2) \right], \tag{11.40}$$

where f_{Tx} is the focal distance of the transmitting lens of the expanded-beam system. Assuming that the focal distances of the transmitting lenses of the unexpanded- and expanded-beam systems are identical, $f_{Tx} = f_T$, one can find that the focussed diameter of the expanded beam will be diminished as

$$d_{fex} = \frac{1}{E_x} d_{fe} \simeq \frac{4 f_T \lambda}{\pi d_{ex}}. \tag{11.41}$$

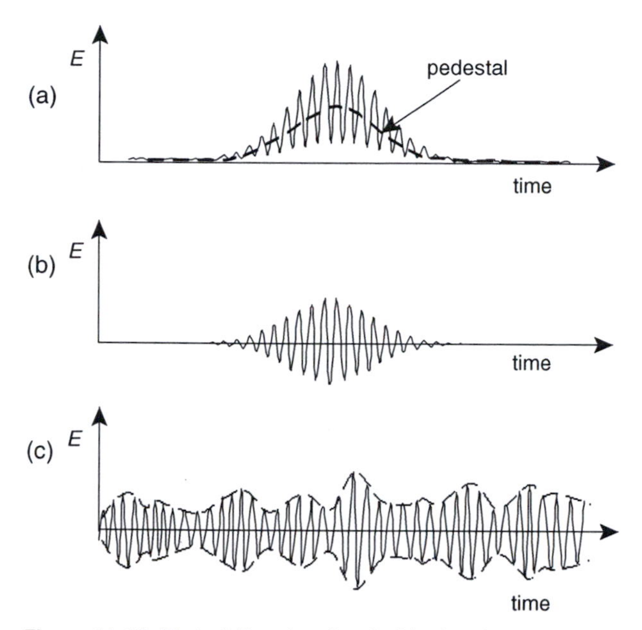

Figure 11.13. Typical Doppler signals (a) of a single particle, (b) of a single particle with the pedestal removed, and (c) of multiple particles with the pedestal removed.

The fringe spacing will also be reduced by a factor equal to E_x, while the number of fringes within the measuring volume will remain unaffected. Most importantly, the width of the measuring volume will decrease by a factor of E_x, the height by a factor somewhat less than E_x [because of the increasing value of $\cos(\theta/2)$], while the length, which is normally by far the longest dimension, will decrease by the factor E_x^2! The volume itself will be reduced by a factor somewhat less than E_x^4, which represents a substantial improvement of spatial resolution.

Doppler signal: First consider that the photodetector is only exposed to light scattered by a single particle traversing the central region of the measuring volume with a constant velocity. The Doppler signal will have a typical appearance shown in Fig. 11.13(a). It consists of a slowly varying component, called the *pedestal*, which is due to the Gaussian distribution of the light intensity within each beam, and a rapidly varying component, called the *burst*, which is generated by fringe crossings by the particle and is an amplitude-modulated oscillatory function with frequency ν'_D. The instantaneous value of the burst signal depends on the size and refractive index of the particle and its position in the measuring volume, whereas the number of cycles within each burst depends on the number of fringes it crosses and thus its path through the measuring volume. Particles crossing the volume near its edges will generate bursts with a smaller number of cycles than those crossing it centrally. Also, particles moving normal to the fringes will cross more fringes than will particles with inclined paths. The pedestal is usually removed by high-pass filtering, which leaves only the bursts, as shown in Fig. 11.13(b). The Doppler signal discussed previously appears to be continuous within the duration of each burst, although it vanishes between bursts. This is based on the assumption that the *photon density*, i.e., the intensity of light received by the photodetector during each burst, is sufficiently high for a continuous output voltage to be generated. At extremely low photon densities, however, the signal consists of a train of pulses corresponding to individual collected photons. This situation is not considered any further, as it is normally to be avoided. When more than one particle is simultaneously present in the measuring volume, the Doppler signal gets more complicated and its appearance at any given time depends on the *burst density*, namely the number of particles crossing the measuring volume. Even if all particles present had the same size and the same velocity, amplitude and phase differences in their bursts would be created by differences in paths and times of entry into the measuring volume. In the extreme case in which the particle concentration is high enough for many particles to reside in the measuring volume at all times, the Doppler signal would be the result of superposition of many bursts (*high burst density*), so that it would be continuous over the entire observation period, with a typical appearance as shown in Fig. 11.13(c). High burst density signals have an average frequency ν'_D, corresponding approximately to the average particle velocity. However, their instantaneous frequency and phase vary randomly, introducing errors in the velocity measurement, called, respectively, *ambiguity noise* and *phase noise*.

Frequency shifting: The relationship between V_θ and ν'_D [Eq. (11.32)] is equally valid for both senses of the direction of V_θ. Because the measured frequency ν'_D is always positive, Doppler signal analysis will always produce positive values of V_θ. This would result in erroneous velocity statistics in regions of reverse flow or very large turbulence intensity. The standard way for removing this ambiguity is by the passing of one of the incident laser beams through one or more *Bragg cell*. A Bragg cell is an acousto-optical device, within which a train of acoustic waves generated by piezoelectric transducers modulates the light frequency, such that the laser beam at its output has a frequency shifted with respect to the incident frequency ν by a fixed amount $\Delta\nu$, equal to about 40 MHz for a single cell. Frequency shifting produces a motion of the interference fringe pattern in the measuring volume with a velocity normal to the fringe plane and equal to

$$V_f = \Delta\nu\delta. \tag{11.42}$$

Then the computed flow velocity would be given by

$$V_\theta = \left(\nu'_D - \Delta\nu\right)\delta, \tag{11.43}$$

which can take zero or negative values and, thus resolve flow velocity sense within a range. To illustrate this concept, consider a particle at rest within the control volume. In the absence of frequency shifting, this particle would produce no Doppler signal, as it would not cross any fringes. With frequency shifting, however, the fringes would be convected past the particle, and thus a Doppler signal with frequency $\nu'_D = \Delta\nu$ would be produced and Eq. (11.43) would give the correct value $V_\theta = 0$. An additional advantage of frequency shifting is that it increases the dynamic range of the LDV system by increasing the number of fringes intersected by each particle crossing the control volume. Frequency shifting is necessary for highly turbulent flows but also improves the amplitude resolution of the system, a feature that is useful in measuring low-amplitude fluctuations in relatively high-speed flows. Frequency shifting by different amounts is also possible by electronic means.

Doppler signal processing: The Doppler signal produced by the photodetector is usually band-pass filtered to remove the pedestal and high-frequency noise, and then processed by a *signal processor*, in order to determine the Doppler frequency. Doppler signal processors have evolved significantly since their early years, following developments in electronics and algorithms. Among the various available options, one could mention the following:

Burst analyzers: At present, these are the most widely used processors, because of their relatively high sensitivity and accuracy and their ability to operate with low particle densities. They require single-particle bursts for their operation and compute the power spectrum of each individual burst with the use of a FFT. Then they determine the Doppler frequency as the frequency at which the spectral peak occurs.

Frequency counters: These also require single-particle bursts. They count the number of zero crossings of the filtered burst over a specified period during which the signal amplitude is relatively large, and determine the particle velocity as the ratio of the fringe spacing over the average time between two successive zero crossings.

Frequency trackers: These are processors suitable for high particle densities, when several particles can be found in the measuring volume at all times. They contain an electronic oscillator, which scans a frequency range and locks at the Doppler frequency, providing an analogue output proportional to it. As the particle velocity fluctuates, the tracker output is continually updated within a fixed number of Doppler cycles. When there are no particles in the measuring volume or when the particle velocity is outside of the specified range of operation, the tracker loses the signal and it necessarily holds the last valid value until it tracks again. An advantage of frequency trackers is that they provide an analogue output, which is readily suitable for spectral analysis. Their main disadvantages are limited dynamic range and the need for heavy seeding.

Photon correlators: These processors detect the emission of individual photons and correlate them with respect to their times of arrival to compute the time delay for peak correlation. Their advantage over other processors is that they can operate with very low light intensity and noisy signals; however, their frequency range is limited, because of the need for time averaging.

Errors and uncertainty in LDV measurements: Previous investigators have identified several sources of measurement error and uncertainty in LDV, which, in many cases, can be avoided or reduced by proper design of the experimental setup. If the beams do not intersect at their waists, the fringes will not be parallel planes, but curved surfaces with spacings that vary across the measuring volume. This would create a measurement uncertainty, called *fringe divergence uncertainty*, as particles with the same speed would appear to have different speeds if they followed different paths through the measuring volume. Another source of uncertainty, which is particularly important for turbulent flows, is due to non-uniform distribution of scattering particles. In such cases, the measured statistical properties would be biased in favour of flow regions that contain more particles.

Even when carefully implemented, LDV would be subjected to a number of inherent systematic (bias) errors [31]. The most serious one is *velocity bias*: When particles cross the measurement volume with different velocities, the arithmetic average of these velocities would be different from the local average velocity. The reason is that faster particles would cross the measurement volume faster than slower ones, with the result that a greater population of faster particles will be encountered and the velocity statistics would be biased towards the higher speeds. This problem is particularly important for high-intensity turbulent flows and would occur even if the particles were uniformly distributed in the stream. An order-of-magnitude estimate of this bias for the mean streamwise velocity is [32]

$$\frac{\overline{U_m}}{\overline{U}} \approx 1 + \frac{\overline{u^2}}{\overline{U}^2}, \tag{11.44}$$

where $\overline{U_m}$ is the mean velocity measured by the LDV system, $\overline{U}$ is the actual mean velocity, and $\overline{u^2}$ is the velocity variance. Among the various methods that have been proposed to correct for this bias, the most popular is the so-called *residence-time* or *transit-time correction* [27], applicable to burst-analysis-type processors. Modern burst processors also measure, besides the particle velocity, the residence time Δt_i of each particle in the measurement volume, determined as the difference between the times of exit and entry. Then the velocity statistics are determined by the weighing of the measured velocity of each particle by the factor $\Delta t_i / \sum_{j=1}^{J} \Delta t_j$, where J is the number of particles considered. A related bias, which is also due to non-uniform velocities, is the *gradient bias*, occurring in shear flows, even laminar ones. In such flows, this bias can be reduced if the measurement volume is kept as small as possible. Near walls, additional errors would be produced by the non-uniform distribution of particles in the flow, as heavier particles would tend to migrate away from the wall. In highly three-dimensional flows, there will also be a *directional bias*, as particles crossing the measurement volume in directions forming small angles with the fringes will not be registered by the processor. Frequency shifting can be used to reduce this bias.

A source of random error, usually called *Doppler ambiguity*, is the broadening of the Doppler frequency f_D that is due to the finite sampling time, as discussed in Section 4.5. Even if the particle had a constant speed within the measurement volume, the measured spectrum of its Doppler burst would not be a delta-like function but a sinc-like function. One may estimate the uncertainty in the measured Doppler frequency by the half-width of its main lobe, which is approximately $1/\Delta t$ (Δt is the residence time), or, in terms of the number of fringes N that the particle intersects, $2f_D/N$ [30]. Obviously this uncertainty would be reduced by an increase in the fringe number that the particle intersects by frequency shifting.

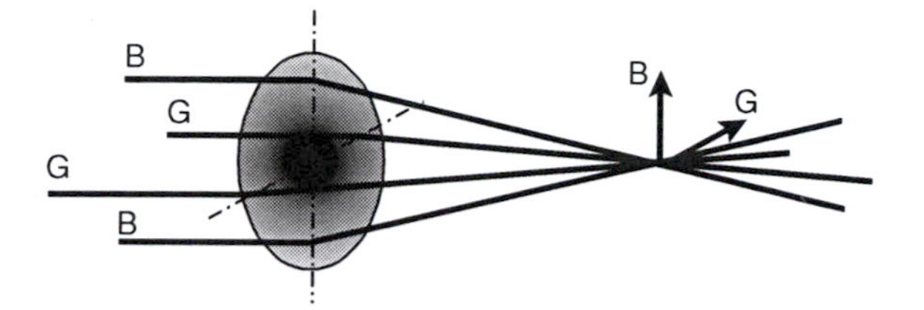

Figure 11.14. Sketch of a two-component LDV configuration with two pairs of beams at orthogonal planes; the letters G and B indicate beams with light at the green and the blue spectral peaks of an Ar-ion laser, as well as the corresponding measured velocity components.

Multicomponent and multipoint systems: The basic LDV system previously described can measure only one velocity component, the one normal to the bisector of the angle between the intersecting beams. A straightforward extension of this configuration to a *two-component system* is to use two pairs of beams, intersecting within the same measuring volume, but such that the axes of each pair form planes perpendicular to each other, as shown in Fig. 11.14. Thus the measuring volume will contain two sets of interference fringes, and it will become necessary to separate the light emitted by a particle as it crosses one set of fringes from light emitted by the same particle as it crosses the other set. This can be achieved by use of light of a different wavelength for each pair of beams and the removal, by an optical filter, of the light of the other pair before it enters the corresponding photodetector. This is called a *two-colour LDV system*. It can be implemented by the combining of two lasers that emit at different wavelengths; however, it is far more common to use a single laser with multiple spectral peaks, mostly the Ar-ion laser at its two strongest spectral peaks in the green ($\lambda = 514.5$ nm) and blue ($\lambda = 488$ nm) ranges. Because the two velocity components of each particle are measured simultaneously, one may also measure their joint statistical properties, such as their covariance, corresponding to a Reynolds shear stress in turbulent flows. A simplified two-component LDV design utilizes three rather than four beams, by combining two beams into one that contains both wavelengths. The two-component system can be extended to a *three-component system* by the addition of a third pair of beams intersecting on the same measuring volume as the other two, but belonging on a third plane. Unless the facility permits optical access from two orthogonal directions (e.g., one side and the top), the third component will not be orthogonal to the other two, and one needs to perform a coordinate transformation to compute the instantaneous velocity vector. The third pair of beams must also be at a different wavelength. This could be the wavelength of a third spectral peak of an Ar-ion laser (purple, at $\lambda = 476.5$ nm) or light from a different type of laser, e.g., a He–Ne laser. In principle, multipoint measurements by LDV can be performed by use of multiple LDV systems with distinct measuring volumes. Compared with multiple hot-wire arrays, they would have the advantage of not interfering with each other. In practice, however, because of the high cost and the need for alignment of many components, this approach is restricted to, at most, two LDV systems. By traversing one system with respect to the other, one can measure spatial correlations and wavenumber spectra in different directions. A more economical alternative is to scan rapidly the measuring volume [33].

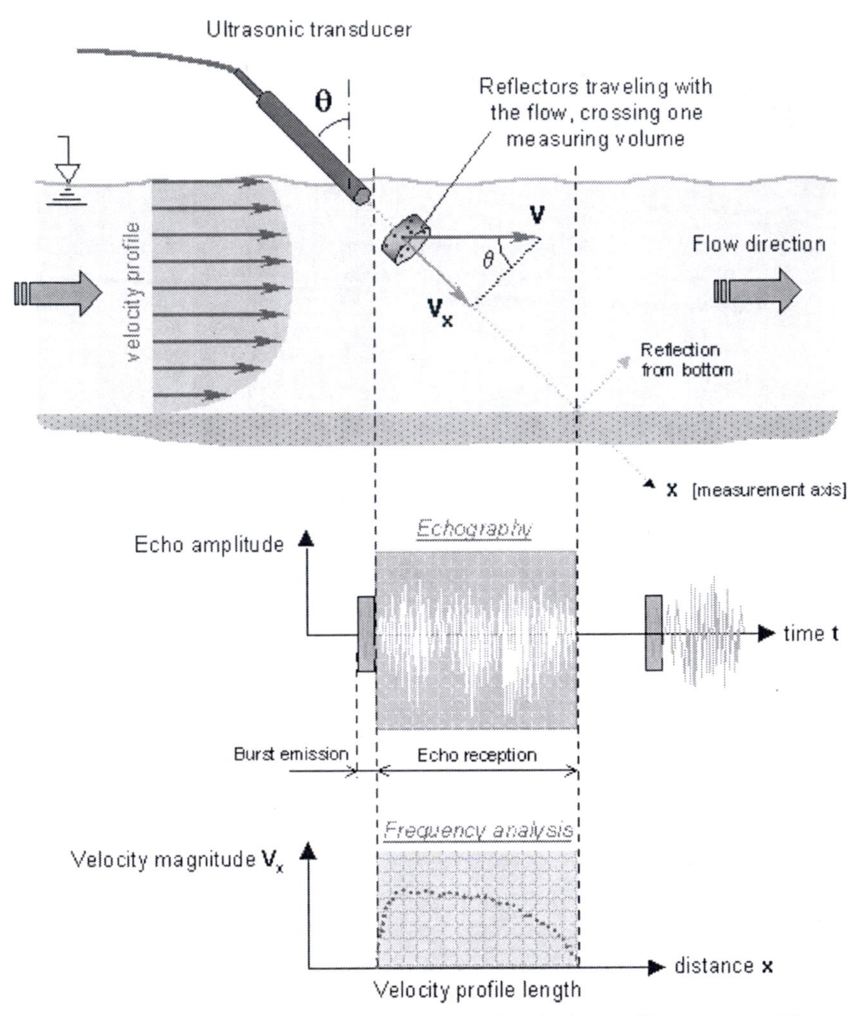

Figure 11.15. Principle of operation of the ultrasonic velocimeter (from www.met-flow.com) with permission of MET–Flow S. A., Lausanne, Switzerland.

11.3 Ultrasonic Doppler velocimetry

This method is based on the Doppler shift of the frequency of ultrasonic waves scattered by moving particles. It has been used widely for the measurement of flow rate in pipes, but, more recently, it has also been extended to the measurement of local flow velocity along the path of an ultrasonic beam (Fig. 11.15). The latter method actually measures the instantaneous velocity of particles at selected locations along the beam path, thus providing instantaneous velocity profiles, which may be time averaged to give average profiles. The velocity component V_x of a particle in the direction of the beam is calculated as

$$V_x = \frac{f_D}{2f} c, \tag{11.45}$$

where f is the frequency of sound emitted by a transmitter, f_D is the Doppler frequency shift that is due to the particle motion, and c is the speed of sound in the medium. The particle position x along the beam path can be calculated as

$$x = \frac{1}{2} c \Delta t, \tag{11.46}$$

where Δt is the time delay between emission and reception of sound by the transducer, which alternates as transmitter and receiver.

For practical reasons, this method can be used only in liquids. Its advantage over optical methods is that it can be applied to opaque fluids, and, for this reason, it has found applications in the food industry, chemical processing, oil refineries, sediment transport, and sewer design. Among its limitations is that it cannot measure directly the streamwise velocity component, although one can estimate this by combining readings of transducers oriented in different directions.

11.4 Particle displacement methods

Flow-marker imaging techniques developed for flow visualization purposes have been refined to record the locations of markers at two or more time instances, from which marker displacement and hence velocity can be computed. The simplest such technique is the direct tracking of the images of individual foreign particles, provided that they can be distinguished from the images of other particles. Methods that distinguish the displacements of many particles simultaneously are more usable, as they can provide the variation of local velocity over a two- or three-dimensional domain. Besides particle imaging by optical means, particle tracing can be accomplished with the use of sound waves and electromagnetic fields. All particle tracing methods are non-intrusive, although, at sufficiently high concentrations, foreign particles may cause some loading of a flow, for example dampen turbulent fluctuations. Another advantage of particle tracing methods is that they provide the velocity directly as the ratio of displacement and time, so that they require no calibration. It is remarkable that the earliest recorded flow velocity measurement, by van Leeuwenhoek in 1689, utilized blood cells as tracers to measure blood flow speed. On the other hand, the accuracy of flow velocity measurement depends on the relative velocity of particle with respect to the surrounding fluid, which may be a complicated function of particle properties and flow conditions (see Section 5.6). Simple conceptually, current particle tracing methods have become quite sophisticated, requiring expensive instrumentation and considerable experimental skill.

Particle tracking: *Particle tracking*, also referred to as *chronophotography*, describes techniques that have been developed to estimate the local flow velocity by measuring the distance between positions of individual particles transported by the flow and observed at distinct and known times [4, 34]. Under certain circumstances, one may be able to actually follow visually the particle's motion, if necessary assisted by a microscopic or telescopic lens or other optical device, and time the passage of the particle between two lines drawn either on the flow apparatus or displayed on the observation lens. This approach is obviously restricted to relatively low speeds and may only be applied to easily visible, isolated particles. In most applications, it would be necessary to record the particle's motion with a camera. The simplest technique is to record all particle positions on the same frame of the film, tape, or other digital storage device. Depending on whether exposure and illumination are continuous or pulsed, the recorded particle trace could be a continuous streak, terminated by the initial and final positions of the particle, or a sequence of images of the particle at different times. A problem that arises in this approach is that the sense of particle motion along its path cannot be resolved. Among the techniques that have been suggested to remove this ambiguity is the use of two or more synchronized light sources with different intensities or colours, such that the particle image at a specific time would be distinguishable from those at other times. Another problem that arises when the concentration of particles is relatively large is the overlapping of pathlines and images of different particles. One method of separating pathlines of different particles is to combine continuous illumination at low intensity and interrupted illumination at a higher intensity, thus being able to record continuous pathlines superimposed on discrete particle positions. A method preferable to single-frame recording is to record the particle images at different times on separate photographic or digital frames. This requires both short-time exposure, achieved with high-speed shutters or pulsed illumination, and rapid advancing of the film, tape, or digital frame of the camera. Image improvement, such as removal of blur that is due to particle movement, can be achieved by image processing methods readily available as part of commercial graphics packages. Specialized methods of illumination may also assist in the process [35]. Manual analysis of particle tracking images is a very tedious process and can be practically achieved for only a small number of sparsely spaced particles. This process can be automated with the use of algorithms that identify particles and their displacement, from which one can estimate the local flow velocity [36, 37]. Sophisticated multi-camera systems and three-dimensional algorithms have also been utilized to reconstruct three-dimensional traces of particles and the three-dimensional velocity field [38–41].

A variant of the particle tracking method is the *line-tracking method*. A line in the fluid transverse to the main flow direction is marked by various means, for example by hydrogen bubbles, wire-generated smoke, or an electrochemical or photochemical reaction. Successive images of the line at different times are analysed to produce displacements of points along the line, from which flow velocity profiles can be determined. This approach would not be effective in highly turbulent flows, in which the line would become quickly untraceable because of mixing, or in strongly three-dimensional flows, containing significant velocity components tangential to the line.

Laser speckle velocimetry [42–44]: *Laser speckle velocimetry* (*LSV*) provides the local flow velocity in very heavily seeded flows, such that the light scattered by each particle would interfere significantly with light scattered by others. When illuminated by planar, coherent light from a laser, this interference would result in patterns, called *speckles*, that would move with the particles that generate them. Thus flow velocity could be computed from successive *specklegrams* (images of speckles). Historically,

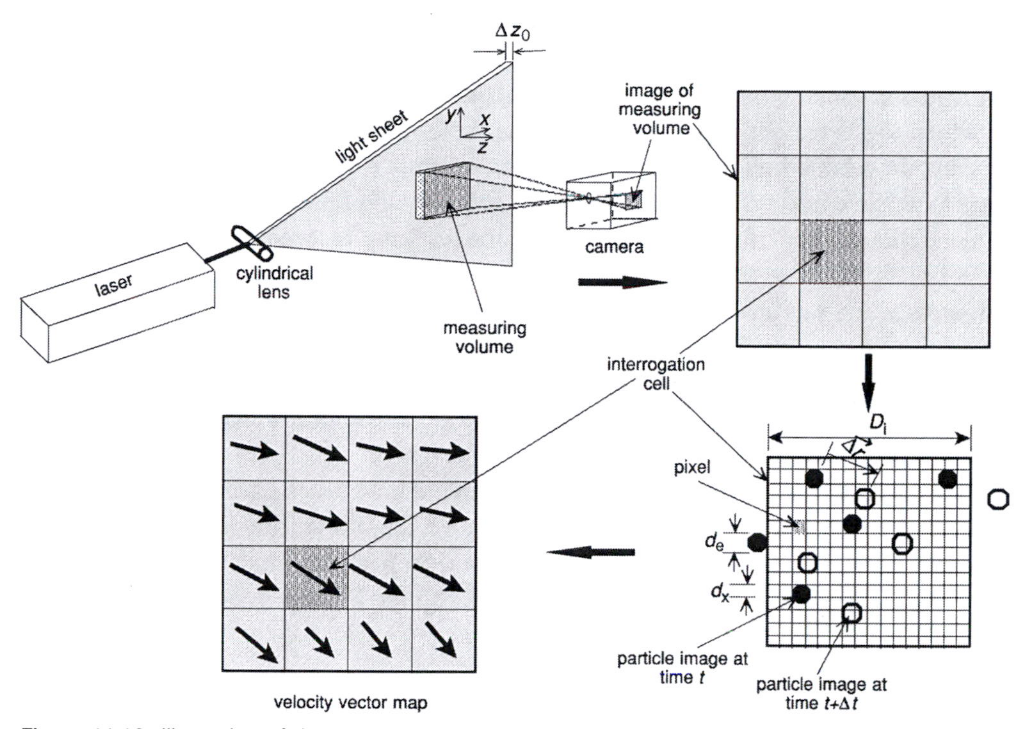

Figure 11.16. Illustration of the components and the principle of operation of a basic particle image velocimeter.

LSV preceded PIV (see next subsection), but it has not gained popularity because of its reliance on impractically high particle densities and will not be discussed further [28].

Particle image velocimetry [28,45–50]: Like LSV, *PIV* provides the velocity of groups of particles illuminated by planar light. Its difference from LSV is that, in PIV, the particle density is sufficiently low for the images of individual particles to be distinct and not to interfere with each other. PIV instrumentation and data processing techniques have undergone rapid development in recent years. Whereas earlier applications utilized photographic recording, current ones are almost exclusively based on digital image recording, to the point that the term *digital particle image velocimetry* (*DPIV*) is now used alternatively with the more general one. PIV setups have been proposed in great variety and the field is still in the developmental stage, with further advances likely to follow in the near future. The following presentation focusses on the general principles of the method, illustrated mainly through the most widely used DPIV configuration.

A basic PIV configuration is illustrated in Fig. 11.16. A laser beam is converted to a planar sheet of light with a thickness Δz_0. A camera lens with a magnification factor M and an f-number $f^{\#}$ is focussed on the light sheet and produces the image of a portion of the illuminated flow, which is the *measuring volume* of the system. The light intensity is assumed to be uniform throughout the entire measuring volume. The method requires the recording of two such images, spaced in time by a time increment Δt. In the most popular current method, a twin-cavity Nd-YAG laser (see Section 5.3) is used to generate two closely spaced, short light pulses, a digital camera is used to record the two images on separate frames, and precise synchronization of the laser firings and the frame advancing is provided by a controller. A technique, called *frame straddling*, in which the first light pulse is triggered just before a frame is advanced to the next one and the second light pulse is triggered immediately following the frame advance, is used to achieve the highest possible recording speed with relatively inexpensive digital cameras. Earlier techniques in which both images were recorded on the same frame have largely become obsolete because of ambiguities in separating the two images. Theoretical analysis has demonstrated that, instead of processing the images of the entire measuring volume simultaneously or the images of individual particles separately, it is more accurate to compute the local flow velocity from subdivisions of the recorded images, called *interrogation cells*, having an area D_i^2 and within which all particles may be assumed to have the same velocity.

First, let us discuss the condition under which the recorded particle images would be distinct from those of other particles. Let C be the number density (i.e., the number of particles per unit volume) of monodisperse spherical particles randomly distributed in the flow. The probability of finding k particles in a volume V obeys a *Poisson distribution*, such that

$$P\,(k \text{ particles in } V) = \frac{(CV)^k}{k!}e^{-CV}, \qquad (11.47)$$

For $CV \ll 1$, this probability can be approximated as $(CV)^k/k!$, which has a very strong peak at $k = 1$. When considering particle images in PIV, one must not use the geometrical image diameter Md_P, but the parameter d_e, defined by Eq. (5.50), which takes into account light diffraction. When projected geometrically back into the flow, the particle image appears to be produced by a source with diameter d_e/M. The images of two or more particles will overlap if their distances on the image plane are less than d_e. To avoid overlap, there should be at most one particle within the cylindrical volume $\Delta z_0 \frac{\pi d_e^2}{4M^2}$, which is projected by the lens onto the particle image area $\frac{\pi d_e^2}{4}$. Thus, to ensure non-overlapping particle images, one must select a *particle source density* N_s such that

$$N_s = C\Delta z_0 \frac{\pi d_e^2}{4M^2} \ll 1. \qquad (11.48)$$

To proceed, one must consider the resolution of the recorded images, which depends on the recording medium. For digital cameras, the resolution is the size d_x of an individual pixel. Optimal utilization of the recording system is achieved when $d_e/d_x \approx$ 1–2. Assuming that the particle diameter d_P is fixed, one must select the camera lens properties such as to satisfy this condition. In selecting D_i, one needs to ensure that the probability that a significant number of particle images can be found within each interrogation cell is high, so that spatial averaging of their displacements would be meaningful. This condition requires one to verify that the number of particles within the *interrogation volume* $V_i = \Delta z_0 D_i^2/M^2$ is significant, expressed by the requirement that the *particle image density* N_i is

$$N_i = C\Delta z_0 \frac{D_i^2}{M^2} \gg 1. \tag{11.49}$$

Optimal values of the ratio D_i/d_e for resolving particle displacement have been found to be in the range 20–30, which points to optimal cells of 16×16 or 32×32 pixels. The length D_i is important as it determines the spatial resolution of the PIV system. Whether such resolution is satisfactory depends on whether the size D_i/M of the cell projection into the flow is comparable with or smaller than the smallest length scale of dynamic interest. For turbulent flows, this scale would be the Kolmogorov microscale [51], which is usually too small to be resolved by PIV systems and so some compromise must be tolerated.

Once the particle images have been recorded and the interrogation cell size fixed, one needs to determine the average particle displacement $\langle\overrightarrow{\Delta r}\rangle$ within each cell, from which the local flow velocity projected on the plane of the light sheet can be estimated as

$$\vec{U}_p \approx \frac{\langle\overrightarrow{\Delta r}\rangle}{M\Delta t}. \tag{11.50}$$

Among the various methods that have been used to estimate this displacement, the most widely used one is the *cross-correlation method*. Each pixel will produce as output a number, chosen from among a small set of permissible values, that represents the intensity of the received light and determines whether a particle image overlaps with the pixel. In the cross-correlation method, two images of each cell at two different times and recorded on separate frames and are processed by an algorithm or special hardware, which correlate all pixels in the first frame with all pixels in the second one. Figure 11.16 shows an example of a cell with both frames superimposed on each other. If the system parameters have been properly selected, the majority of the particles would have both images recorded within the same cell and all particle image displacements would be identical within the cell. Then the cross correlation would have a peak whose location with respect to the zero-separation point would correspond to the average particle image displacement on the plane of the image. Interpolation methods are available to determine the peak location at subpixel resolution. Shear in the interrogation volume would introduce particle velocity differences, which would cause broadening of the correlation peak and even change of its shape. For broadening effects on the estimate of $\langle\overrightarrow{\Delta r}\rangle$ to be negligible, the flow velocity differences $|\overrightarrow{\Delta U}_p|$ within the interrogation volume must satisfy the condition

$$\left|\overrightarrow{\Delta U}_p\right| \ll \frac{d_e}{M\Delta t}. \tag{11.51}$$

It is also evident that correlation distortion would occur by the fact that some particles may only have a single image recorded within a cell, because their motion is such that they enter or exit the interrogation volume during Δt. Such errors will be negligible if the in-plane $\vec{U}_p$ and out-of-plane $\vec{U}_n$ flow velocity components are such that

$$\frac{\left|\vec{U}_p\right| M \Delta t}{D_i}, \frac{\left|\vec{U}_n\right| \Delta t}{\Delta z_0} \ll 1. \tag{11.52}$$

As an example, when $N_i = 12$, the value $1/4$ may be considered as acceptable for both preceding ratios [49]. Another source of error in PIV, known as *velocity bias*, is that the broadening of the correlation peak results in a systematic underestimation of the average image displacement, which would be relatively small if all particle speeds were identical, but worsen with increasing shear. Displacement errors that are due to velocity bias are typically of subpixel magnitude and can be corrected by use of methods available in the literature [49, 52].

Once the velocity $\vec{U}_p$ is determined at all interrogation cells, it can be displayed in the form of a *velocity vector map* (see Fig. 11.16), in which each vector is displayed on the centre of each cell, in a manner similar to vector plots in computational fluid dynamics (CFD) simulations. These velocity values can be used to estimate the vorticity component normal to the light-sheet plane and streamline projections on this plane. The measurement of flow velocity variation over a plane section, rather than at a single point, is the great advantage of PIV over LDV, HWA, and other probe methods that require multiple systems for simultaneously measuring multipoint velocities. This is the main reason why PIV has gained popularity over the other methods. On the other hand, it is also well understood that the spatial and temporal resolutions of PIV are markedly lower than those of LDV and HWA. Whereas spatial resolution of PIV is mainly associated with pixel size, temporal resolution is associated with the ability to produce strong light pulses at a high repetition rate. Conventional PIV, which uses flash-pumped Nd-YAG lasers, can produce measurements at the maximum rate of 30 Hz, which is grossly insufficient to resolve many unsteady and turbulent flow phenomena. Faster data rates, exceeding 1 kHz, have been achieved with the use of diode-pumped Nd-YAG and copper-vapour lasers, combined with fast digital cameras, in configurations referred to as *time-resolved PIV*. A limitation of these methods is the relatively low light energy per pulse, which reduces their range of applications.

Many other improvements and extensions of the basic PIV configuration have been presented by various authors. An interesting example has been the replacement of the camera lens with a microscope lens, which has allowed the measurement of flow velocity in microdevices [53]. One of the main concerns in experimental fluid mechanics has been the measurement of the three-dimensional velocity vector. A classical approach with several variations is three-dimensional particle tracking in fully illuminated flows with the use of multiple cameras (see earlier subsection) or by holographic techniques [41,54]. A more popular method, known as *stereoscopic PIV*, is illustrated in Fig. 11.17(a). This configuration resembles conventional PIV as it measures velocity on the plane of a planar light sheet, but, in addition to measuring the in-plane particle displacements, it also measures the out-of-plane displacements, from which the three-dimensional velocity vector variation on the plane of the sheet is reconstructed. Two cameras are positioned symmetrically with respect to the normal light sheet, with their lenses inclined by

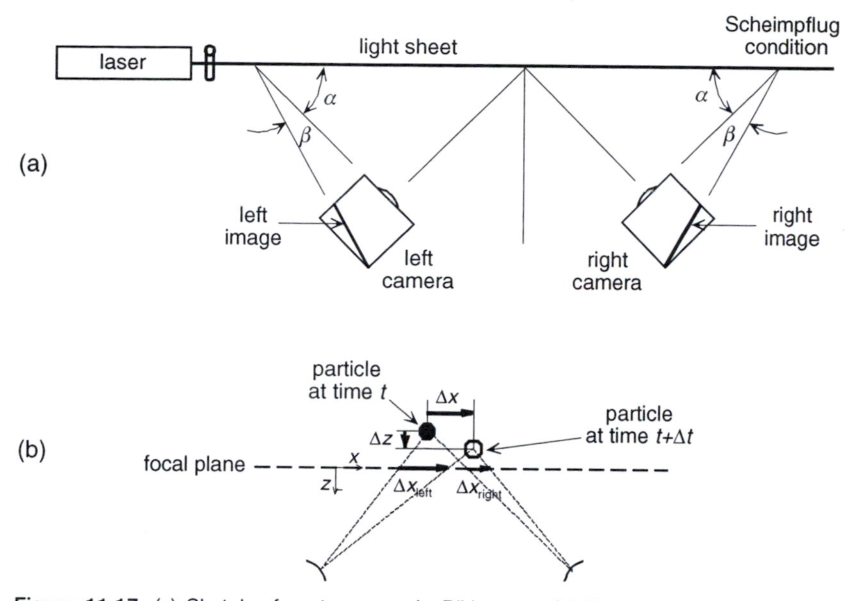

Figure 11.17. (a) Sketch of a stereoscopic PIV setup; (b) illustration of stereoscopic view of particle displacement.

an angle α with respect to the sheet plane. If the depth of field of the lens is relatively narrow, lens inclination would produce images that would be partially out of focus. One corrects this by tilting the CCD chips of the cameras by an angle β, such that the lens plane and the image plane intersect on the object plane, an arrangement known as the *Scheimpflug condition*. One corrects distortion of the object proportions that is due to image inclination by calibrating the cameras by using a target marked by an array of spots at known positions. Because the motion of particles in the measuring volume is viewed by the two cameras from different angles, the two images would be different, in a manner similar to stereoscopic vision. For example [Fig. 11.17(b)], let us assume that the true in-plane displacement of a particle in the measuring volume during time Δt is Δx. Assuming that both lenses are focussed on the midplane of the light sheet, the left and right cameras will record displacements Δx_{left} and Δx_{right}, which are different from each other and Δx. Analysis of these values can provide both Δx and the out-of-plane displacement Δz [45]. The error in the determination of Δz can be related to the error in Δx by the approximate expression [55]

$$\varepsilon_{\Delta z} = \frac{\varepsilon_{\Delta x}}{\sqrt{2}M \sin\beta}. \tag{11.53}$$

Molecular tagging velocimetry [56]: *Molecular tagging velocimetry* (*MTV*), also referred to as *laser-induced photochemical anemometry* and *flow tagging velocimetry*, includes methods that trace the motion of photosensitive molecules in the fluid. Its advantage, compared with that of PIV and seed-particle tracking methods, is that it requires no seeding, thus being suitable for both single-phase and multiphase flows. It does require, however, the presence of special materials, whose molecules, when activated, will

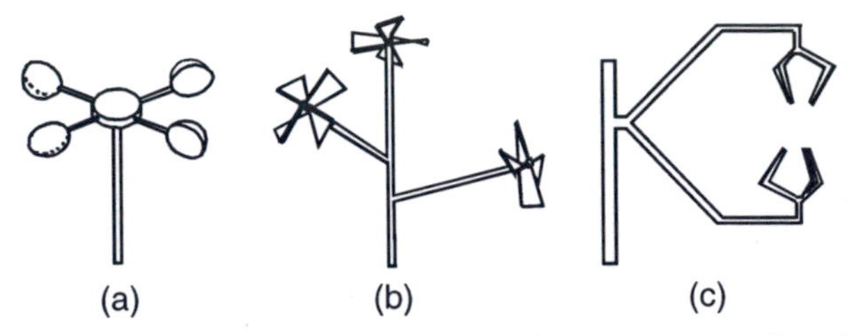

Figure 11.18. Sketches of representative anemometers for wind velocity and turbulence measurement: (a) cup anemometer, (b) three-component Gill anemometer, and (c) three-component sonic anemometer.

become distinguishable from the background fluid. Three types of such materials have been used in MTV:

Photochromic materials: These have been briefly discussed in Section 10.2. Their disadvantages compared with more popular MTV methods are that their absorption-activated visibility has a relatively low contrast and they are not water soluble.

Phosphorescent supramolecules: These are molecules that, as a result of photon absorption, undergo transition to metastable electronic states and emit radiation at a different wavelength, following a relatively long delay time, which exceeds 0.1 ms (see discussion on phosphorescence in Section 5.4). The emitted radiation is collected by a camera during subsequent 'interrogation' times to locate the tagged fluid and compute its displacement. Recipes for such molecules that are water soluble have been proposed in recent literature [56]. Tagging in the form of single and multiple lines or in meshes of intersecting lines has been employed to map one-, two-, and three-dimensional displacement fields [57, 58].

Caged dyes: These are fluorescent dyes, such as fluorescein disodium, whose fluorescence has been 'caged' by the attachment of a chemical group. Fluorescence is recovered locally following absorption by the caging group of an ultraviolet photon. This is a permanent, irreversible change and allows the tracing of the tagged fluorescent molecules indefinitely, provided that they get illuminated by suitable laser light. A disadvantage of this method is the high cost of the caged dye material.

A related method utilizes fluorescence of ions of strontium atoms seeded into a gas mixture [59]. It utilizes two different lasers: one to produce a beam along which ionization occurs and a second one, fired following some specified time delay, to produce a light sheet that induces ionic fluorescence. This method has been applied to the measurement of velocity in low-speed burner flames.

11.5 Measurement of wind velocity

The following subsections describe instruments designed to measure mean wind magnitude, wind direction, and, in some cases, atmospheric turbulence statistics [60].

Cup anemometers: These consist of two, three, or more cups, usually of hemispherical shape, mounted on a shaft that is free to rotate, as shown in Fig. 11.18(a). A net torque is generated by flow on planes normal to the shaft axis, as a result of higher drag on the concave than the convex side of each cup. The speed of rotation ω of the anemometer in steady laminar flow is related to the flow speed V by the linear expression

$$\omega = \frac{k_c V}{r}, \tag{11.54}$$

where r is the mean radius of the cup with respect to the axis of rotation and k_c is an empirical numerical coefficient with a typical value of about 0.3. In turbulent flow, the cup anemometer tends to overspeed, showing velocities higher than values calculated from expression (11.54). For very small turbulence intensities, the cup anemometer dynamic response can be approximated by a first-order system with a time constant

$$\tau = \frac{I}{\rho r^2 A V}, \tag{11.55}$$

where A is the frontal area and I is the moment of inertia of the cup. More sophisticated, non-linear models have also been formulated. Cup anemometers are usually oriented with their axes of rotation vertical and are meant to measure horizontal wind. When a significant wind component outside the plane of rotation is present, they tend to indicate a speed that is larger than the actual component on that plane, with the error increasing significantly at flow inclinations higher than 10° with respect to the rotation plane.

Propeller anemometers: These consist of one or more propeller-type rotors, meant to measure the wind velocity parallel to their axes. The most common type is called the *Gill anemometer* [Fig. 11.18(b)], which has four helicoidal blades. Gill anemometers are often mounted on rotating vanes, which orient them into the wind direction. Their speed of rotation ω is, within limits, linearly related to axial wind speed V as

$$\omega = k_p V, \tag{11.56}$$

where k_p is a calibration coefficient with units of inverse length. When misaligned with the flow direction, Gill anemometers tend to measure axial velocity components that are lower than the actual ones. The dynamic response of propeller anemometers is non-linear. For rough purposes, one may approximate it by a first-order response with a time constant

$$\tau = \frac{L}{V}, \tag{11.57}$$

where L is a length depending on the instrument geometry and flow conditions. Gill anemometers also tend to overspeed in turbulent and gusty flows, but much less than cup anemometers do. For this reason, combinations of three Gill anemometers, together with temperature and humidity sensors, are commonly used for measuring atmospheric turbulence. One must keep in mind, however, that, because of their limited frequency response (at most extending to a few hundred hertz), they cannot resolve the turbulence fine structure and may even be missing part of the turbulence kinetic energy.

Vane anemometers: Two types of vanes are used in wind measurement: *rotating vanes* and *fixed vanes*. A main function of rotating vanes is to orient other instruments into the wind direction as well as to indicate this direction by their instantaneous position.

In high-turbulence flows, vane dynamic response is non-linear. A simplified model gives a second-order response, which means that underdamped vanes are subjected to resonance, with possible implications on the measurement of turbulent spectra. Fixed vanes are mounted on strain gauges and measure flow direction from measurements of the applied torque.

Sonic anemometers: Although sonic Doppler anemometers have also been used for atmospheric turbulence measurement, the most common sonic anemometers operate based on the difference in the speed of sound waves propagating in the direction of the flow and against it. A three-axis sonic anemometer is shown in Fig. 11.18(c). It consists of three sets of transducer–receiver combinations. Each consists of a pair of transmitters and a pair of receivers, such that one transmitter and one receiver are on one side, facing the other transmitter and receiver on the opposite side. Both transmitters are triggered simultaneously, and the produced sound waves are picked up by the opposite receivers. Assuming that the wind velocity component parallel to the axis of these transducers is V_x, there will be a time delay between the reception of the two waves, approximately equal to

$$\Delta t \simeq \frac{2w}{c} V_x, \tag{11.58}$$

where w is the distance between the opposing transmitter–receiver pair and c is the speed of sound in the air. Because the speed of sound depends on temperature, it is necessary to measure the local temperature, as well. In order to maintain a reasonable sensitivity, it is necessary to keep the distance w relatively large (e.g., 0.20 m), which limits the spatial resolution of this instrument. Additional errors are introduced by flow blockage by the transducers and humidity effects. Moreover, sonic anemometers cannot be used in dusty or foggy air or when there is precipitation. For these reasons, sonic anemometers have not been utilized in routine atmospheric monitoring and their main application has been in atmospheric research.

Lidar: The term *lidar* is an acronym for light detection and ranging, sometimes also referred to as *ladar* (laser radar) or *CLR* (coherent laser radar). It uses a coherent pulsed laser light beam to scan the lower atmosphere and detects amplitude attenuation and phase shift in the light collected back at the source to detect the location of reflecting objects. Such systems are utilized to provide wind velocity profiles and to identify air disturbances, including clear air turbulence, aircraft vortices, and thermal plumes [61,62]. The range of lidars may extend to tens of kilometres, and their spatial resolution would at best be of the order of tens of meters. Lidar technology is emerging, with advances anticipated to be made in the near future.

QUESTIONS AND PROBLEMS

1. The thermal resistivity coefficient of annealed tungsten at 20 °C is 0.0048 K^{-1}. A hot-wire sensor made of tungsten has a 'cold' resistance of 5.65 Ω, measured at 25 °C. Determine the maximum allowable operating resistance R_w of this sensor

to avoid oxidization. The sensor, operating at the resistance R_w as found previously, is inserted in a flow that has a temperature of (a) 25 °C and (b) 130 °C. Determine the corresponding resistance and temperature overheat ratios. Compare qualitatively the anticipated velocity sensitivities of this sensor in the two flows.

2. Consider the constant current anemometer shown in Fig. 11.3, for which $R_c' = 100\ \Omega$. The time constant of the sensor was determined by the square-wave test to be 7.5 ms. Determine the values of the resistor and the capacitor in the R–C circuit that would give a compensated output time constant of 0.05 ms. Plot, in logarithmic coordinates, the amplitude ratios of the uncompensated and compensated outputs vs. frequency. Specify the 3-dB upper cutoff frequencies of the uncompensated and the compensated systems.
3. Consider that the response of a hot-wire anemometer follows the modified King's law, Eq. (11.8), in which the calibration coefficients and their 95% confidence limits were found as $A = 0.0555 \pm 1\%$, $B = 0.00250 \pm 2\%$, and $n = 0.48 \pm 0.01$, all in SI units. During an experiment, the following values and their 95% confidence limits were measured: $E = 3.673 \pm 0.001$ V; $T_w = 230 \pm 1$ °C; $T = 24 \pm 0.2$ °C. Determine the value of the flow velocity and its uncertainty. Explain whether this uncertainty would be acceptable and identify the parameters that contribute most to it. If possible, suggest means to reduce the uncertainty.
4. A hot-wire probe will be used for measuring the turbulent flow velocity profile at the outlet of a heating air duct located on the ceiling of the room. The probe will be calibrated in a heated jet.
   a. Describe how you would measure the instantaneous local flow velocity in the operating duct, whose temperature fluctuates.
   b. Consider describing the hot-wire response by a fitted fourth-order polynomial rather than King's law. Would you recommend this approach for this experiment? Justify your answer.
5. The response of a hot-wire, made of platinum-plated tungsten, 5 μm in diameter and 1.25 mm long, is tested by calibration in a heated air jet.
   a. First, the jet speed is kept constant at 8.00 m/s and the resistance of the unheated wire is measured at different flow temperatures, according to the following table:

T (°C)	25.81	27.71	29.89	32.80
R_w (Ω)	3.44	3.47	3.50	3.54

   Determine the thermal resistivity coefficient of the wire at a reference temperature and compare it with values from the literature.
   b. A next test consists of varying the overheat ratio of the wire by setting its operating resistance R_w by use of a constant-temperature anemometer circuit and measuring the circuit output voltage E, which is related to the current i through the wire by the expression

$$E = i\,(R_w + R_b),$$

where $R_b = 50.30\ \Omega$ is a resistor in series with the hot-wire. During these tests the flow temperature is kept at 25.80 °C and the flow speed is kept at 8.00 m/s. The results are given in the following table:

R_w (Ω)	4.13	4.82	5.50	6.19	5.16	3.78
E (V)	2.448	3.232	3.731	4.098	3.500	1.791

Determine the corresponding resistance and temperature overheat ratios. Then, compute the experimental values of the Nusselt number and compare them with estimates based on the Collis and Williams expressions. Discuss your findings.

6. A constant-temperature hot-wire anemometer is calibrated in an air jet. The following table provides the anemometer output voltage at different air speeds

V (m/s)	7.06	8.76	9.48	10.61	11.91	14.06	14.73	15.33	17.08
E (V)	2.894	2.977	3.008	3.054	3.104	3.177	3.197	3.218	3.269

Absorbing the temperature difference into the coefficients of the modified King's law, determine the optimum value of the exponent n in that expression by minimizing the mean-squared difference between the measured velocities and those calculated from the fitted expression. Further evaluate the sensitivity of this difference to the selected exponent value. Discuss your findings.

7. A laser Doppler velocimeter is used to measure water flow velocity in a water tunnel having a test section with a width of 200 mm. The LDV system includes a He–Ne laser, a Bragg cell with a frequency shift of ± 40 MHz, and three transmitting lenses with focal distances of 130, 450, and 800 mm. The beam separation is 30 mm and the beam diameter is 0.8 mm. (a) Determine the beam intersection half-angle in the water; compare the calibration factors in air and in water. (b) Select the lens that is best suited for measuring the flow velocity in the entire test section; explain your reasoning. (c) Determine the dimensions of the measuring volume and the number of fringes in the control volume. (d) Assuming that frequency shifting is used to measure the flow velocity in a recirculating flow region, determine the highest negative velocity that could be measured unambiguously; would you use positive or negative frequency shift?

8. Two popular instruments for measuring local flow velocity are the hot-wire anemometer and the laser Doppler velocimeter. Explain which of the two would be preferable in the following situations:
   a. When the highest possible spatial resolution is required.
   b. When the highest possible temporal resolution is required.
   c. When the turbulence intensity exceeds 30%.
   d. When the flow frequently reverses direction.
   e. When the flow is three dimensional.
   f. When measurements in both gases and liquids are required.
   g. When cost is a determining factor.

h. When calibration must be avoided.
i. When multipoint measurements must be conducted simultaneously.

9. Compare the ranges of application of LDV and PIV. Describe comparatively the advantages and disadvantages of these methods. Provide an example of flow measurement for which only LDV would be suitable, and another for which only PIV would be suitable.
10. Describe an experiment for which both LDV and pulsed-wire anemometry would be suitable but HWA would not. Also give an example of an experiment for which pulsed-wire anemometry would be preferable to LDV.

REFERENCES

[1] H. H. Bruun. *Hot-Wire Anemometry*. Oxford University Press, Oxford, UK, 1995.
[2] C. G. Lomas. *Fundamentals of Hot Wire Anemometry*. Cambridge University Press, Cambridge, UK, 1986.
[3] A. E. Perry. *Hot-Wire Anemometry*. Clarendon, Oxford, UK, 1982.
[4] R. J. Emrich (Editor). Fluid dynamics. In L. Marton and C. Marton (Editors-in-Chief), *Methods of Experimental Physics*, Vols. 18A and B. Academic, New York, 1981.
[5] R. J. Goldstein. *Fluid Mechanics Measurements* (2nd Ed.). Taylor & Francis, Washington, 1996.
[6] G. Comte-Bellot. Hot-wire anemometry. *Annu. Rev. Fluid Mech.*, 8:209–231, 1976.
[7] L. V. King. On the convection of heat from small cylinders in a stream of fluid; determination of the convection constants of small platinum wires with application to hot-wire anemometry. *Philos. Trans. R. Soc. Lond. Ser. A*, 214:373–432, 1914.
[8] H. L. Dryden and A. M. Kuethe. The measurement of fluctuations of air speed by the hot-wire anemometer. Tech. Rep. 581, National Advisory Committee for Aeronautics Washington, DC, 1929.
[9] L. S. G. Kovasznay. Simple analysis of the constant temperature feedback hot-wire anemometer. Tech. Rep. AERO/JHU CM-478, Johns Hopkins University, Baltimore, MD, 1948.
[10] E. Ossofsky. Constant temperature operation of the hot-wire anemometer at high frequency. *Rev. Sci. Instrum.*, 19:881–889, 1948.
[11] D. C. Collis and M. J. Williams. Two-dimensional convection from heated wires at low Reynolds numbers. *J. Fluid Mech.*, 6:357–384, 1959.
[12] J. D. Li, B. J. McKeon, W. Jiang, J. F. Morrison, and A. J. Smits. The response of hot wires in high Reynolds-number turbulent pipe flow. *Meas. Sci. Technol.*, 15:789–798, 2004.
[13] F. DeSouza and S. Tavoularis. Hot-wire response in high-subsonic flow. In *Proceedings of the 37th AIAA Aerospace Sciences Meeting*, AIAA Paper 99-0310, pp. 1–11, American Institute of Aeronautics and Astronautics, Reston, VA, 1999.
[14] G. D. Huffman. Calibration of tri-axial hot-wire probes using a numerical search algorithm. *J. Phys. E*, 13:1177–1182, 1980.
[15] T. L. Butler and J. W. Wagner. Application of a three-sensor hot-wire probe for incompressible flow. *AIAA J.*, 21:726–732, 1983.
[16] T. J. Gieseke and Y. G. Guezennec. An experimental approach to the calibration and use of triple hot-wire probes. *Exp. Fluids*, 14:305–315, 1993.
[17] M. Samet and S. Einav. A hot-wire technique for simultaneous measurement of instantaneous velocities in 3d flows. *J. Phys. E*, 20:683–690, 1987.

[18] K. Döbbeling, B. Lenze, and W. Leuchel. Basic considerations concerning the construction and usage of multiple hot-wire probes for highly turbulent three-dimensional flows. *Meas. Sci. Technol.*, 1:924–933, 1990.
[19] K. S. Wittmer, W. J. Devenport, and J. S. Zsoldos. A four-sensor hot-wire probe system for three-component velocity measurement. *Exp. Fluids*, 24:416–423, 1998.
[20] J. M. Wallace and J. F. Foss. The measurement of vorticity in turbulent flows. *Annu. Rev. Fluid Mech.*, 27:469–514, 1995.
[21] P. Vukoslavcevic and J. M. Wallace. A 12-sensor hot-wire probe to measure the velocity and vorticity vectors in turbulent flow. *Meas. Sci. Technol.*, 7:1451–1461, 1996.
[22] P. M. Handford and P. Bradshaw. The pulsed-wire anemometer. *Exp. Fluids*, 7:125–132, 1989.
[23] I. P. Castro and M. Dianat. Pulsed wire velocity anemometry near walls. *Exp. Fluids*, 8:343–352, 1990.
[24] W. J. Devenport, G. P. Evans, and E. P. Sutton. A traversing pulsed-wire probe for velocity measurements near a wall. *Exp. Fluids*, 8:336–342, 1990.
[25] H. Yeh and H. Z. Cummins. Localized fluid flow measurements with a He–Ne spectrometer. *Appl. Phys. Lett.*, 4:176–179, 1964.
[26] F. Durst, A. Melling, and J. H. Whitelaw. *Principles and Practice of Laser Doppler Anemometry*. Academic, New York, 1976.
[27] P. Buchhave, W. K. George, and J. L. Lumley. The measurement of turbulence with the laser Doppler anemometer. *Annu. Rev. Fluid Mech.*, 11:443–503, 1979.
[28] R. J. Adrian. Laser velocimetry. In R. J. Goldstein, editor, *Fluid Mechanics Measurements*, pp. 175–299. Taylor & Francis, Washington, DC, 1996.
[29] Anonymous. Laser Doppler anemometry: An introduction to the basic principles. Tech. Rep. 1203E, DISA Elektronik A/S, Skovlunde, Denmark, 1981.
[30] B. M. Watrasiewitz and M. J. Rudd. *Laser Doppler Measurements*. Butterworths, Sydney, Australia, 1976.
[31] R. V. Edwards. Report of the special panel on statistical particle bias problems in laser anemometry. *J. Fluids Eng.*, 109:89–93, 1987.
[32] D. K. McLaughlin and W. G. Tiedermann. Biasing correction for individual realization of laser anemometer measurements in turbulent flow. *Phys. Fluids*, 16:2082–2088, 1973.
[33] K. A. Shinpaugh and R. L. Simpson. A rapidly scanning two-velocity component laser Doppler velocimeter. *Meas. Sci. Technol.*, 6:690–701, 1995.
[34] E. F. C. Somerscales. Fluid velocity measurement by particle tracking. In R. B. Dowdell, editor, *Flow – Its Measurement and Control in Science and Industry*, pp. 795–808. Instrument Society of America, Pittsburgh, PA, 1974.
[35] G. J. Kostrzewsky and R. D. Flack. Simple system for fluid flow visualization and measurement using a chronophotographic technique. *Rev. Sci. Instrum.*, 57:3066–3074, 1986.
[36] P. R. Jonas and P. M. Kent. Two-dimensional velocity measurement by automatic analysis of trace particle motion. *J. Phys. E*, 12:604–609, 1979.
[37] T. Kobayashi, T. Ishihara, and N. Sasaki. Automatic analysis of photographs of trace particles by microcomputer system. In W. J. Yang, editor, *Flow Visualization III*, pp. 231–235. Hemisphere, Washington, DC, 1985.
[38] J. Doi, T. Miyake, and T. Asanuma. Three-dimensional flow analysis by on-line particle tracking. In W. J. Yang, editor, *Flow Visualization III*, pp. 14–18. Hemisphere, Washington, DC, 1985.
[39] K. Nishino, N. Kasagi, and M. Hirata. Three-dimensional particle tracking velocimetry based on automated digital image processing. *J. Fluids Eng.*, 111:384–391, 1989.

[40] N. Kasagi and K. Nishino. Probing turbulence with three-dimensional particle-tracking velocimetry. *Exp. Therm. Fluid Sci.*, 4:601–612, 1991.

[41] K. D. Hinsch. Three-dimensional particle velocimetry. *Meas. Sci. Technol.*, 6:742–753, 1995.

[42] W. Merzkirch. *Flow Visualization* (2nd Ed.). Academic, New York, 1987.

[43] J. C. Dainty (Editor). *Laser Speckle and Related Phenomena*. Springer-Verlag, New York, 1975.

[44] R. J. Adrian. Scattering particle characteristics and their effect on pulsed laser measurements of fluid flow: Speckle velocimetry vs. particle image velocimetry. *Appl. Opt.*, 23:1690–1691, 1984.

[45] M. Raffel, C. E. Willert, and J. Kompenhans. *Particle Image Velocimetry, A Practical Guide*. Springer, Berlin, 1998.

[46] C. Mercer (Editor). *Optical Metrology for Fluids, Combustion and Solids*. Kluwer Academic, Dordrecht, The Netherlands, 2003.

[47] F. Mayinger and O. Feldmann (Editors). *Optical Measurements* (2nd Ed.). Springer, Berlin, 2001.

[48] R. Adrian. Particle-imaging techniques for experimental fluid mechanics. *Annu. Rev. Fluid Mech.*, 23:261–304, 1991.

[49] J. Westerweel. Fundamentals of digital particle image velocimetry. *Meas. Sci. Technol.*, 8:1379–1392, 1997.

[50] I. Grant. Particle image velocimetry: A review. *Proc. Inst. Mech. Eng. Part C*, 211:55–76, 1997.

[51] J. O. Hinze. *Turbulence* (2nd Ed.). McGraw-Hill, New York, 1975.

[52] R. D. Keane and R. J. Adrian. Theory of cross-correlation analysis of PIV images. *Appl. Sci. Res.*, 49:191–215, 1992.

[53] C. D. Meinhart, S. T. Wereley, and M. H. B. Gray. Volume illumination for two-dimensional particle image velocimetry. *Meas. Sci. Technol.*, 11:809–814, 2000.

[54] D. H. Barnhart, R. J. Adrian, and G. C. Papen. Phase-conjugate holographic system for high-resolution particle image velocimetry. *Appl. Opt.*, 33:7159–7170, 1995.

[55] M. P. Wernet. Stereo viewing 3-component, planar PIV utilizing fuzzy inference, Paper AIAA-96-2268. American Institute of Aeronautics and Astronautics, Reston, VA, 1996.

[56] A. J. Smits and T. T. Lim (Editors). *Flow Visualization Techniques and Examples*. Imperial College Press, London, 2000.

[57] R. B. Hill and J. C. Klewicki. Data reduction methods for flow tagging velocity measurements. *Exp. Fluids*, 20:142–152, 1996.

[58] R. Sadr and J. C. Klewicki. A spline-based technique for estimating flow velocities using two-camera multi-line MTV. *Exp. Fluids*, 35:257–261, 2003.

[59] H. Rubinsztein-Dunlop, B. Littleton, P. Barker, P. Ljungberg, and Y. Malmsten. Ionic strontium fluorescence as a method for flow tagging velocimetry. *Exp. Fluids*, 30:36–42, 2001.

[60] J. C. Wyngaard. Cup, propeller, vane and sonic anemometers in turbulence research. *Annu. Rev. Fluid Mech.*, 13:399–423, 1981.

[61] J. A. Zak. Atmospheric boundary layer sensors for application in a wake vortex advisory system. Tech. Rep. NASA/CR-2003-212175, ViGYAN, Inc., Hampton, VA, 2003.

[62] M. Keane, D. Buckton, M. Redfern, C. Bollig, C. Wedekind, and F. Koepp. Axial detection of aircraft wake vortices using doppler lidar. *J. Aircraft*, 39:850–861, 2002.

12 Measurement of temperature

Unlike velocity, temperature cannot be defined scientifically, although it is commonly understood to be a measure of 'hotness' or 'coldness' of an object. The value of temperature is defined with respect to continuous temperature scales, of which two types exist: *thermodynamic scales*, based entirely on thermodynamic considerations, such as the perfect-gas law or thermal noise on electric components; and *practical* or *laboratory scales*, constructed by the definition of discrete reference values and interpolation in the intervals between consecutive reference values by use of consistent instruments [1]. Following a brief review of the currently accepted practical temperature scale, this chapter describes the most common thermometers and other temperature measurement methods that are used in fluid mechanics research. More details on specific methods can be found in a few general references [2–8].

12.1 A practical temperature scale

The International Practical Temperature Scale of 1968, amended in 1975, defines 13 *fixed points* of temperature, corresponding to phase equilibria or phase transitions of water, gases, and metals, which can be reproduced consistently [1]. For example, the value 273.16 K, or 0.0100 °C, is assigned to the temperature of the *triple point of water*, namely the unique state at which vapour, liquid, and ice coexist in equilibrium, which occurs when the pressure is 611 Pa; this state can be readily achieved with the use of a water triple-point cell. The highest temperature defined by a fixed point in this scale is $T_{Au} = 1337.58$ K (1064.43 °C), corresponding to the solid–liquid phase equilibrium of gold at a pressure of 101,325 Pa. Between fixed points, the temperature scale is defined continuously by the response of the following standard thermometers: between 13.81 and 273.15 K, a capsule platinum resistance thermometer; between 273 and 903 K, a long-stem platinum resistance thermometer; and between 903 and 1337 K, a platinum/10% rhodium–platinum thermocouple. Temperatures above 1337 K are defined by reference to T_{Au}, such that the spectral radiance of a blackbody at the wavelength λ (in metres) and at temperature T (in degrees Kelvin) in vacuum satisfies *Planck's radiation law* [Eq. (5.27)]:

$$\frac{L_{e\lambda}(T)}{L_{e\lambda}(T_{\text{Au}})} = \frac{e^{C_2/(\lambda T_{\text{Au}})} - 1}{e^{C_2/(\lambda T)} - 1}, \tag{12.1}$$

where $C_2 = hc/k_B = 0.014388$ m K and $L_{e\lambda}(T_{\text{Au}})$ is provided by a special blackbody operating at T_{Au} and used as a standard [9]. Such temperatures are determined with the use of an optical pyrometer (see Section 12.5), but they are not very reliable for values above about 4000 °C. Higher temperatures, up to 10^9 K, such as those that occur in atomic explosions or in the interior of stars, are defined with spectroscopic methods. Temperatures lower than 30 K and approaching absolute zero have been defined with the rhodium–iron resistance thermometer and various semiconductor thermometers.

12.2 Thermometers

A variety of thermometers is available. They can be generally classified into the following types, depending on the principle of their operation:

Thermal expansion thermometers, whose operation utilizes the relationship between temperature and volume or length of a material. Common types are the *liquid-in-glass thermometers* and the *bimaterial thermometers*.

Thermocouples, whose operation is based on the thermoelectric effect.

Resistance thermometers, whose operation is based on the relationship between temperature and electric resistance. These include *RTDs*, which are metallic resistance sensors, and *thermistors*, which are semiconductor resistance sensors.

Liquid-in-glass thermometers [4, 7, 10]: These contain a liquid (commonly mercury or alcohol), which partially fills a sealed and evacuated glass container, consisting of a bulb-type reservoir and a graduated capillary tube (stem). As the temperature of the liquid in the bulb increases, the liquid expands to fill a longer length of the capillary tube. Laboratory-grade thermometers have typical resolutions of about 0.05–1 K and can be classified as *complete immersion*, *total immersion*, and *partial immersion* types. Complete immersion thermometers must be entirely immersed in the fluid of interest, which may make their reading difficult, at least for liquid flows; total immersion thermometers must be inserted in the fluid of interest up to the end of the liquid column; and partial immersion thermometers must be inserted in the fluid up to a marked line. Properly used, total immersion thermometers are free of errors that are due to temperature differences between the immersed and non-immersed sections, whereas partial immersion thermometers suffer from such errors, which may significantly exceed their resolution. The readings of total immersion thermometers used in the partial immersion mode can be corrected by the addition of a *stem correction*. For mercury-in-glass thermometers calibrated in degrees Kelvin, this correction is $0.00016N(T_{\text{imm}} - T_{\text{non-imm}})$, where N is the number of degrees Kelvin corresponding to the non-immersed portion of the liquid column, T_{imm} is the temperature of the immersed portion, and $T_{\text{non-imm}}$ is the temperature of the non-immersed portion, measured by another thermometer. Following extensive use, the volume of the bulb might change slightly because of accumulating thermal stresses, and it is advisable to check periodically the reading of accurate liquid-in-glass thermometers versus an ice-point bath or other reference temperature. When the thermometer is inserted to flowing fluids through holes or wells in pipe walls, one must also consider possible errors that are due to heat conduction through the exposed stem or the well wall. Thermal insulation can be use to reduce such errors. Liquid-in-glass thermometers have poor spatial and temporal resolutions and are unsuitable for measuring temperatures that change rapidly with location or time; however, they are excellent laboratory standards for the calibration of other instruments.

Bimaterial thermometers [2–4]: These consist of thin plates of two different materials A and B, bonded together tightly. The materials are selected so as to have vastly different thermal expansion coefficients α_A and α_B (assume that $\alpha_A < \alpha_B$). When the temperature of the assembly changes, the side with the lower α will pull the other side, thus changing the curvature of the assembly. It has been shown that, if an assembly with an initial radius of curvature R_0 at temperature T_0 is exposed to a temperature T, it will be deformed to a radius of curvature R, given by

$$\frac{1}{R} - \frac{1}{R_0} = \frac{6(1+m)^2(\alpha_B - \alpha_A)(T - T_0)}{\delta\left[3(1+m)^2 + (1+mn)(m^2 + 1/mn)\right]}, \tag{12.2}$$

where $m = \delta_A/\delta_B$ is the thickness ratio, $\delta = \delta_A + \delta_B$ is the total thickness, and $n = E_A/E_B$ is the ratio of the Young's moduli of elasticity of the two materials. If the two sides have equal thicknesses, expression (12.2) can be simplified to

$$\frac{1}{R} - \frac{1}{R_0} \approx \frac{3(\alpha_B - \alpha_A)(T - T_0)}{2\delta} \tag{12.3}$$

Because of such deformation, the free end of a bimaterial cantilever beam will be displaced, and the beam will become an actuator. The actuation can be amplified by use of a spiral or a helically coiled assembly. If the free end is linked to a pointer moving across a calibrated scale, the assembly will act as a thermometer. Bimaterial assemblies are also used in thermostatic controls. In most applications, metals are used for both sides, although non-metallic materials, including carbon and semiconductors, have also been used. The relevant properties of materials used in bimaterial thermometers are listed in Table 12.1 [3]. A favourite choice for side A is Invar, a 36% nickel alloy of steel, which, among the known metals and alloys, is the one with the lowest expansion coefficient. Typical resolution of bimetallic thermometers is about 1% of full scale, and their maximum operating temperature is about 500 °C.

Thermocouples [2–4, 10]: Thermocouples are the most common temperature sensors in engineering applications, being inexpensive, simple to use, and robust and having the widest range than any other thermometer. Their operation is based on the *Seebeck effect*, which describes the phenomenon that any electrical conductor will develop a potential difference (*thermoelectric voltage*) between two of its points that have a temperature difference. Although the Seebeck effect applies even to single, homogeneous materials,

Table 12.1. Properties of materials used in bimetallic elements

Material	α (K^{-1})	E (GPa)
Cu	383.1×10^{-6}	129.8
Al	24×10^{-6}	61 to 71
Ni	13.3×10^{-6}	199.5
Fe	12.1×10^{-6}	211.4
Ti	8.9×10^{-6}	120.2
Invar	1.7 to 2.0×10^{-6}	140 to 150
Si	4.7 to 7.6×10^{-6}	113
SiO_2	0.50×10^{-6}	57 to 85

the practical thermocouple configuration is to use two dissimilar metallic wires joined firmly at two junctions by welding, soldering, or pressing. One junction is exposed to the temperature of interest, while the other one, called the *reference junction*, is kept at a known, constant temperature. In open circuit, the voltage difference between the two junctions depends on only the materials and the temperature difference and so can be used for temperature measurement. In closed circuit, a current will flow, which also depends on the resistance of the wires and reduces the thermoelectric voltage. Modern thermocouple circuits are essentially free of current effects. In the not-too-distant past, the reference junction was immersed in an ice bath, conveniently establishing the reference temperature of 0 °C. Currently, thermocouples are usually connected to a heated block, maintained at a constant temperature by electronic control. Commercially available thermocouples are designated by a letter, such as J, E, and K. The properties of some common types are listed in Table 12.2. The sensitivities shown in this table are approximate, intended to serve for comparative purposes only, as thermocouple response is non-linear. Typical resolutions of thermocouples are of the order of 1 °C. For highest accuracy, it is advisable to calibrate individual sensors separately, fitting the response curve by a low-order polynomial. Good-quality commercial thermocouples are within 0.5 to 3 °C from tabulated values supplied by the manufacturer. Several types of common thermocouples have ranges exceeding 1000 °C, with the highest temperature of 2930 °C reached by the tungsten–rhenium type. For measuring higher temperatures, up to about 4000 °C, cooled sensor arrangements have been developed, in which two thermocouples are used to measure temperature differences in a cooling stream inside a tube exposed to

Table 12.2. Common types of thermocouples and their properties

Type	Materials	Range (°C)	Sensitivity (μV/°C)
J	Iron – constantan	0–760	55
E	Chromel – constantan	−200–870	70
K	Chromel – alumel	−200–1260	40
S	Platinum/10% rhodium – platinum	0–1480	8
T	Copper – constantan	−200–370	50

the hot fluid. Typical size of thermocouple junctions is of the order of 1 mm, with time constants of the order of 1 s. Much smaller microjunctions are also offered commercially, with time constants of the order of 1 ms, and frequency response may be further boosted by electronic compensation [11]. The reader is referred to manufacturers' catalogues for specialized thermocouple probe and sensor designs and arrangements.

Resistance temperature detectors (RTDs) [2–4,7,10]: The RTD is a commonly used name for pure metal thermometers. The most popular material is platinum, but nickel, copper, and different alloys are also used. The sensor is wound over a ceramic substrate in the form of a coil or deposited on the substrate as a thin film. RTDs are very stable and accurate. Common types have a typical resolution of about 0.1 K, and are used extensively in the industry and many technological applications, while the Standard Platinum Resistance Thermometer (SPRT), with a resolution of 0.0001 K, is used as a laboratory standard. RTDs are connected in a dc or ac bridge configuration, allowing the measurement of their resistance by passing a small amount of current through them. A limitation of RTDs is their relatively low sensitivity (for platinum this is about 0.4% of its resistance per degree Kelvin), which necessitates the use of signal conditioning and high-gain amplification. The resistance temperature relationships of platinum and other metals are non-linear and are given in the form of fitted low-order polynomials. For example, the resistance R_{pt} of platinum RTD in the range between -100 and 700 °C is commonly described by the *Callendar–Van Dusen equation*

$$\frac{R_{pt}(T)}{R_{pt0}} = 1 + 3.81 \times 10^{-3}T - 6.02 \times 10^{-7}T^2 + 6.0 \times 10^{-10}T^3 - 6.0 \times 10^{-12}T^4, \tag{12.4}$$

where R_{pt0} is the resistance at 0 °C, T is in degrees centigrade, and the numerical coefficients may vary somewhat from one unit to another.

Cold-wires [12]: *Cold-wires* are thin metallic resistance thermometers used for measuring temperature fluctuations in air flows. Their name was established by contrast to hot-wires, which are similar sensors used for velocity measurement. Like hot-wires, cold-wire sensors are typically mounted on two prongs (see Fig. 11.1), either by spot-welding a bare wire or by soldering a Wollaston wire and then etching the silver coating of a central segment. The sensor is powered by a constant-current source [13–15], which permits the measurement of resistance through voltage measurement, following calibration. The current must be sufficiently low to avoid self-heating and sensitivity to velocity fluctuations. Typical current values are 0.5 mA or less, depending on the flow velocity and the intensity of temperature fluctuations. Considering that cold-wire signals are generally very weak, one should strive to keep the current as high as possible to reduce the need for high-gain amplification, which is accompanied by electronic noise. Common sensor materials are platinum, which has the highest sensitivity to temperature fluctuations, and platinum alloys or tungsten, which have higher mechanical strength. Most researchers have used cold-wires with a diameter of 1 μm, but wires with diameters 2.5, 0.6, and 0.25 μm have also been utilized. It is obvious that the finer the sensor is, the better its frequency response would be; however, choice of diameter is also influenced by the need for mechanical strength and ease in handling. As a rough indication, one may consider that clean cold-wires with 1-μm diameter have a high-frequency 3-dB cutoff of about 3 kHz, which increases to about 10 kHz for sensors with 0.25-μm diameters [16–19]. The frequency response generally increases with increasing flow speed but deteriorates with dirt accumulation on the sensor. Frequent dipping of cold-wires in acetone or other solvents to remove coatings is recommended. Frequency compensation or use of more elaborate response relationships can be employed to further improve the temporal resolution of cold-wires [19–21]; such approaches require dynamic calibration of individual sensors, discussed separately later in this section. Cold-wires could be subjected to spatial resolution limitations, in addition to temporal ones, especially when used to measure fine structure properties in high-Reynolds-number turbulence. Theoretical estimates of sensor finite-length effects are available and could be used for estimating the measurement uncertainty and for devising corrections [22–24]. Two additional problems are related to the interference of the prongs [25, 26]. One problem is heat conduction from the sensor to the much larger prongs, which results in distorting the sensor's transfer function in the low-frequency range such that its magnitude attains an intermediate plateau, rather than smoothly approaching a maximum. Another problem may arise from immersion of the sensor in the thermal boundary layers of the prongs; this problem would be accentuated if the sensor were operated in a high-conductivity gas, such as helium. Both adverse effects would be reduced, at the expense of spatial resolution, by an increase in the sensor's length-to-diameter ratio.

Thermistors [2–4, 27]: Thermistors are semiconductor elements whose resistance is a very strong function of temperature. Although thermistors with a positive temperature coefficient (PTC) of resistance, namely increasing resistance with increasing temperature, are also available, it is metal oxide thermistors with a negative temperature coefficient (NTC) of resistance that are used for temperature measurement in the laboratory. Within their range of operation, thermistors offer several advantages over other thermometers; however, they are also subjected to important limitations. A main advantage is their extremely high sensitivity to temperature. As an example, the resistance of a metal oxide thermistor will drop by 7 orders of magnitude within the temperature range between -100 and 400 °C, whereas in the same range, the resistance of a platinum element will increase by 1 order of magnitude. In the room-temperature range, the sensitivity of a thermistor is typically of the order of 10^5 Ω/K, dropping significantly as temperature increases. Another advantage is their insensitivity to aging, electromagnetic fields, and radiation. A calibrated thermistor is likely to remain faithful to its calibration relationship over many years. One disadvantage is the strong non-linearity of its response, which one may reduce by operating it in special bridge configurations. Among the various proposed relationships between thermistor resistance R_{th} and temperature, a fairly accurate one is

$$\frac{1}{T} = A + B \ln R_{\mathrm{th}} + C (\ln R_{\mathrm{th}})^3 \tag{12.5}$$

where A, B, and C are empirical coefficients specific to each unit and T is the absolute temperature. Thermistors are generally not interchangeable and need to be calibrated individually. Their range of operation is limited to a few hundred degrees centigrade and their time constant typically exceeds 1 s. To measure its resistance, one has to supply the thermistor with some current i, supplied by a constant-current source. If the current exceeds a certain level, the dissipated electric power is sufficient to raise the thermistor temperature above the ambient level (self-heating effect). This will result in a decrease of its resistance and an error in the voltage output $i R_{\mathrm{th}}$. In fact, if one keeps increasing the current, there will be a threshold above which the voltage output will start decreasing rather than increasing. Thermistors are available commercially in different forms and shapes, including bare and glass-coated beads, disks, rods, washers, and different types of assemblies. Beside their use as thermometers, they find extensive application as parts of control circuits.

12.3 Dynamic response of thermometers

Simple models: When a thermometer is exposed to a time-dependent temperature field, it will indicate a measured temperature T_m, which, in general, would be different from the actual local temperature T. To determine the relationship between these two temperatures, one needs to devise a model of the thermometer's operation. Before proceeding with some examples, it seems useful to recall that heat transfer always takes place from a warmer to a colder object. Therefore there is no thermal property analogous to mechanical inertia or electrical inductance, and a thermometer will not overshoot above the maximum temperature it has been exposed to. The commonly used term 'thermal inertia' actually indicates a lag in thermal response, and it is unrelated to mechanical inertia. Of course, when thermal sensors are parts of electric or electronic circuits, one would have to consider the dynamic response of the entire system and not the thermal response of the sensor alone.

Let us first use the lumped parameter approach to model a simple thermometer as having a volume V, an area A in contact with the surrounding fluid, a density ρ, and a specific heat c. Further assume that the fluid has a uniform temperature T, the thermometer has a uniform temperature T_m, and the overall convective heat transfer coefficient from the fluid to the thermometer is h. Then the energy equation, combined with Newton's law of convection, gives the heat transfer during time δt as

$$hA(T - T_m)\,\delta t = c\rho V \delta T_m, \tag{12.6}$$

which, in the limit $\delta t \to 0$, leads to the first-order differential equation

$$\frac{c\rho V}{hA}\frac{\mathrm{d}T_m}{\mathrm{d}t} + T_m = T. \tag{12.7}$$

Thus the thermometer would respond to flow-temperature changes as a first-order system with a time constant

$$\tau = \frac{c\rho V}{hA}. \tag{12.8}$$

Clearly the time constant depends on the heat transfer coefficient, which generally increases with increasing flow speed. Thus the dynamic response of a thermometer would be better in fast streams than in slow streams or still fluids. The assumption of uniform temperature within the thermometer would be valid only if its thermal conductivity k were sufficiently high for the *Biot number* to be [28]

$$\frac{hV}{kA} \ll 1. \tag{12.9}$$

Further deviations from first-order response would occur if the sensing element were surrounded by a coating, sheath, well, or the like, which is a common occurrence in temperature measurement. To illustrate this case, consider a thermistor bead with properties as described in the previous example and enclosed within a glass coating, having a volume V_g, an area A_g exposed to the surrounding fluid, a density ρ_g, and a specific heat c_g. Using the lumped parameter approach, assume that the temperature T_g of the glass coating is uniform and that the overall heat transfer coefficient from the fluid to the glass coating is h_g, whereas the overall heat transfer coefficient from the glass coating to the thermistor is h. Then the heat transfer from the flow to the glass–thermistor assembly and from the glass coating to the thermistor can be respectively modelled as

$$\frac{c_g \rho_g V_g}{h_g A_g} \frac{\mathrm{d}T_g}{\mathrm{d}t} + T_g + \frac{c\rho V}{h_g A_g} \frac{\mathrm{d}T_m}{\mathrm{d}t} = T, \tag{12.10}$$

$$\frac{c\rho V}{hA} \frac{\mathrm{d}T_m}{\mathrm{d}t} + T_m = T_g. \tag{12.11}$$

Eliminating T_g between these equations, one gets the second-order equation [4]

$$\tau \tau_g \frac{\mathrm{d}^2 T_m}{\mathrm{d}t^2} + \left(\tau + \tau_g + \tau_c\right) \frac{\mathrm{d}T_m}{\mathrm{d}t} + T_m = T, \tag{12.12}$$

which, in addition to the thermistor time constant τ, contains the coating time constant

$$\tau_g = \frac{c_g \rho_g V_g}{h_g A_g} \tag{12.13}$$

and a coupling time

$$\tau_c = \frac{c\rho V}{h_g A_g}. \tag{12.14}$$

Although the dynamic response of the coated sensor is of second order, its step response would never be oscillatory, because, as can be easily proved, any system described by Eq. (12.12) would be overdamped. Third-order and non-linear models of thermometers have also been proposed in the literature [4].

Dynamic calibration: Although analytical expressions like (12.8), (12.13), etc., are useful for rough estimates of the dynamic response of thermometers as well as a guide for its improvement, precise determination of the time constant of a particular sensor can be achieved only by dynamic calibration. Furthermore, dynamic calibration can also reveal possible deviations from first-order response and the need for a more complex model. Among the various sensor heating methods that have been used for dynamic calibration purposes, one could mention these:

Self-heating technique. This is the easiest method to implement, but it is suitable only for thermometers sensing temperature through a change in their electric resistance, which includes metallic resistance thermometers and thermistors. The sensor is exposed to a flow stream, and a low-amplitude square-wave current is superimposed to the steady current that is used to measure its resistance. A sudden change in the Joule heating of the sensor would result in a gradual change of the thermometer output, from which the time constant can be determined. Alternatively, a sinusoidal current is applied to the sensor and its frequency is varied, while its amplitude is kept constant. The frequency response of the thermometer (in amplitude only) can be determined as the ratio of the standard deviation of its output at a given frequency and that at extremely low frequencies. Either process can be repeated at different flow speeds to determine their possible effect on the dynamic characteristics.

External heating technique. This is the most straightforward method and is suitable for all types of thermometers. One approach is to immerse the thermometer, suddenly or periodically, into a steady heated stream, such as a jet, wake, or buoyant plume, by use of a fast-action linear or rotating mechanism, and then to determine its step response or frequency response. An alternative approach is to keep the sensor fixed and interrupt the flow (e.g., by interrupting a heated jet with the use of a rotating perforated disk) or its heating (e.g., by putting the sensor in the wake of a thin heating wire, supplied with a pulsed current [18]). Obviously, *in situ* dynamic calibration would be desirable, but this requires that the time-dependent temperature field be measured by an accurate, fast-response method that would serve as calibration standard. An example is the dynamic calibration of fine thermocouples in turbulent flows, with a cold-wire serving as the standard [11]. In such cases, dynamic calibration may be assisted by the use of cross spectra, cross correlations, and joint probability densities of the two temperature signals.

Laser heating technique. In this method the thermometer is heated by the beam of a pulsed laser or the chopped beam of a CW laser [20, 29]. Visible or infrared laser radiation can be used, and the beam may be focussed to increase its intensity. Because only part of the sensor may be heated, one would have to determine whether complications may arise because of heat conduction within the sensor or to its supports.

Compensation: Quite commonly, the dynamic response of thermometers is inadequate to measure accurately temperature variations in unsteady and turbulent flows. Improved accuracy can be achieved with the use of compensation. For example, if the time constant of a first-order temperature sensor is known, one can estimate the actual flow temperature T from the sensor's reading T_m by using Eq. (12.7). This can be done by analogue components, which include a differentiator to provide the temperature derivative, an amplifier, and an adder [20]. Better yet, one can perform these operations digitally, from discrete-time histories of the temperature signal, by estimating the derivative as a finite difference, preferably by a second- or higher-order difference method [30].

12.4 Thermochromic materials

Thermochromic materials are materials that, when illuminated appropriately, emit visible radiation that changes colour at a certain temperature. Such materials are used for visualization purposes (see Section 10.2), but, with proper calibration, can also be used for mapping isotherms and measuring local temperature as well. In most applications, the thermochromic material is applied on a surface in the form of a coating, but thermochromic suspensions have also been used for in-flow temperature mapping in liquids. Two main types of thermochromic materials are available: *temperature-sensitive paints*, having temperature ranges up to several hundred degrees centigrade, and *liquid crystals*, usable in ranges near room or body temperatures.

Temperature-sensitive paints [6, 31]: The term *TSP* describes coatings that contain fluorescent substances whose luminescence is sensitive to temperature. Their operating principle is the same as that for PSPs and was discussed in Section 8.3. To minimize sensitivity to pressure by oxygen quenching, TSPs utilize coatings that are impenetrable to oxygen. Furthermore, according to approximation (8.15), for increased temperature sensitivity, materials with a relatively low activation energy ΔE would be desirable. Typical variations in luminescence of TSPs are in the range of a few percentage units per degree centigrade. A number of recipes for luminophore compounds and binder combinations have been proposed in the literature.

A different type of thermochromic paint has been used as a temperature indicator in aerodynamics research [32] as well as in a variety of industrial and commercial applications [3]. These paints consist of metallic compound powders dispersed in an acrylic lacquer, that is applied as a coating to the heated surface. When the paint is exposed to a triggering temperature for a certain time, a chemical reaction occurs that causes it to change colour, and some materials may undergo multiple colour changes as temperature rises. The temperature value can be determined from appropriate calibration in a temperature-controlled oven. Several such materials are available commercially, and their properties are regulated by national and international standards. For some paints the colour change is permanent, whereas for others it is reversible. The temperature range of some paints reaches 1300 °C and their uncertainty ranges between 1 and 5 °C. Besides paints, crayons are also available, which may be easily applied on surfaces. Temperature indicators also include materials that undergo phase change or deform at certain temperatures and are available in a variety of shapes [3, 33].

Liquid crystals [34, 35]: *Liquid crystals* (*LCs*), or *mesophases*, are certain organic compounds whose molecular structure and optical properties are intermediate between those of conventional liquids, which are optically isotropic, and crystalline solids, which are anisotropic. Among the different types available, the ones suitable for temperature measurement are the *thermotropic* or *thermochromic* LCs. These have both a pure crystalline state and an amorphous liquid state, but, within a range of temperatures, they also have an anisotropic liquid state. Mixtures of LCs with other LCs or other materials may also be optically anisotropic. Within their range of operation, LCs have a colour that changes with temperature and could therefore be used for temperature mapping, either applied to a surface as a coating, or suspended in a liquid medium, in which case they can also serve as flow markers. Compared with TSPs, LCs have higher temperature, spatial, and temporal resolutions.

Among the three general categories of LC materials, known as *smectic* (soap-like), *nematic* (thread-like), and *cholesteric* (containing cholesterol) or *chiral nematic*, it is the latter that are thermochromic. Nematic LCs have long molecules that are oriented in approximately the same direction, although not as perfectly structured as crystals. Cholesteric materials consist of parallel nematic layers, about 3 Å thick, with each layer containing molecules oriented at a slightly different angle with respect to the previous one, in a helical fashion. A linearly polarized beam of light propagating through such a material would rotate its plane by as much as 50 rotations/mm thickness. Cholesteric LCs also decompose white light into clockwise- and counterclockwise-rotating components (*dichroism*) and transmit one component while reflecting the other, with the reflection properties changing with temperature, within a certain range. Therefore their colour appears to change, at a certain temperature, but the process also depends on electromagnetic field strength, shear stress, pressure, and angles of illumination and observation.

To avoid chemical reactions with the surrounding fluids, LCs are usually enclosed in microcapsules (5–30 μm in diameter), having a gelatinous shell and suspended in a liquid in the form of a slurry. A common approach is to apply the slurry on a surface by a brush or other means and to let it dry. Single layers should have a thickness of about 50 μm, but multiple layers of different LCs can also been applied consecutively. The surface must be painted black before application to ensure contrast. LCs are also available in sheet form, about 100 μm thick, which can be bonded to a surface. Finally, LC slurries can be also suspended in water or other liquids, usually at very small concentrations, typically less than 0.1%.

Each LC is characterized by its *event temperature range*, namely the temperature range at which it is optically active. The event temperature ranges of different LCs cover temperatures between 0 and 100 °C. The width of the event temperature range of a given LC is called the *ambiguity bandwidth*, typically ranging from 1 to 15 °C. As the temperature increases and a LC spans its event temperature range, its colour changes from the infrared to red, through the visible colours to blue, and then to ultraviolet. To obtain a sharp isotherm, one may use monochromatic illumination or white light and band-pass optical filters. The use of multiple filters or the mixing of LCs with different, non-overlapping, event temperature ranges could provide several isotherms at different temperatures. Continuous variation of temperature can be resolved by use of LCs with a wide ambiguity band, colour photography, and digital image processing. Although colours appear in brilliant forms, their succession is highly non-linear with temperature. LCs may be calibrated vs. the readings of thermocouples embedded in the surface underneath them or by exposing the surface to a uniform, constant, or slowly varying, temperature [36]. Although the time constant for colour change of common LCs is relatively small, typically of the order of 0.1 s, the response of applied layers is considerably slower and depends on the layer thickness, the thermophysical properties of the materials, and the heat transfer coefficient. Typical cutoff frequencies in practical applications are of the order of 0.1 Hz. When proper care is taken for their calibration and use, the temperature resolution of LCs is very good, reaching values as small as 0.1 K. Another advantage of liquid crystals is that their operation is reversible and can be used repeatedly to map temperature fields. In combination with in-flow velocity measurement methods, surface-coated LCs can be used to measure convective heat transfer [37].

12.5 Radiation emission methods

Pyrometry: Literally, the term *pyrometry* means measurement in fires. *Optical pyrometers* [2–8] are radiation thermometers, which measure temperature of objects through thermal radiation laws (see Section 5.3). Compared with other thermometers, they have the advantages of non-contact operation and application to much higher temperatures. A variety of instruments is offered by different manufacturers, often employing proprietary designs.

The classical and most accurate pyrometer design is the *disappearing-filament optical pyrometer*, also known as *monochromatic-brightness pyrometer*. In its original version, the instrument required manual adjustments by the user. First, the user observed a small area of the 'target', whose temperature was to be measured, through an eyepiece. The image of an electrically heated, incandescent tungsten filament was superimposed on the target image and the combined image was viewed through an optical band-pass filter, which removed all radiation but a narrow band in the red part of the spectrum ($\lambda_p = 0.655\ \mu$m). The user adjusted the current through the filament until its image was no longer distinguishable from the background, which indicated that the spectral radiance of the filament matched that of the target. The instrument was calibrated to provide as output the *brightness temperature* T_B, defined as the temperature of a blackbody with the same spectral radiance as that of the filament. If the target was not a blackbody, its temperature T would have to be higher, to compensate for a lower emissivity. Wien's radiation law [Eq. (5.28)] provides the equality

$$L_{e\lambda} = \frac{2\pi hc^2}{\lambda_p^5} e^{-C_2/(\lambda_p T_B)} = \frac{2\pi hc^2 \varepsilon_\lambda}{\lambda_p^5} e^{-C_2/(\lambda_p T)}, \tag{12.15}$$

which leads to a relationship between the two temperatures as

$$\frac{1}{T} = \frac{1}{T_B} + \frac{\lambda_p}{C_2} \ln \varepsilon_\lambda, \tag{12.16}$$

where ε_λ is the monochromatic emissivity of the target for $\lambda_p = 0.655\ \mu$m and $C_2 = 0.01439$ m K. Notice that the temperature sensitivity to ε_λ is relatively low because of the latter's appearance in the logarithm. For this reason, an uncertainty in the determination

Table 12.3. Monochromatic emissivity of some materials

Material	ε_λ
C	0.80–0.93
Cu	0.10 (solid), 0.15 (liquid)
Fe or steel	0.35 (solid), 0.37 (liquid)
Pt	0.30 (solid), 0.38 (liquid)
Ag	0.07 (solid or liquid)

of ε_λ would result in a much lower relative uncertainty in T. Representative values of ε_λ for common materials with unoxidized surfaces are listed in Table 12.3 [7]. For many materials, however, or for different surface conditions, the emissivity is not known and the preceding correction method cannot be applied.

The manual pyrometer has been superseded by automated versions, which utilize a PMT to sense differences in the brightness of the filament and target images and to adjust the current to enforce equality of these brightnesses. A more recent version is the *two-colour pyrometer*, which measures the ratio of spectral radiances $L_{e\lambda 1}$ and $L_{e\lambda 2}$ at two wavelengths λ_1 and λ_2. Assuming that the corresponding monochromatic emissivities are $\varepsilon_{\lambda 1}$ and $\varepsilon_{\lambda 2}$, one may use Wien's radiation law to estimate the target temperature as

$$\frac{1}{T} = \frac{1}{C_2\left(\frac{1}{\lambda_1} - \frac{1}{\lambda_2}\right)}\left(\ln\frac{L_{e\lambda 1}}{L_{e\lambda 2}} + 5\ln\frac{\lambda_1}{\lambda_2} + \ln\frac{\varepsilon_{\lambda 1}}{\varepsilon_{\lambda 2}}\right). \tag{12.17}$$

If the monochromatic emissivities at the two wavelengths can be assumed to be equal, as in the case of grey bodies, then the preceding expression would become independent of the emissivity value. Even if this is not the case, one would need to estimate only approximately the emissivity ratio, as the uncertainty in this ratio plays a minor role in the accuracy of the temperature measurement.

When the absorption of radiation by the medium between the pyrometer and the target is not negligible, it would be necessary to replace the monochromatic emissivity in the preceding expressions with its product with the *monochromatic transmissivity* $\tau_\lambda \leq 1$, defined as the ratio of the radiation transmitted through the medium at a given wavelength and the emitted radiation.

Infrared thermography: *Infrared thermography* [2,3,5,8,38,39] is a non-contact technique for the measurement of surface temperature of an object by measurement of the infrared radiation emitted by it. Infrared radiation detectors providing point measurements of temperature are available; however line- and area-scanning instruments are more usable. A common device is the *infrared scanning radiometer* (*IRSR*), which provides two-dimensional images of the surface temperature distribution, called *thermograms*. A typical IRSR is a camera containing two mirrors or prisms, which oscillate on planes normal to each other to scan the image, and an infrared radiation detector, which converts the radiation to an electronic video signal. *Staring-type infrared cameras* are also available. These have no moving parts but contain a two-dimensional array of detectors that integrate the oncoming radiation over a small interval and provide this information to a CCD array. To convert the received radiation energy to surface temperature, one must know the emissivity of the surface. Further errors would be produced if radiation is absorbed or reflected along its path from the target to the detector, which includes losses in the intervening fluids, test-section windows, and camera windows. The transmissivity of air is a strong function of radiation wavelength. Two spectral bands have been identified as matching a high air transmissivity and strong response of available infrared detectors: the 3.5–5-μm range, called the 'short-wave band' and the 8–12-μm range, called the 'long-wave band' . These are the two bands that are utilized in most infrared cameras. Glass and quartz have a low transmissivity to infrared radiation and are not suitable as camera windows. Materials with high transmissivity in the ranges of interest include sapphire, germanium, and zinc selenide. The temperature resolution of infrared cameras can be characterized by two parameters: the *noise-equivalent temperature difference*, which is the temperature difference that would create a change in output voltage equal to the peak-to-peak noise voltage, and the minimum resolvable temperature difference, which is the lower limit of temperature difference between a background surface and an array of strips for which the strips can be discriminated visually from the background. Infrared cameras are calibrated vs. a blackbody radiation source at a known temperature. Because of the uncertainty in the values of the surface emissivity and the transmissivity of the radiation path, it is a sound practice to perform *in situ* calibration, for example by comparing the camera output to the reading of surface mounted thermocouples. Infrared thermography is used extensively in many industrial applications. In the laboratory, it is a valuable tool for heat transfer experiments and for high-speed flows, especially in the hypersonic range, in which temperature differences on immersed objects are generated spontaneously. For low-speed flows, heating of the surface electrically or by other means can be used to provide thermograms that illustrate surface flow patterns, such as flow separation and transition to turbulence, through their effect on the local heat transfer coefficient.

Laser-induced fluorescence (LIF): Although LIF methods are mainly used for the measurement of concentration in liquids and gases (see Section 13.5), the dependence of fluorescence on temperature can be used for temperature measurement. This approach has been applied to the measurement of temperature fluctuations in water by use of Rhodamine B, whose fluorescence quantum efficiency decreases, almost linearly, with temperature at a relatively strong rate (2.3%/K at 20 °C). The dye is mixed with the water at a uniform concentration, a laser light sheet is used to activate the process, and images are recorded with a digital camera or by other suitable means. If the dye concentration is kept at very low levels and the water is free of other substances, light absorption in its path through the water can be neglected and the radiation received by each element of the camera would be proportional to the local quantum efficiency. Because of spatial, and possibly temporal, variation of light intensity on the light sheet, and other disturbances along the light path, pixel-by-pixel calibration would be required. An alternative approach is to monitor undesirable local light intensity variations by measuring simultaneously the local emission by a second fluorescent dye (e.g., Rhodamine 110), whose fluorescence quantum efficiency sensitivity to temperature is negligible [40].

12.6 Optical techniques

The optical techniques described in Section 10.3 for flow visualization purposes are based on the variation of the refractive index of the fluid medium, which is directly related to the fluid density. Under experimental conditions such as density variation that is caused by temperature variation alone, one could, in principle, use these optical techniques to measure the local fluid temperature. In practice, however, the resolution of shadowgraphs is insufficient to permit the accurate double integration required for determining temperature from the measured light intensity variation, so that the shadowgraph technique remains a strictly qualitative one. The resolution of high-quality Schlieren images could be adequate to determine temperature by single integration, following proper calibration. On the other hand, interferograms would directly provide isotherms, from which one could easily extract temperature values, provided that temperature at a reference location is known by other means. An example of this procedure was illustrated by Eq. (10.27). For details and references, the reader is referred to Section 10.3.

QUESTIONS AND PROBLEMS

1. A liquid-in-glass thermometer is calibrated in a dual-jet calibration facility such that it is alternately immersed in either of two air streams with different temperatures. Assume that the air temperature surrounding the thermometer during calibration varies sinusoidally between a minimum of 30 °C and a maximum of 50 °C with a period of 8 s. The reading of the thermometer fluctuates between a minimum of 35 °C and a maximum of 45 °C.
   a. Determine the thermometer's time constant.
   b. Determine the time difference between the occurrence of maximum thermometer reading and the occurrence of maximum airstream temperature.
2. Compute the slope $\mathrm{d}R_{\mathrm{pt}}/\mathrm{d}T$ of Callendar–Van Dusen equation (12.4) at $T = 0$ °C. Assuming that a linear expression with the same slope describes the entire resistance temperature relationship for a platinum RTD, compute the error in temperature computed from the linear expression and plot it vs. T for the range between -100 and 700 °C.
3. Calibration of a thermistor probe provided the following data:

T (°C)	20.0	40.0	60.0
R (Ω)	3440	1280	512

Determine the coefficients A, B, and C in Eq. (12.5). In semi-logarithmic axes, plot the thermistor resistance vs. temperature for the extended range between -20 and 100 °C. Then compute and plot the sensitivities $\mathrm{d}R/\mathrm{d}T$ and $\mathrm{d}T/\mathrm{d}R$ over the extended range. Compare the values of these sensitivities for $T = -20$, 20, and

100 °C. Compute the maximum error that would be produced in the range of the calibration data if a linear expression were fitted to these data.

4. Consider a bimetallic element, 30 mm long and consisting of a plate of copper, 2 mm thick, bonded on a plate of Invar, 1 mm thick. The element is clamped at one end and straight when the ambient temperature is 20 °C. Determine the deflection of its free end when the temperature reaches 40 °C.
5. Prove that the system described by Eq. (12.12) is overdamped.
6. Compare the ranges of applications of thermistors, thermocouples, and cold-wires. Give examples of temperature measurements for which each of the three methods would be suitable, but the other two would not.
7. Provide an example of an experiment for which LCs would be preferable to a thermocouple array and another example for the opposite case.
8. Provide an example of an experiment for which thermocouples would be preferable to thermistors and another example for the opposite case.

REFERENCES

[1] J. F. Schooley. *Thermometry*. CRC Press, Boca Raton, FL, 1986.

[2] J. G. Webster (Editor-in-Chief). *The Measurement, Instrumentation and Sensors Handbook*. CRC Press, Boca Raton, FL, 1999.

[3] J. G. Webster (Editor). *Mechanical Variables Measurement: Solid, Fluid and Thermal*. CRC Press, Boca Raton, FL, 2000.

[4] E. O. Doebelin. *Measurement Systems Application and Design* (5th Ed.). McGraw-Hill, New York, 2004.

[5] F. Mayinger and O. Feldmann (Editors). *Optical Measurements* (2nd Ed.). Springer, Berlin, 2001.

[6] C. Mercer (Editor). *Optical Metrology for Fluids, Combustion and Solids*. Kluwer Academic, Dordrecht, The Netherlands, 2003.

[7] R. P. Benedict. *Fundamentals of Temperature, Pressure and Flow Measurements* (2nd Ed.). Wiley-Interscience, New York, 1977.

[8] T. Arts et al. *Measurement Techniques in Fluid Dynamics* (2nd Ed.). Von Karman Institute for Fluid Dynamics, Rhode-Saint Genese, Belgium, 2001.

[9] M. W. Zemansky. *Temperatures Very Low and Very High*. Dover, New York, 1964.

[10] Omega Engineering. *The Temperature Handbook*. Omega Engineering, Inc., Stamford, CT, 2001.

[11] R. Talby, F. Anselmet, and L. Fulachier. Temperature fluctuation measurements with fine thermocouples. *Exp. Fluids*, 9:115–118, 1990.

[12] H. H. Bruun. *Hot-Wire Anemometry*. Oxford University Press, Oxford, UK, 1995.

[13] S. Tavoularis. A circuit for the measurement of instantaneous temperature in heated turbulent flows. *J. Phys. E*, 11:21–23, 1977.

[14] R. Peattie. A simple, low-drift circuit for measuring temperatures in fluids. *J. Phys. E*, 20:565–567, 1987.

[15] J. Haugdahl and V. Lienhard. A low-cost, high performance DC cold-wire bridge. *J. Phys. E*, 21:167–170, 1988.

[16] J. Hojstrup, K. Rasmussen, and S. E. Larsen. Dynamic calibration of temperature wires in still air. *DISA Inf.*, 20:22–30, 1976.

[17] J. C. LaRue, T. Deaton, and C. H. Gibson. Measurement of high frequency turbulent temperature. *Rev. Sci. Instrum.*, 46:757–764, 1975.

[18] R. A. Antonia, L. W. B. Browne, and A. J. Chambers. Determination of time constants of cold wires. *Rev. Sci. Instrum.*, 52:1382–1385, 1981.

[19] J. Lemay and A. Benaissa. Improvement of cold-wire response for measurement of temperature dissipation. *Exp. Fluids*, 31:347–356, 2001.

[20] A. R. Weeks, J. K. Beck, and M. L. Joshi. Response and compensation of temperature sensors. *J. Phys. E*, 21:989–993, 1988.

[21] P. V. Vukoslavcevic and J. M. Wallace. The simultaneous measurement of velocity and temperature in heated turbulent air flow using thermal anemometry. *Meas. Sci. Technol.*, 13:1615–1624, 2002.

[22] J. C. Wyngaard. Spatial resolution of a resistance wire temperature sensor. *Phys. Fluids*, 14:2052–2054, 1971.

[23] S. Larsen and J. Hojstrup. Spatial and temporal resolution of a thin-wire resistance thermometer. *J. Phys. E,* 15:471–477, 1982.

[24] J. C. Lecordier, A. Dupont, P. Gajan, and P. Paranthoën. Correction of temperature fluctuation measurements using cold wires. *J. Phys. E,* 17:307–311, 1984.

[25] A. E. Perry, A. J. Smits, and M. S. Chong. The effects of certain low frequency phenomena on the calibration of hot wires. *J. Fluid Mech.*, 90:415–431, 1979.

[26] P. Paranthoën, C. Petit, and J.C. Lecordier. The effect of thermal prong-wire interaction on the response of a cold wire in gaseous flows (air, argon and helium). *J. Fluid Mech.*, 124:457–473, 1982.

[27] Fenwal Electronics. *Thermistor Manual*. Fenwal Electronics, Inc., Milford, MA, 1974.

[28] J. P. Holman. *Heat Transfer* (9th Ed.). McGraw-Hill, New York, 2002.

[29] R. Budwig and C. Quijano. A new method for in situ dynamic calibration of temperature sensors. *Rev. Sci. Instrum.*, 60:3717–3720, 1989.

[30] M. J. Downs, D. H. Ferriss, and R. E. Ward. Improving the accuracy of the temperature measurement of gases by correction for the response delays in the thermal sensors. *Meas. Sci. Technol.*, 1:717–719, 1990.

[31] T. Liu, T. Campbell, S. Burns, and J. Sullivan. Temperature and pressure sensitive luminescent paints in aerodynamics. *Appl. Mech. Rev.*, 50:227–246, 1997.

[32] W. Merzkirch. *Flow Visualization* (2nd Ed.). Academic, New York, 1987.

[33] E. Kimmel. Temperature sensitive materials. *Meas. Control*, October 1979:98–103, 1979.

[34] N. Kasagi, R.J. Moffat, and M. Hirata. Liquid crystals. In W.-J. Yang, editor, *Handbook of Flow Visualization*, pp. 105–116. Taylor & Francis, Washington, DC, 1989.

[35] A. J. Smits and T. T. Lim (Editors). *Flow Visualization Techniques and Examples*. Imperial College Press, London, 2000.

[36] D. R. Sabatino, T. J. Praisner, and C. R. Smith. A high-accuracy calibration technique for thermochromic liquid crystal temperature measurements. *Exp. Fluids*, 28:497–505, 2000.

[37] T. J. Praisner, D. R. Sabatino, and C. R. Smith. Simultaneously combined liquid crystal surface heat transfer and PIV flow-field measurements. *Exp. Fluids*, 30:1–10, 2000.

[38] T. Astarita, G. Cardone, G. M. Carlomagno, and C. Meola. A survey of infrared thermography for convective heat transfer measurements. *Opt. Laser Technol.*, 32:693–610, 2000.

[39] W.-J. Yang (Editor). *Handbook of Flow Visualization*. Taylor & Francis, Washington, DC, 1989.

[40] J. Sakakibara and R. J. Adrian. Whole field measurement of temperature in water using two-color laser induced fluorescence. *Exp. Fluids*, 26:7–15, 1999.

13 Measurement of composition

This chapter describes methods of identifying the components of a fluid mixture and their relative proportions (concentrations). Much of the interest focusses on the products of combustion and other chemical reactions because of our concern for their efficiency and the production of pollutants. Additional topics include the measurement of size and number density (particles per unit volume) of solid and liquid particles suspended in fluids and the measurement of the percentage of volume occupied by the gas phase (void fraction) in gas–liquid flows. Only an introductory treatment of composition measurement methods can be afforded here, as in-depth understanding of some of these methods requires an extensive background in chemistry, physics, and other fields, which is beyond the scope of the present exposition. Many composition measurement methods, especially those based on radiation absorption and emission, were developed for application in chemistry, physics, biochemistry, biology, and medical sciences. We are not concerned with such aspects here; instead, we shall focus on only aspects relevant to fluid mechanics research.

13.1 Sample analysis

Sampling is the removal of a volume of the fluid through a tube or sampling probe for subsequent analysis. It is necessarily intrusive and subjected to several limitations. First, excessive or inadequate suction applied during sampling may bias the sample in favour of heavier or lighter components, thus distorting the measured composition of mixtures and multiphase flows. For this reason, it is advisable for sampling to be applied *isokinetically*, which means that the sampling tube should be aligned with the flow direction and the sampling speed inside the tube, corrected for viscous effects and probe wall thickness, should be regulated to match the flow speed. In the case of reactive flows, one must consider possible changes of composition within the sampling instrumentation, as a result of continuing reactions. Similarly, errors in composition measurement may result from condensation or evaporation of some components within the sampling tubing. When this is a possibility, it may be necessary to control the temperature of the sampling tube and subsequent parts of the apparatus. Other precautions and controls may also be necessary, depending on the particular experimental conditions. In general, sampling is

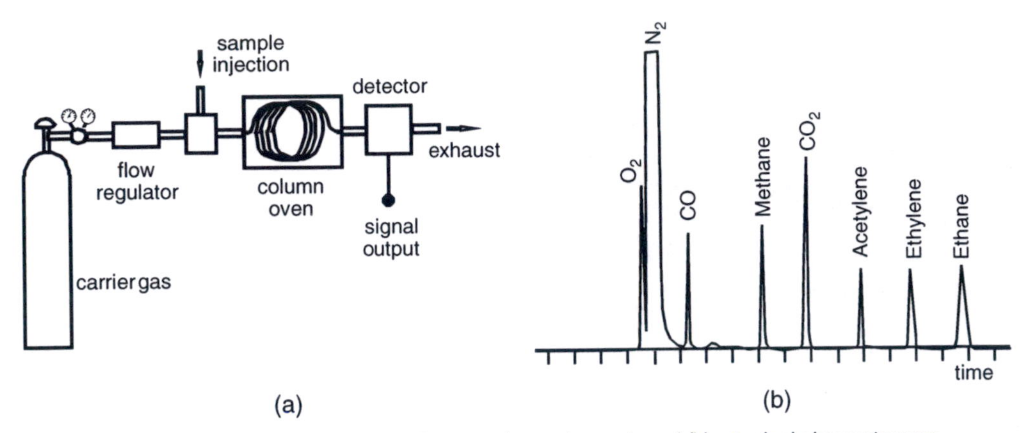

Figure 13.1. Sketches of (a) a representative gas chromatograph and (b) a typical chromatogram.

a relatively slow process, with typical response times of the order of 1 s or more. Much faster sampling rates have been achieved with the use of sampling probes utilizing solenoid valves or piezoelectric transducers [1]. A few representative sample analysis methods for composition measurement are subsequently described.

Orsat analyser: This is a classical, rather old-fashioned, device used for measuring the combustion efficiency of fossil fuels. A sample of the exhaust gas is injected into a series of graduated glass containers, partly filled with liquid reagents, each of which reacts with a particular gas, for example O_2, CO_2, or CO. Absorption of each gas produces an easily measurable change in volume of the corresponding reagent. This method is simple and inexpensive, but also manual, and therefore slow and tedious.

Gas chromatography: *Gas chromatography* [2–5] is a method for the separation and analysis of gas and volatile liquid mixtures by *eluting* (i.e., removing by dissolving) its components from a heated column packed with a liquid or granular solid sorbent. The main components of a simple gas chromatograph are illustrated in Fig. 13.1(a). A small gas sample is injected into a stream of an inert carrier gas, usually helium or nitrogen, which transports it through the column. Each component of the mixture is retained by the sorbent for a certain time and exits the column with a certain time lag, depending on the affinity of the component to the sorbent and on the carrier flow rate. CO_2, unsaturated hydrocarbons, O_2, and CO can be eluted sequentially from a certain sorbent. A practical gas chromatograph also contains a detector, which identifies the components. A common detector type is the thermal conductivity detector. It contains two heated elements, one in the analysed stream and one in the pure carrier gas. The presence of a component is detected from the change in the electric resistance of the element that is due to its temperature change caused by a change in thermal conductivity. The resistance change is sensed by a bridge, which produces an electric signal output, which is recorded or displayed on a potentiometric strip chart recorder. Other types of detectors have also been utilized, including flame ionization and photoionization detectors. A typical *chromatogram*, shown in Fig. 13.1(b), appears as a sequence of peaks, each associated with a gas component and separated by specific time intervals. The components of the mixture are identified by the timing of the peaks, and their concentrations can be found by measurement of the areas under each peak.

Electronic testers: These are specialized instruments measuring the concentrations of various gases, particularly products of combustion. They contain *electrochemical gas sensors*, each measuring the concentration of a particular type of molecule. For example, a zirconium oxide cell is used to measure oxygen concentration because of its capacity to develop a voltage difference across its two sides when exposed to differences in oxygen concentration. Testers of CO and other gases utilize disposable chemical capsules, which change colour when the particular gas is drawn through them.

Continuous-emission monitors: Current pollutant emission regulations require continuous monitoring of exhaust gases from the stacks of industrial plants. This is achieved with a variety of electronic instruments, referred to as *continuous-emission monitors* (*CEMs*). Both extractive and *in situ* CEMs are available. The three main methods of sample analysis are absorption spectroscopy, luminescence techniques, and electroanalytical methods. Detailed descriptions of CEMs can be found in specialized references, for example Ref. 6. In the following subsections, some basic principles of absorption spectroscopy and spectrophotometry are briefly discussed.

Absorption spectrophotometry: When radiation passes through a gas, it gets absorbed according to Beer's law (see Section 5.2). If the gas consists of a single chemical species, this law can be written in the form [2]

$$A = \log \frac{I_0}{I} = \varepsilon l C \tag{13.1}$$

where A is the *absorbance*, I_0 and I are the incident and transmitted radiant intensities (see Section 7.3), respectively, ε is the *molar absorptivity* coefficient of the species, whose value depends on the radiation wavelength, l is the length of the absorption path in the gas, and C is the concentration of the molecules, expressed as moles per unit volume. Thus, by measuring A, it is possible to compute C. In gases containing two or more types of molecules, the total absorbance would be the sum of the individual absorbances, as

$$A = l \sum_{i=1}^{N} \varepsilon_i C_i. \tag{13.2}$$

Because each type of molecule preferentially absorbs radiation at particular wavelengths corresponding to its vibrational and rotational energy levels [2, 7–9] (see Section 5.4), one may analyse the spectrum of the absorbed radiation to identify a particular molecule. Furthermore, by using radiation of different wavelengths, one would, at least in principle, be able to compute the concentrations of the components of a gas mixture by solving a system of equations of the type shown by Eq. (13.2). In practice, the applicability of this method is restricted by various effects, such as deviation from Beer's law that is due to large species concentrations [2]. A variety of *spectrophotometers* are available, operating in the ultraviolet, visible, or infrared ranges.

Mass spectrometry: *Mass spectrometry* refers to methods that utilize electric or magnetic fields to separate ions according to their mass or charge. These methods may also be applied to neutral species, provided that they are first ionized. A *mass spectrometer* consists of a sampling system, an ion source, a mass analyser, and an ion detector. Mass spectroscopy is used extensively in several fields of chemistry, physics, and biology. Its main application in fluid mechanics is in the study of shock waves and flames. Details on related methods and references can be found in Ref. 2. Mass spectroscopy has also been utilized in association with *molecular-beam sampling* to measure species concentrations in rarefied gas flows [1]. In these applications, a gas sample is converted to a molecular beam, which is ionized before been analysed by a mass spectrometer.

13.2 Thermal probes

Convective heat transfer from a heated object immersed in a fluid mixture depends on the thermal conductivity of the fluid, which, in turn, depends on the mass fractions of its constituents. Thus the composition of a binary mixture of fluids can, at least in principle, be estimated from heat transfer measurements by use of thermal sensors. Nevertheless, heat transfer also depends on flow velocity and temperature, among other factors, and one would have to separate the sensor's sensitivity to concentration from other sensitivities. Furthermore, to achieve high spatial and temporal resolution, one would have to use fine thermal sensors, such as hot-wire–hot-film anemometers. This led to the suggestion [10, 11] of a probe consisting of two closely spaced but non-interfering hot-wire sensors with substantially different physical characteristics, so that their sensitivities to velocity and concentration would be significantly different. Such probes have been applied [12, 13] to the simultaneous measurement of the fluctuating velocity and concentration in isothermal binary gas mixtures. This approach is, however, difficult to implement, mostly because of inadequate separations of the different sensitivities. More practical and more extensively utilized are two somewhat different thermal techniques: the *interfering thermal probe* and the *aspirating probe*.

The interfering thermal probe also utilizes two hot-wire–hot-film sensors with differing materials, sizes, and overheats, but positioned such that one of them would be slightly upstream and within the thermal field of the other [Fig. 13.2(a)]. The extent of this thermal field would depend on the thermal conductivity, and thus on the concentration, of the surrounding fluid. As a result, the sensitivity of the upstream probe to concentration would increase dramatically and, following calibration of the two-sensor probe in flows of varying but known velocity and concentration, it would be possible to measure both the flow velocity and the concentration. Addition of a third sensor would permit measurement of a second velocity component or the temperature. Although tedious and subject to drift and various interferences, this method has been applied successfully to the study of turbulent mixing [14–19].

The ideal concentration measurement probe would be one that would be entirely insensitive to velocity variations. This has been achieved by enclosure of the thermal sensor

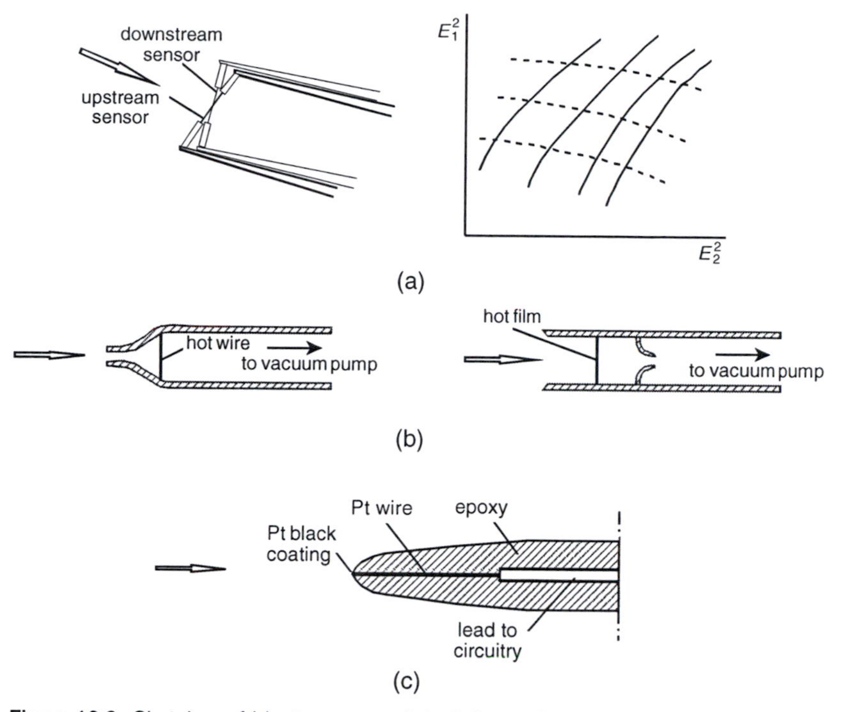

Figure 13.2. Sketches of (a) a two-sensor interfering probe and its calibration plot (solid curves represent constant concentration lines, and dashed curves represent constant velocity lines; E_1 and E_2 are the hot-wire voltage outputs), (b) two aspirating probe tips, and (c) a conductivity probe tip.

inside a small sampling tube, upstream or downstream of a sonic nozzle [Fig. 13.2(b)]. Suction is applied to this tube at a level sufficient for the flow to be 'choked', in which case the flow velocity inside the tube would remain constant, independently of velocity fluctuations in the flow from which the fluid is sampled. This type of probe, sometimes referred to as a *catharometer* or an aspirating probe, has a strong sensitivity to concentration fluctuations and provides reliable fluctuating concentration measurements [20–24]. Their outputs are also sensitive to temperature fluctuations; however, the two sensitivities are decoupled, so that one may calibrate it separately for the two effects and apply corrections to the concentration measurement if temperature is measured by other means [25]. Finally, these probes also exhibit sensitivity to pressure variations, which would always be present in turbulent flows and may express themselves as an increased output noise level. A variation of the aspirating probe utilizing mass transfer has recently been applied to the measurement of hydrocarbon concentration in a spark ignition engine [26].

13.3 Electric conductivity probes

Electric conductivity probes [27–34] measure the local electric resistivity of aqueous solutions of electrolytes, such as common salt, from which, through calibration, one can estimate the local concentration of the electrolyte. Such probes have been utilized in the study of turbulent mixing of passive scalars as well in studies of chemical reactions. Their use for measurement of the void fraction is discussed in Section 13.7.

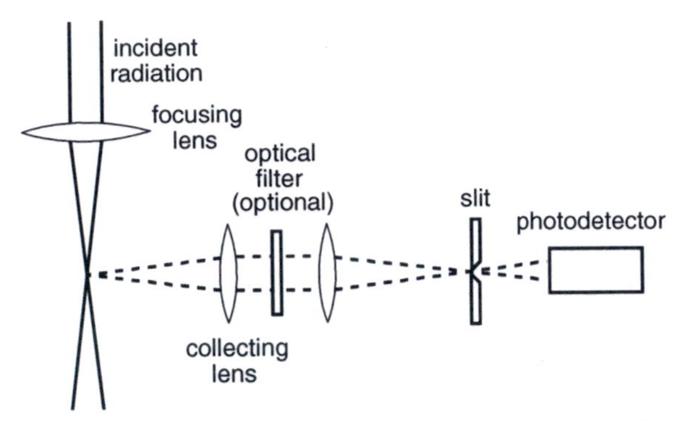

Figure 13.3. Sketch of a typical optical arrangement for the measurement of scattered light.

Their principle of operation is that the electric resistance of a path between two electrodes in the liquid solution decreases with increasing concentration of the electrolyte. Although probes with two closely spaced electrodes have been utilized, the most common configuration is the single-electrode type, in which the probe tip is the active electrode, and the walls of the apparatus, or a distant metallic piece, act as the second electrode. The probe tip [Fig. 13.2(c)] consists of a short, fine, platinum wire, encased in epoxy, except for the exposed active length, which is coated with platinum black to reduce surface polarization and noise. The wire is supplied with a relatively high-frequency ac carrier voltage, and the output consists of the low-pass-filtered voltage across a fixed resistor in series with the probe. Frequent cleaning, replating with platinum black, and calibration of the probe in liquid solutions with known uniform concentrations are necessary. The spatial resolution of conductivity probes is remarkably good (typically with characteristic lengths of the order of tens of micrometres) and adequate for the measurement of the fine structure of salinity fluctuations in moderate-Reynolds-number flows.

13.4 Light-scattering methods

A variety of light-scattering methods have been developed and are becoming increasingly popular in fluid mechanics research. Such methods utilize the light-scattering phenomena introduced in earlier chapters (Section 5.4 and elsewhere), which include Mie scattering, Rayleigh scattering, Raman scattering, and fluorescence. A basic experimental arrangement for many light-scattering methods is illustrated in Fig. 13.3. A laser beam or other collimated beam of light is focussed on a small measuring volume, which can be as small as 1 mm^3 or less. The light scattered from particles within this volume is collected by a collecting lens, separated from other collected radiation with the use of a slit or pinhole acting as a spatial filter, and then projected on a photodetector for subsequent analysis. The scattered light is normally collected at a right angle to the incident beam, in contrast with usual *in situ absorption* and *emission spectroscopic methods* [35], in which the photodetector is aligned with the incident beam (line-of-sight methods). As a result, the latter methods produce a path-average measurement, whereas light-scattering methods measure 'local' values. *Planar light-scattering methods*, in which light scattered within a volume bounded by two closely spaced planes is collected by photodetector arrays or cameras, have also become standard tools. This section briefly discusses some of the major light-scattering methods, and LIF techniques are discussed separately in the following section.

Mie-scattering methods: The intensity of light scattered from suspended tracer particles, such as smoke or mists contained in gases, can be used to measure their local concentration; this approach is sometimes referred to as *nephelometry* [36–38]. In principle, this method is quite straightforward. It requires a strong source of light focussed in a small measurement volume and a light collection system focussed on the same volume and leading to a photodetector. PMT outputs are linearly related to light intensity over many orders of magnitude, and so the theoretical dynamic range of this method is very wide. In practice, the accuracy of the method could be compromised by variations in particle size (polydispersity) and refractive index, particle coagulation, and absorption. Mie scattering has also been utilized for planar measurements (*tomography*) of particle concentrations with the use of laser light sheets [39,40].

Molecular radiation emission: Far more common than Mie scattering is the variety of techniques based on radiation scattering at the molecular level. Like the corresponding radiation absorption methods, radiation emission methods are capable of identifying both the species in a mixture as well as their concentrations. These methods utilize the transition of molecules at an energy state n to a state of lower energy m, by emission of photons with energy $h\nu_{nm}$. If N_n is the number of such molecules within the emitting volume, the intensity of spontaneously emitted radiation would be [2]

$$I_{nm} = A_{nm} N_n h \nu_{nm}, \tag{13.3}$$

where A_{nm} is the *Einstein transition probability*, specific to each transition and available from specialized sources or by calibration. Thus, in principle, one can measure the concentration of such molecules by measuring the intensity of radiation at the specific frequency ν_{nm}. Clearly, radiation emission methods require the use of high-power radiation sources and a careful experimental setting.

Rayleigh-scattering methods: Consider a pure gas consisting of a single chemical species and such that the mean free path between molecular collisions is relatively large. When illuminated by monochromatic light of frequency much lower than the lowest resonant frequency of its molecules, the gas would emit radiation at the same frequency as that of the incident radiation, which is precisely what we call Rayleigh scattering. If molecule aggregates comparable in size with the wavelength of incident radiation were also present, they would generate Mie scattering at the same frequency, which is potentially much stronger than Rayleigh scattering. Lenses, mirrors, etc., and the walls of the apparatus would also scatter light at the incident frequency. Thus measurement based on Rayleigh scattering would be possible only following careful elimination or drastic reduction of other scattering effects and the application of corrections to remove residual influences. Moreover, even for single-species gases, the emitted radiation would not exactly be monochromatic, but its frequency spectrum would have a nominally

Gaussian shape because of Doppler shifting caused by thermal motions of the molecules. The width of the peak depends on temperature, which introduces the possibility of temperature measurement by examination of the spectral shape of the scattered light. To make matters worse, when the molecular mean free path is relatively short, the process is contaminated by Doppler shifting caused by the propagation of sound waves in the gas. Then the frequency spectrum of the scattered light has two peaks, called *Brillouin peaks*, on either side of the incident frequency. This effect and other interferences make spectral shapes difficult to interpret [2]. In practice, composition measurements based on Rayleigh scattering mainly rely on light intensity monitoring. For a single species, the collected light intensity would be

$$I = AN\sigma_R, \tag{13.4}$$

where A is a calibration constant, N is the concentration of the species, and σ_R is the Rayleigh-scattering cross section (see Section 5.4), which has been tabulated for common species and light wavelengths [41,42]. Therefore the gas density can be found as proportional to the scattered-light intensity. For a mixture of molecules at relatively low densities, the scattering intensities are cumulative, as

$$I = A\sum N_i\sigma_{\mathrm{R}i}. \tag{13.5}$$

This approach has been applied successfully to binary mixtures of gases with substantially different scattering cross sections, such as methane and air (molecular oxygen and nitrogen have nearly the same σ_R) or carbon monoxide and nitrogen, in which the degree of mixedness has been measured with a relatively high-frequency response, in some cases exceeding 10 kHz [43–45]. For multispecies mixtures, or when the scattering cross sections are not sufficiently separated, Rayleigh-scattered radiation cannot be easily converted to species concentrations. When this method works, its advantage over competing methods is a nearly continuous signal output and a good temporal resolution.

Spontaneous Raman spectroscopy: *Spontaneous Raman spectroscopy* (*SRS*) utilizes the spontaneous Raman-scattering effect to identify species in fluid mixtures as well as to measure their concentrations. The great advantage of Raman-scattering methods, compared with Rayleigh-scattering methods, is that the former are not contaminated by incident light or light scattered by impurities in the flow and the walls of the apparatus: All these, as well as Rayleigh-scattered light, can be separated by a spectral filter (e.g., a spectrometer). Its disadvantage is the low intensity of Raman-scattered radiation, which limits the resolution of the method. The collected light intensity obeys the relationship

$$I = AN\sigma_{\mathrm{Rm}}, \tag{13.6}$$

where σ_{Rm} is the Raman-scattering cross section, available in the literature for common species [46]. This method has been applied successfully to measurements of concentration in high-speed gas flows and in flames [1,2,35,47]. Because each type of molecule has unique vibrational energy levels, it is possible to identify the presence of many different types of molecules, for example the multiple products of combustion of complex fuels. One can find the number of densities and concentrations of different species in the mixture by measuring the areas of the corresponding spectral bands and summing the contributions of bands that belong to the same species. It must be noted that, as the flow temperature increases, the number of vibrational energy levels that contribute to Raman scattering increases. Similarly, the number of rotational energy levels increases, with the result of broadening the bands of vibrational levels. Both of these effects increase the probability of spectral band overlapping for different molecules, thus reducing the usability of the method. In general, Raman spectroscopy is relatively easy for mixtures of simple molecules (e.g., O_2, CO_2, etc.), but much more difficult when higher hydrocarbons or other complex molecules are present. High-power laser sources are required to overcome the low sensitivity of the method. Pulsed lasers, with typical pulse lengths of 10 ns, are used for measurements in high-speed and rapidly reacting flows, whereas high-power CW lasers can be applied to the study of steady phenomena. Although light in the visible range has been used by some researchers, ultraviolet light is more usable, because some molecules (O_2, N_2, and others) have no visible resonance absorption energy bands. It is also preferable to use radiation with as long wavelength as possible, because the intensity of scattered light increases as the fourth power of the wavelength [see Eq. (5.39)]. The most commonly used laser for SRS in combustion research is the KrF excimer laser. Radiation is measured either through a number of PMTs, each restricted to a particular spectral band, or with the use of an ICCD camera, which measures the full spectrum simultaneously. The typical spatial resolution of SRS systems is about 0.5 mm and temporal response is of the order of 20 ns [42]. Raman scattering can also be used for measuring temperature, thus permitting the simultaneous measurement of temperature and concentration. Moreover, in combination with LDV or PIV, or it can be used for simultaneously measuring concentration and flow velocity.

Coherent anti-Stokes Raman spectroscopy: *Coherent anti-Stokes Raman spectroscopy* (*CARS*) *methods* are specialized variants of Raman-scattering methods [42, 48, 49]. They are generally preferable to SRS for combustion research, because they provide a much stronger signal and they are not as sensitive to the presence of soot. The basic CARS configuration utilizes two monochromatic beams with frequencies ν_1 and $\nu_2 < \nu_1$. Both beams are focussed colinearly on the same measuring volume within the fluid [50] (see Fig. 13.4). If some molecules within this volume have a vibrational energy state with a frequency $\Delta\nu = \nu_1 - \nu_2$, they will be driven to vibration coherently (namely in phase), scattering the incoming light into two new colinear and coherent beams with wavelengths $\nu_1 + \Delta\nu$ (anti-Stokes sideband) and $\nu_2 - \Delta\nu$ (Stokes sideband), respectively. Practical systems utilize the anti-Stokes band as easier to detect spectroscopically. To match the vibrational energies of different species, one of the lasers has to be tunable (i.e., a dye laser), whereas the other could have a fixed wavelength (commonly used is the frequency-doubled Nd:YAG laser). By scanning the tunable laser wavelength, one will be able to detect different types of molecules, each of which has distinct vibrational energy levels. Furthermore, one can find the concentration of each species by measuring the intensity of light scattered at a particular spectral band, taking into account that the intensity of scattered radiation is proportional to the square of concentration. Because of the low sensitivity of the method, it is necessary to use

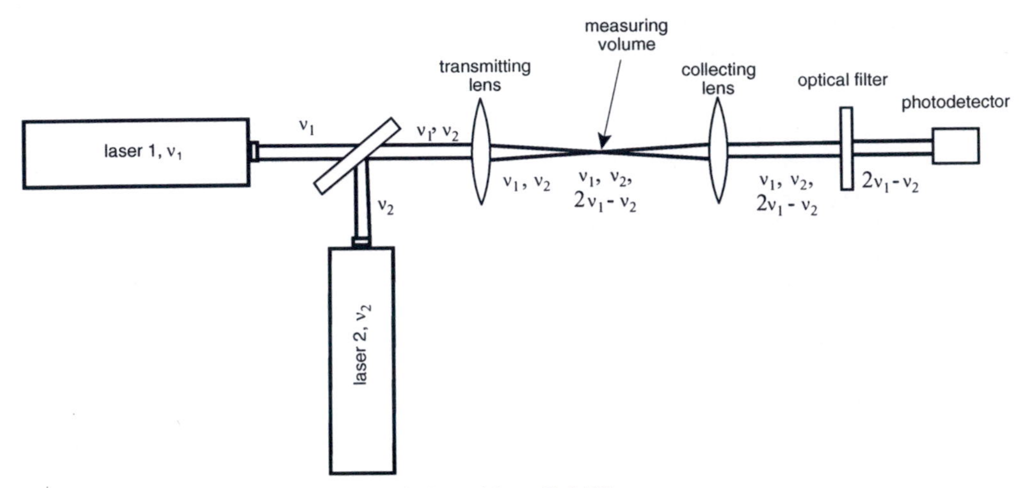

Figure 13.4. Sketch of a CARS system (adapted from Ref. 50).

high-power pulsed lasers. Several different variations of CARS have been applied to the measurement of concentration in combustion processes and high-speed aerodynamics. A variant, referred to as *CARS thermometry*, has also been applied to the measurement of temperature in flames.

13.5 Laser-induced fluorescence

The term *laser-induced fluorescence* (*LIF*) describes a host of experimental methods utilizing the phenomenon of fluorescence of certain molecules (see Section 5.4) when excited by laser light. In all cases, for the method to be effective, the collected light must be passed through an optical band-pass filter, centred around the peak of the fluorescence emission, thus eliminating all interference from the incident radiation, reflections and stray light. Two types of LIF measuring configurations are possible:

local LIF, in which the fluid is illuminated by a laser beam; light is collected by a lens focussed on a small control volume along the beam and recorded by a photodetector [Fig 13.5(a)]

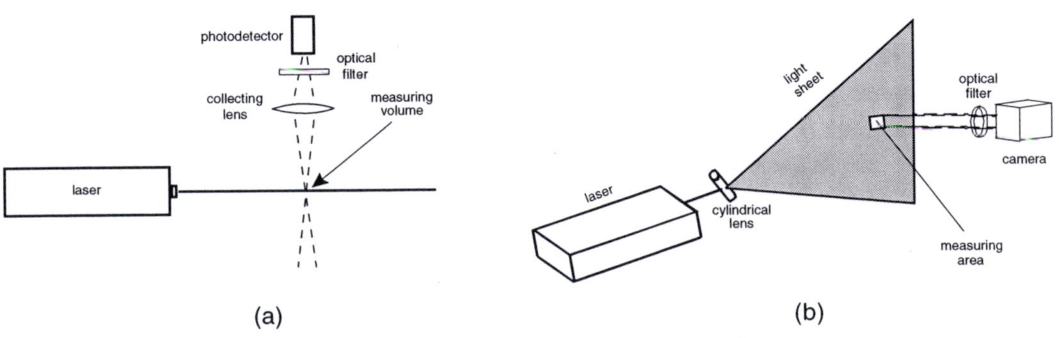

Figure 13.5. Sketches of optical arrangements for (a) local LIF and (b) PLIF.

Table 13.1. Compounds commonly used in LIF research

Compound	Absorption peak (nm)	Fluorescence peak (nm)
Fluorescein disodium	488	514.5
Rhodamine 6G	530	556
Rhodamine B	542.75	575

planar LIF (*PLIF*), in which the laser beam is converted into a thin sheet and a two-dimensional image of the concentration field is recorded by a conventional or a CCD camera [Fig 13.5(b)]; three-dimensional information can be obtained by use of rotating mirrors to rapidly scan the laser sheet in a direction normal to it.

In both cases, the light is usually collected at right angles with respect to the transmission path. LIF methods can be further subdivided to two classes: LIF in liquids and LIF in gases.

LIF in liquids: The most common compounds used for LIF in water flows, together with representative peaks in their absorption and fluorescence spectra, are listed in Table 13.1.

A buffered solution of the dye is injected into the fluid stream, and laser light is directed into the region of interest. Qualitative LIF for flow visualization purposes is relatively easy to implement and does not require particular attention to details. On the other hand, the quantitative application of LIF for measuring dye concentration is subjected to the following limitations [51, 52]:

- *Depletion* (bleaching): Fluorescence of molecules will diminish with time, when the same fluid volume is steadily subjected to laser radiation; this could happen during calibration of the technique, if a small container is used without mixing, but would not be a problem in a turbulent flow of a large volume of water, in which adequate mixing is provided.
- *Extinction*: for a fixed, relatively low dye concentration, fluorescence intensity is proportional to the incident-light intensity, which indicates that absorption effects over the path of the light would be negligible; this persists for at least 5 orders of magnitude of concentration; however, as the concentration increases, the attenuation of the laser beam or sheet along its path will inevitably become significant, causing measuring errors.
- *Temperature sensitivity*: the fluorescence quantum efficiency of the substances listed in Table 13.1 decreases with increasing temperature, with the steepest slope exhibited by Rhodamine B (see Section 12.5); thus, temperature variations in the flow would introduce uncertainty in concentration measurements; on the other hand, this effect makes LIF suitable also for the measurement of temperature fluctuations in liquids having a uniform dye concentration.
- *pH sensitivity*: Within a range, pH affects significantly the extinction of light in the fluorescent medium; for fluorescein disodium, the pH effect is negligible for $pH \geq 8.5$.

- *Spatial resolution*: fluorescent dyes have particularly small molecular diffusivities γ in water, with Schmidt numbers Sc ($Sc = \nu/\gamma$; ν is the kinematic viscosity of the solution) roughly equal to 2000; this means that, in a turbulent flow, the scalar fine structure would be much finer than the fine structure of the turbulence itself, thus requiring extremely fine spatial resolution of the measuring system, which may not be achievable for relatively large-Reynolds-number flows; as an illustration, let us assume that the Kolmogorov microscale (i.e., the length scale of the smallest dynamically significant motions) in a flow is $\eta = 0.2$ mm; then the Batchelor microscale (i.e., the length scale of the smallest significant concentration differences) would be $\eta_B = \eta Sc^{-1/2} \simeq 4\ \mu$m; this value would most likely much smaller than the typical spatial resolution of LIF systems (for local LIF, this would be the focal volume of the collecting lens, whereas for PLIF, this would depend on the laser-sheet thickness, the pixel size–film resolution, and the image magnification factor)
- *Non-uniformity of illumination*: As mentioned in Section 5.3, laser sheets produced by cylindrical lenses have non-uniform intensities; Powell lenses can be used to produced a nominally uniform intensity across the sheet width, but these cannot eliminate the appearance of dark stripes, which are the expanded images of dust particles in the laser cavity; this effect would create erroneous concentration variations, even in cases with perfectly uniform concentration; the reader is referred to Section 5.3 for methods to improve the illumination uniformity and calibration methods of the optical system to correct for such non-uniformity.

Common sources of illumination for LIF in water are CW Ar-ion and He–Ne lasers and pulsed Nd:YAG lasers, which conveniently emit light at wavelengths near the absorption peaks of the compounds listed in Table 13.1.

LIF in gases: LIF has been applied in non-reacting gas flows, utilizing passive fluorescent seeding materials, such as biacetyl, which emits radiation in the visible range. However, its most extensive application has been in flames, in which fluorescence is usually produced by radicals, such as OH, CH, CO, and NO, thus making it possible to map chemical reactions, even in transient states. The hydroxyl radical OH is the most commonly utilized species, for both quantitative and qualitative purposes, as it provides a strong signal and its concentration jump can be used to map the interface between burned and unburned parts of flames. Flame extinction is usually detected by the concentration of CH. By comparison with other spectroscopic methods, LIF produces the strongest radiation signal from a single species. Together with other optical methods, LIF has become an indispensable tool in combustion research. Excitation of fluorescence is normally provided by tunable, pulsed, dye lasers and excimer lasers, operating in the visible or ultraviolet ranges and pumped by various means. Details of the associated techniques may be found in several excellent reviews [42, 49, 53–55].

13.6 Particulate measurement

The term *particulate* covers all sorts of airborne particles, both solid and liquid. Their sources can be natural, as in the case of dust, pollen and haze, or industrial–commercial–residential, as in the case of soot and smoke. Many types of particulates constitute health and environmental hazards and contribute to pollution. Sprays and aerosols, which are liquid droplet suspensions in gases, are encountered in many industrial systems, such as combustors and paint injectors. From the fluid mechanical viewpoint, the characterization of particulates and aerosols includes the description of their mass flux, number density, mean size, size distribution, active surface area, and shape. Among the common particulate measurement methods are these:

Gross gravimetric analysis: The simplest approach for obtaining the average mass flux of suspended matter in a fluid stream would be to pass it through a filter of sufficient solidity (i.e., ratio of blocked to total area), which will capture all particles or those larger than a certain size. The average mass flow rate would be equal to the increase in weight of the filter, compared with its clean weight, over a measured time interval. Besides using filters, one may capture particles by hitting the stream on surfaces coated with solid or liquid films [56]. When passing the entire stream through a filter is inconvenient or undesirable, one may extract a sample of the fluid through a tube. In such cases, it would be advisable to apply *isokinetic sampling* (i.e., to sample at a speed equal to the local flow speed) to avoid biasing the composition of the sample towards heavier or lighter components. Obviously these approaches cannot separate the different sizes or densities of particles and would be unsuitable for non-stationary flows and for reacting flows, in which the composition of the particulate may vary.

Separation and measurement of sizes: Mixtures of solid particles with linear dimensions greater than a few micrometres can be separated by size with the use of *sieves*. Samples of captured particles, solid or liquid, can be analysed by use of optical microscopes, for sizes larger than about 0.5 μm, or electron microscopes, for sizes as small as 10 nm [2, 57]. Separation of particles in a fluid stream according to their inertia can be achieved by the passing of a fluid sample through a *cascade sampler*, consisting of a set of different nozzles and baffles, or by the exposure of the sample to oscillatory flow [56]. The characterization of polydispersed particles, particles with different densities, and particles with non-spherical shapes requires an unambiguous definition of the *average* (or, more appropriately, *equivalent*) *diameter*. This parameter may be defined in different ways, depending on the measuring method and the intended use of the measurements: the *Sauter mean diameter* (*SMD*) is defined as the diameter of spheres that have the same surface area per unit volume of the suspension, whereas the *mean aerodynamic diameter* takes into account the drag on the particles, irrespective of their size, shape, and density. When analysing particle samples, one must take care to avoid coalescence, deposition, condensation, or evaporation in the sampling tubes; this may require heating or cooling of some components. A number of *particle sizers*, based on different principles, are available commercially and one is advised to contact the manufacturers for details.

Photographic, visual, and optical methods: Unlike sampling methods, on-site photographic, visual, and optical methods are non-intrusive, thus avoiding many of the problems previously mentioned. Conventional photography is not very suitable for particle imaging, because it cannot resolve depth of field; three-dimensional views of particles

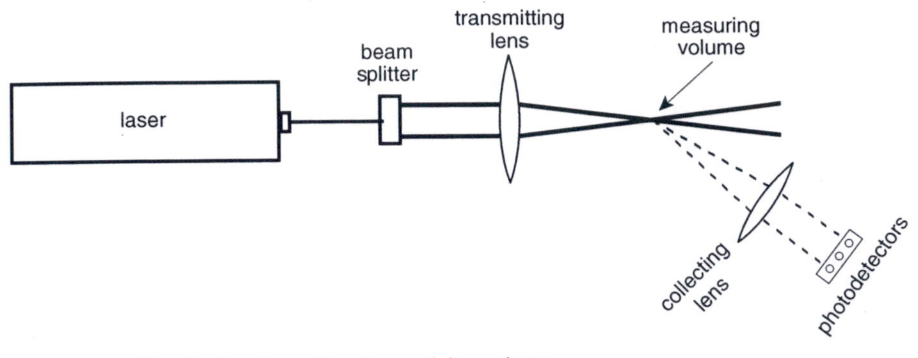

Figure 13.6. Sketch of a phase Doppler particle analyser.

can be recorded by *stereo photography* and holographic methods [56]. The various *light-scattering methods* described in Section 5.4 may also be used for measuring the local concentration of particulate. Related to these are *obscuration techniques*, which are based on the absorption of light by particles in the path of a light beam, usually transmitted and collected by optical fibres.

Phase Doppler analysis and related techniques: Among the variety of available optical methods, of particular interest is *phase Doppler particle analysis* (*PDPA* or *PDA*), also known as *phase Doppler interferometry* (*PDI*) [58,59], because it can be combined with LDV (see Section 11.2) to provide simultaneous measurements of flow velocity and size distribution of droplets or other spherical particles. A simplified illustration of the technique is shown in Fig. 13.6. A laser beam is split into two parallel beams, which are made to intersect within a small measuring volume, in which they form interference fringes. A particle passing through this volume scatters light, according to Mie-scattering theory. The scattered light is collected by a lens located off axis and projected on an array of two or more photodetectors. The output of each photodetector is a Doppler burst, whose frequency is proportional to the particle velocity, according to LDV analysis. At the same time, light reflected on the surface of the particle or refracted through it will acquire a phase shift, which would depend on the geometrical arrangement of the incident-beam–particle-detector system and the size of the particle. By determining the phase shifts between the Doppler bursts of different detectors and taking into account the wavelength of light and some optical parameters, one can calculate the particle diameter without a need for calibration. The phase Doppler technique is not affected by multiple scattering and absorption along the light path. However, it requires the presence of a single particle in the measuring volume, thus limiting its range to relatively low particle concentrations. It is also suitable only for optically homogeneous, spherical particles. Among the techniques that can characterize non-spherical and inhomogeneous particles [60] is the *shadow Doppler technique* [61, 62], which essentially maps the projections of individual moving particles on a large (up to 35) array of photodiodes, from which both the shapes and the velocities of the particles can be extracted. Another recent method of interest is *global phase Doppler* (*GDP*) technique [58]. This technique is similar to PDA, the main difference being that the measuring volume is much larger, as the collecting lens is off-focussed on the intersection of two light sheets. Particles crossing this volume appear to have interference fringes and the size of particles is determined by counting the number of fringes.

Laser-induced incandescence: The exhaust gases of internal combustion engines, furnaces, and other combustion systems contain particulate that is the result of incomplete oxidization of the fuel and is known as *soot*. Soot particles are irregularly shaped aggregates of near-spherical carbon particles of nanoscale dimensions, called the *primary particles*. In addition, the gases may contain liquid droplets, which are the product of vapour condensation, and particles composed of metal sulfates or oxides, which constitute the *ash*. *Laser-induced incandescence* (*LII*) methods [63] utilize high-power, short-duration laser radiation pulses to heat the airborne soot particles to sublimation temperatures, causing them to emit radiation, which is collected by photodetectors and analysed to provide the particle size and volume fraction. Because carbon absorbs laser radiation at a much higher rate than liquids or ash, this method is insensitive to the presence of the latter. The size of the soot particles is much smaller than the laser-light wavelength, so that the light emission is in the Rayleigh regime. Heat transfer from the hot particles to the surrounding cooler gas takes place mainly by conduction, and the particle temperature, following the initial heating stage, decays exponentially, with a time constant that is proportional to the particle size. Thus, assuming that the primary particle distribution is monodisperse, one can estimate their size by analysing the temporal decay of the emitted radiation signal [64, 65]. Mathematical analysis of the LII signal, band-pass-filtered at two wavelengths, may also provide the size distribution of polydisperse particulates [66]. The volume fraction of soot particles can be determined from the intensity of the LII signal, calibrated in a laminar flame containing particles of known size, which has been measured independently by other means; a self-calibrating method providing soot volume fraction has also been proposed [63]. A recent LII method, combined with two-colour pyrometry, is capable of both measuring particle size and determining its composition. This method applies not only to soot particles, but also to tungsten, silicon, and other materials, and it is based on measuring the boiling temperature of the particles [67]. LII is still in its development stage and has good potential for further growth. Compared with other optical methods, it has the advantages of high spatial and temporal resolutions and requires relatively simple means.

13.7 Void measurement

An important property of gas–liquid two-phase flows is the *void fraction*, defined as the ratio of the volume occupied by the gaseous phase over the total volume of interest. Depending on the volume considered, one may distinguish between a *bulk void fraction*, which is spatially averaged over the entire channel or other large volume of the fluid, a cross-section, or a line (e.g., a diameter of a pipe), and a *local void fraction*, which is (usually) time averaged at a location in the flow [68]. One may also be interested in determining, in addition to the void fraction, which is an average property, the shape of the interface between the gas and liquid phases, and, for bubbly flows, the size distribution of bubbles. There is significant interest in such parameters in many applications ranging from nuclear power generation to physical oceanography. A great variety of methods has been developed for gas–liquid flow measurement, widely varying in resolution and accuracy, and often specialized to the needs and conditions of a particular application. The following list includes a sample of representative techniques.

Quick-closing valve method: This method is suitable for bulk void fraction measurement in stationary flows through closed channels, which can be temporarily interrupted without affecting the operation of the system. A part of the mixture is trapped in a section of the channel when two solenoid valves are simultaneously closed at its inlet and outlet. Then one measures the liquid volume by draining it into a volumetric container [68].

Sampling methods: A number of sampling methods have been proposed, all resulting in continuously removing samples of one or both phases of the two-phase mixture, which can be separated and analysed externally. The *wall scoop method* and the *porous wall method* have been used for sampling the liquid film in annular-type flows, whereas *isokinetic sampling probes* have been used to measure the mass flux of liquid droplets in vapour streams [69].

Electrical impedance meters: These methods are based on the dependence of the electrical impedance of a multiphase flow on the concentration of the different phases. Although various improved designs have been introduced by different authors, the basic *impedance void fraction meter* consists of a number of concentric ring-type electrodes, powered by a high-frequency ac and surrounding a non-conducting section of a pipe through which the fluid mixture is passing. This method provides time-dependent cross-sectional or annular averages of void fraction with a high-frequency response. However, because the outputs of impedance meters are very sensitive to the phase pattern (i.e., whether it is in the bubbly, annular, slug, or mist flow regime), they can be used only if the flow regime is known [68, 70]. Local void fraction impedance probes are also available [69], whose operation is also based on the electrical impedance dependence on the phase distribution. Commonly used types are the various *conductivity probes* (see Section 13.3), which have a thin electrode in contact with the fluid. Their output signals discriminate between the liquid and the gas phases, so that they can be used to measure not only the time-averaged local void fraction, but also bubble size and speed. A variety of designs has been proposed for specific use in studies of boiling, cavitation, fluidized beds, and oceanic physics [70–73].

Thermal methods: The constant-temperature *hot-film anemometer*, which is normally used as a velocity sensor, can be also used to discriminate between liquid and gas phases in contact with the sensor. The instrument output is proportional to the electrical power required for maintaining the sensor in constant temperature; because of the reduced heat transfer in the gas compared with that in the liquid, the anemometer output voltage would be much lower for the gas phase than for the liquid phase. By using an adjustable threshold to separate the signal into low-level and high-level segments, one can measure the fraction of time that the sensor would be immersed in gas, which, ideally, would be equal to the void fraction. With calibration of the probe and application of various corrections, it may be possible to measure both void fraction and local fluid velocity in one or both phases [69]. A different approach has been to use *microthermocouples* both as local temperature sensors and conductivity sensors to separate the gas and liquid phases [74].

Optical methods: A variety of optical methods have been suggested, mostly utilizing the difference in refractive index between the gas and liquid phases. In the *glass-rod method* [69], the tip of a thin glass rod is polished to a 45° angle with respect to its axis. The probe is introduced into the flow, and light is transmitted through the glass rod. According to Snell's law (see Section 5.2), when the tip is in contact with gas, the light beam will exit the rod, whereas when the tip is in contact with liquid, total internal reflection will occur (see Section 5.2) and the light will be reflected back into the rod. By collection of the reflected light, it is possible to measure the percentage of time that the rod is in contact with the gas, and thus the void fraction. Besides the basic glass-rod method, a number of variations and improvements by use of fibre optics have been developed. A separate approach has been to measure the attenuation or extinction of light transmitted through bubbly flows through reflections, refractions, and diffraction on the bubbles. Methods suitable for both relatively high and relatively low concentrations of bubbles are available [75, 76].

Acoustical methods: Void fraction and bubble size distribution have been measured in bubbly flows by use of the changes in sound (usually ultrasound) propagating through a liquid, when it encounters a bubble. Acoustical methods have been used in both industrial [69] and oceanic applications [77].

Radiation methods: Some widely used bulk and local void fraction methods are based on the difference in absorption of gamma and x rays propagating through the gas and liquid phases [68]. Absorption of a single collimated ray would provide the bulk void fraction along its path. By traversing the ray across the flow or using broad-beam or multibeam instruments, one may obtain the cross-section averaged void fraction. One can measure the local void fraction by positioning the radiation detector at an angle with the beam (sidescattering) so that it would detect radiation scattered by a small volume on the beam's path. A related method, based on neutron scattering, has been used in applications in which a particle accelerator is available.

QUESTIONS AND PROBLEMS

1. Consider a homogeneous mixture of two neutral gases with molecular weights M_1 and M_2 and Gladstone–Dale constants K_1 and K_2. Would it be possible to determine its composition from a measurement of its Gladstone–Dale constant? If so, describe the procedure and provide appropriate relationships.

2. Consider a stationary turbulent mixing layer between a stream of air and a stream of helium with different speeds and temperatures. Describe in detail the instrumentation and measuring procedures required for the following measurements:
   a. The measurement of the local mean velocity and turbulence statistics across the mixing layer.
   b. The measurement of the local mean and rms concentrations across the mixing layer.
   c. The determination of instantaneous concentration maps over a cross section of the flow.
   d. The simultaneous measurement of local velocity, temperature, and concentration fluctuations with a temporal resolution of at least 1 ms.

REFERENCES

[1] B. E. Richards (Editor). *Measurement of Unsteady Fluid Dynamic Phenomena*. Hemisphere, Washington, DC, 1977.

[2] R. J. Emrich (Editor). Fluid dynamics. In L. Marton and C. Marton (Editors-in-Chief), *Methods of Experimental Physics,* Vols. 18A and B. Academic, New York, 1981.

[3] J. G. Webster (Editor-in-Chief). *The Measurement, Instrumentation and Sensors Handbook*. CRC Press, Boca Raton, FL, 1999.

[4] Anonymous. *Elementary Theory of Gas Chromatography with Bibliography and Experiments*. Gow-Mac Instrument Co., Bridgewater, NJ, 1978.

[5] J. M. Bobbitt, A. E. Schwarting, and R. J. Gritter. *Introduction to Chromatography*. Reinhold, New York, 1968.

[6] J. A. Jahnke. *Continuous Emission Monitoring* (2nd Ed.). Wiley, New York, 2000.

[7] P. F. Bernath. *Spectra of Atoms and Molecules*. Oxford University Press, New York, 1995.

[8] C. N. Banwell and E. M. McCash. *Fundamentals of Molecular Spectroscopy* (4th Ed.). McGraw-Hill, New York, 1994.

[9] E. R. Cohen, D. R. Lide, and G. L. Trigg. *AIP Physics Desk Reference* (3rd Ed.). Springer, New York, 2003.

[10] S. Corrsin. Extended applications of the hot-wire anemometer. *Rev. Sci. Instrum.*, 18:469–471, 1947.

[11] S. Corrsin. Extended applications of the hot-wire anemometer. Tech. Rep. Tech. Note 1864, National Advisory Committee for Aeronautics, Washington, DC, April 1949.

[12] J. McQuaid and W. Wright. The response of a hot-wire anemometer in flows of gas mixtures. *Int. J. Heat Mass Transfer*, 16:819–828, 1973.

[13] J. McQuaid and W. Wright. Turbulence measurements with hot-wire anemometry in non-homogeneous jets. *Int. J. Heat Mass Transfer*, 17:341–349, 1974.

[14] J. Way and P. A. Libby. Hot-wire probes for measuring velocity and concentration in helium–air mixtures. *AIAA J.*, 8:976–978, 1970.

[15] J. Way and P. A. Libby. Application of hot-wire anemometry and digital techniques to measurements in a turbulent helium jet. *AIAA J.*, 9:1567–1573, 1971.

[16] R. A. Stanford and P. A. Libby. Further applications of hot-wire anemometry to turbulence measurements in helium-air mixtures. *Phys. Fluids*, 17:1353–1361, 1974.

[17] P. A. Libby. Studies in variable-density and reacting turbulent shear flows. In B.E. Launder, editor, *Studies in Convection*. Academic, New York, 1977.

[18] A. Sirivat and Z. Warhaft. The mixing of passive helium and temperature fluctuations in grid turbulence. *J. Fluid Mech.*, 120:475–504, 1982.

[19] J. L. Harion, M. Favre-Marinet, and B. Camano. An improved method for measuring velocity and concentration by thermo-anemometry in turbulent helium-air mixtures. *Exp. Fluids*, 22:174–182, 1996.

[20] G. L. Brown and M. R. Rebollo. A small, fast-response probe to measure composition of a binary gas mixture. *AIAA J.*, 10:649–652, 1972.

[21] D. Adler. A hot wire technique for continuous measurement in unsteady concentration fields of binary gaseous mixtures. *J. Phys. E*, 5:163–169, 1972.

[22] D. Adler and Y. Zvirin. The time response of a hot wire concentration transducer. *J. Phys. E*, 8:185–188, 1975.

[23] A. D. Birch, D. R. Brown, M. G. Dodson, and F. Swaffield. Aspects of design and calibration of hot-film aspirating probes used for the measurement of gas concentration. *J. Phys. E*, 9:59–63, 1986.

[24] D. J. Wilson and D. D. J. Netterville. A fast-response, heated-element concentration detector for wind-tunnel applications. *J. Wind Eng. Ind. Aerodyn.*, 7:55–64, 1981.

[25] M. Cabannes, M. Ferchichi, and S. Tavoularis. Temperature variation correction for aspirating concentration probes. *Meas. Sci. Tech.*, 15:1211–1215, 2004.

[26] P. Guilbert and E. Dicocco. Development of a local continuous sampling probe for the equivalence air–fuel ratio measurement. application to spark ignition engine. *Exp. Fluids*, 32:494–505, 2002.

[27] P. Danckwerts. The definition and measurement of some characteristics of mixtures. *Appl. Sci. Res. A*, 3:279–296, 1952.

[28] C. H. Gibson and W. H. Schwarz. Detection of conductivity fluctuations in a turbulent flow field. *J. Fluid Mech.*, 16:357–364, 1963.

[29] C. V. Alonso. Comparative study of electrical conductivity probes. *J. Hydraul. Res.*, 9:1–10, 1971.

[30] S. K. Chua, J. W. Cleaver, and A. Millard. The measurement of salt concentration in a plume using a conductivity probe. *J. Hydraul. Res.*, 24:171–178, 1986.

[31] S. Komori, T. Kanzaki, and Y. Murakami. Simultaneous measurement of instantaneous concentrations of two reacting species in a turbulent flow with a rapid reaction. *Phys. Fluids A*, 3:507–510, 1991.

[32] M. Mahouast. Concentration fluctuations in a stirred reactor. *Exp. Fluids*, 11:153–160, 1991.

[33] F. Ncube, E. G. Kastrinakis, S. G. Nychas, and K. E. Lavdakis. Drifting behavior of a conductivity probe. *J. Hydraul. Res.*, 29:643–654, 1991.

[34] K. Voloudakis, P. Vrahliotis, E. G. Kastrinakis, and S. G. Nychas. The behaviour of a conductivity probe in electrolytic liquid/solid suspensions. *Meas. Sci. Technol.*, 10:100–105, 1999.

[35] T. Arts et al. *Measurement Techniques in Fluid Dynamics* (2nd Ed.). Von Karman Institute for Fluid Dynamics, Rhode-Saint Genese, Belgium, 2001.

[36] H. A. Becker, H. C. Hottel, and G. C. Williams. On the light-scatter technique for the study of turbulence and mixing. *J. Fluid Mech.*, 30:259–284, 1967.

[37] H. A. Becker. Mixing, concentration fluctuations and marker nephelometry. In B.E. Launder, editor, *Studies in Convection*. Academic, New York, 1977.

[38] E. J. Shaughnessy and J. B. Morton. Laser light-scattering measurements of particle concentration in a turbulent jet. *J. Fluid Mech.*, 80:129–148, 1977.

[39] M. B. Long, B. F. Webber, and R. K. Chang. Instantaneous two-dimensional gas concentration measurements in a jet flow by Mie scattering. *Appl. Phys. Lett.*, 34:22–24, 1979.

[40] P. Guilbert, M. Durget, and M. Murat. Concentration fields in a confined two-gas mixture and engine in-cylinder flow: Laser tomography measurement by mie scattering. *Exp. Fluids*, 31:630–642, 2001.

[41] W. C. Gardiner, Y. Hidaka, and T. Tanzawa. Refractivity of combustion gases. *Combust. Flame*, 40:213–219, 1981.

[42] E. P. Hassel and S. Linow. Laser diagnostics for studies of turbulent combustion. *Meas. Sci. Technol.*, 11:R37–R57, 2000.

[43] S. C. Graham, A. J. Grant, and J. M. Jones. Transient molecular concentration measurements in turbulent flows using Rayleigh light scattering. *AIAA J.*, 12:1140–1142, 1974.

[44] W. M. Pitts and T. Kashiwagi. The application of laser-induced Rayleigh light scattering to the study of turbulent mixing. *J. Fluid Mech.*, 141:391–429, 1984.

[45] J. Haumann, G. Wu, and A. Leipertz. Low power laser Rayleigh probe for flow mixing studies. *Exp. Fluids*, 5:230–234, 1987.

[46] A. Weber (Editor). *Raman Spectroscopy of Gases and Liquids*. Springer, Berlin, 1979.

[47] G. F. Widhoff and S. Lederman. Specie concentration measurements utilizing Raman scattering of a laser beam. *AIAA J.*, 9:309–316, 1971.

[48] D. A. Greenhalgh. Quantitative CARS spectroscopy. In R. J. H. Clarc and R. E. Hester, editors, *Advances in Nonlinear Spectroscopy*, pp. 193–252. Wiley, New York, 1987.

[49] A. C. Eckbreth. *Laser Diagnostics for Combustion Temperature and Species* (2nd Ed.). Gordon & Breach Publishers, Amsterdam, 1996.

[50] J. P. Taran. Molecular diagnostics for rarefied flows. In *AGARD Conference Proceedings CP-601*, Advisory Group for Aerospace Research and Development, Neuilly-sur-Seine, France, 1997. AGARD.

[51] G. G. Guibault. *Practical Fluorescence: Theory, Methods and Techniques*. Dekker, New York, 1973.

[52] D. A. Walker. A fluorescence technique for measurement of concentration in mixing liquids. *J. Phys. E*, 20:217–224, 1987.

[53] W.-J. Yang (Editor). *Handbook of Flow Visualization*. Taylor & Francis, Washington, DC, 1989.

[54] J. W. Daily. Laser induced fluorescence spectroscopy in flames. *Prog. Energy Combust. Sci.*, 23:133–199, 1997.

[55] R. K. Hanson, J. M. Seitzman, and P. H. Paul. Planar laser-fluorescence imaging of combustion gases. *Appl. Phys. B*, 50:441–454, 1990.

[56] B. J. Azzopardi. Measurement of drop sizes. *Int. J. Heat Mass Transfer*, 22:1245–1279, 1979.

[57] R. R. Irani and C. F. Callis. *Particle Size: Measurement, Interpretation, and Application*. Wiley, New York, 1973.

[58] H.-E. Albrecht, M. Borys, N. Damaschke, and C. Tropea. *Laser Doppler and Phase Doppler Measurement Techniques*. Springer-Verlag, Berlin, 2003.

[59] F. Mayinger and O. Feldmann (Editors). *Optical Measurements* (2nd Ed.). Springer, Berlin, 2001.

[60] N. Damaschke, G. Gouesbet, G. Grehan, and C. Tropea. Optical techniques for the characterization of non-spherical and non-homogeneous particles. *Meas. Sci. Technol.*, 9:137–140, 1998.

[61] Y. Hardalupas, K. Hishida, M. Maeda, H. Morikita, A. M. K. P. Taylor, and J. H. Whitelaw. Shadow Doppler technique for sizing particles of arbitrary shape. *Appl. Opt.*, 33:8417–8426, 1994.

[62] H. Morikita and A. M. K. P. Taylor. Application of shadow Doppler velocimetry to paint spray: Potential and limitations in sizing optically inhomogeneous droplets. *Meas. Sci. Technol.*, 9:221–231, 1998.

[63] C. Mercer (Editor). *Optical Metrology for Fluids, Combustion and Solids*. Kluwer Academic, Dordrecht, The Netherlands, 2003.

[64] R. L. Vander Wal, T. M. Ticich, and A. B. Stephens. Can soot primary particle size be determined using laser-induced incandescence? *Combust. Flame*, 116:291–296, 1999.

[65] D. R. Snelling, G. J. Smallwood, I. G. Cambell, J. E. Medlock, and O. Gulder. Development and application of laser-induced incandescence (LII) as a diagnostic for soot particulate measurements. In *Proceedings of the AGARD 90th Symposium of the Propulsion and Energetics Panel on Advanced Non-Intrusive Instrumentation for Propulsion Engines*, pp. 1–9, Advisory Group for Aerospace Research and Development, Brussels, Belgium, 1997.

[66] P. Roth and A. V. Filippov. In situ ultrafile particle sizing by a combination of pulsed laser heatup and particle thermal emission. *J. Aerosol Sci.*, 27:95–104, 1996.

[67] M. Stephens, N. Turner, and J. Sandberg. Particle identification by laser-induced incandescence in a solid-state laser cavity. *Appl. Optics*, 42:3726–3736, 2003.

[68] G. F. Hewitt. Void fraction. In G. Hestroni, editor, *Handbook of Multiphase Systems*, pp. 10.21–10.33. Hemisphere, Washington, DC, 1982.

[69] O. C. Jones. Two-phase flow measurement techniques in gas-liquid systems. In R. J. Goldstein, editor, *Fluid Mechanics Measurements* (1st Ed.), pp. 479–558. Hemisphere, Washington, DC, 1983.

[70] S. L. Ceccio and D. L. George. A review of electrical impedance techniques for the measurement of multiphase flows. *J. Fluids Eng.*, 118:391–399, 1996.

[71] L. G. Neal and S. G. Bankoff. A high resolution resistivity probe for determination of local void properties in gas-liquid flow. *AIChE J.*, 9:490–494, 1963.

[72] C. Wolff, F. U. Briegleb, J. Bader, K. Hektor, and H. Hammer. Effect of suspended solids on the hydrodynamics of bubble columns for application in chemical and biotechnological processes. *Chem. Eng. Technol.*, 13:172–184, 1990.

[73] N. P. Cheremisinoff. *Instrumentation for Complex Fluid Flows*. Technomic Publishing, Lancaster, PA, 1986.

[74] J. M. Delhay, R. Semeria, and J. C. Flamand. Void fraction, vapor and liquid temperatures: Local measurements in two-phase flow using a microthermocouple. *J. Heat Transfer*, 95C:365–370, 1973.

[75] D. M. Leppinen and S. B. Dalziel. A light attenuation technique for void fraction measurement of microbubbles. *Exp. Fluids*, 30:214–220, 2001.

[76] B. Shamoun, M. El Beshbeeshy, and R. Bonazza. Light extinction technique for void fraction measurements in bubbly flow. *Exp. Fluids*, 26:16–26, 1999.

[77] S. Vagle and D. M. Farmer. A comparison of four methods for bubble size and void fraction measurements. *IEEE J. Ocean. Eng.*, 23:211–222, 1998.

14 Measurement of wall shear stress

The value of the wall shear stress is of great interest in fluid mechanics research, as it represents the local tangential force by the fluid on a surface in contact with it. By integrating the wall shear stress along the surface, one can compute its contributions to the lift and drag on immersed objects and the pressure drop in pipes and other internal flows. Sometimes the interest in wall shear stress is qualitative only, for example, when shear stress is used as an indicator of flow separation or transition to turbulence. In engineering applications involving stationary turbulent flows, it is the time-averaged wall shear stress that is usually of interest, and so the temporal resolution requirement is often relaxed. There are other occasions, however, when both temporal and spatial resolutions of the wall shear stress measurement method must be high. In particular, measurements of wall shear stress fluctuations are required for the validation of turbulence theories and numerical models. In addition to a variety of classical methods, a number of new wall shear stress measurement techniques have been proposed in recent years. The established methods and new developments can be found in several thorough reviews [1–6].

14.1 Estimates from measured velocity profiles

For Newtonian fluids, the wall shear stress τ_w is related to the velocity derivative normal to the wall as

$$\tau_w = \mu \frac{\partial U}{\partial y}\bigg|_{y=0}, \tag{14.1}$$

where μ is the fluid viscosity. In laminar boundary layers, the velocity changes gradually, and it is usually possible to fit a smooth curve to near-wall measurements, from which τ_w can be obtained through Eq. (14.1). Unfortunately, this is not the case in turbulent boundary layers, in which the velocity changes dramatically over a short distance. The resolution and accuracy of conventional instrumentation, such as Pitot tubes, hot-wires, and LDV and PIV systems, become inadequate as the wall is approached, and even the distance from the wall cannot always be determined very precisely.

In the near-wall region of turbulent boundary layers, the relevant velocity scale is not the *free-stream velocity* U_e, but the *friction velocity*

$$u_\tau = \sqrt{\frac{\tau_w}{\rho}}, \tag{14.2}$$

and the mean velocity is presented in dimensionless form as

$$u^+ = \frac{\overline{U}}{u_\tau}. \tag{14.3}$$

Similarly, the relevant length scale is the *viscous length* ν/u_τ, and the dimensionless distance from the wall is expressed as

$$y^+ = \frac{y}{\nu/u_\tau} \tag{14.4}$$

In turbulent flows over smooth walls, the flow region closest to the wall, called the *viscous* or *laminar sublayer* and usually assumed to extend in the range $0 \leq y^+ \lesssim 5$–7, is dominated by viscous forces. Within this sublayer, the mean velocity varies linearly as

$$u^+ = y^+. \tag{14.5}$$

Thus, if reliable mean velocity measurements within the viscous sublayer were available, linear least-squares fitting, combined with Eq. (14.1), would easily provide the mean wall shear stress. In practice, however, such measurements may be subject to significant systematic and random errors, which may be apparent in the form of profile non-linearity or scatter, but may also include subtle systematic effects [7,8]. Thus it is often necessary to estimate wall shear stress from velocity measurements outside the viscous sublayer, which are subjected to much lower uncertainty.

A suitable region beyond the viscous sublayer is the *inertial or logarithmic sublayer*, in which the velocity is commonly described by the *law of the wall*,

$$u^+ = \frac{1}{\kappa} \ln y^+ + B, \tag{14.6}$$

where κ and B are empirical constants (κ is called the *von Kármán constant*). This layer starts at $y^+ \approx 30$, but its outer bound cannot be specified, because it increases with increasing Reynolds number. The region 5–$7 \lesssim y^+ \lesssim 30$ is called the *buffer sublayer*, where the velocity is described by an interpolation expression. Although the universality of both expression (14.6) and the values of the constants remain topics of argument among researchers, one may plausibly adopt this expression with the typical values $\kappa \approx 0.39$ and $B \approx 5.0$. Equation (14.6) is often written as

$$u^+ = A \log y^+ + B, \tag{14.7}$$

where, obviously,

$$A = \frac{\ln 10}{\kappa} \approx 5.9. \tag{14.8}$$

A traditional approach to estimating u_τ from velocity measurements within the logarithmic sublayer is to use the *Clauser chart* [9], which consists of plots of the

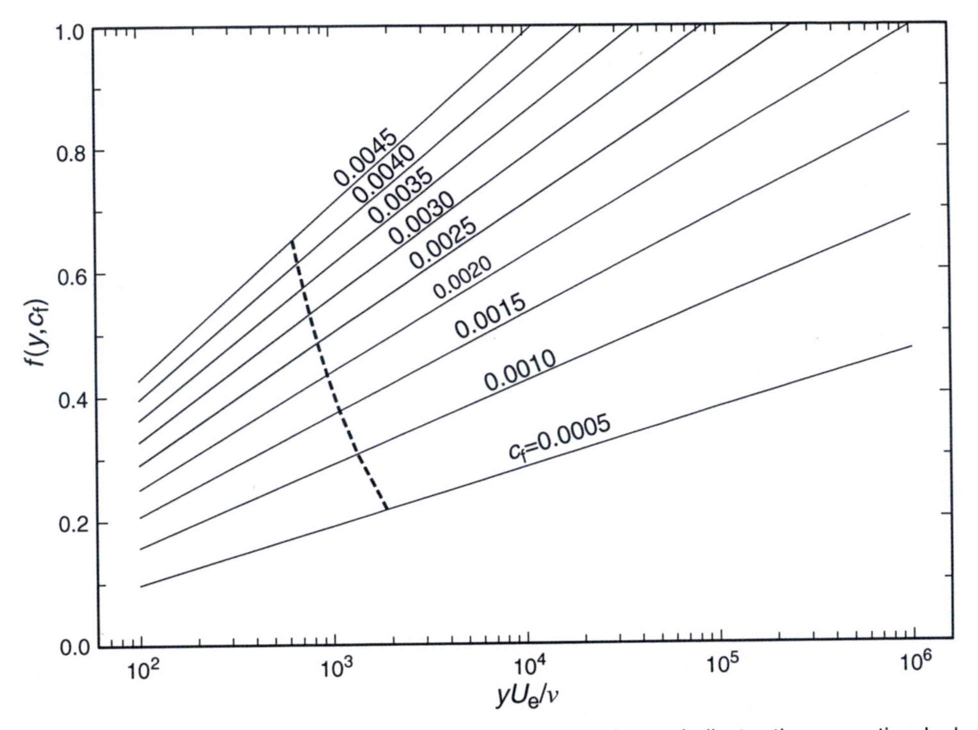

Figure 14.1. Clauser chart for $A = 5.9$ and $B = 5.0$; the dashed curve indicates the conventional edge of the logarithmic sublayer, at $y^+ = 30$.

function

$$f(y, c_f) = \sqrt{\frac{c_f}{2}}\left(A \log \frac{yU_e}{\nu} + A \log \sqrt{\frac{c_f}{2}} + B\right), \tag{14.9}$$

vs. $\log \frac{yU_e}{\nu}$, with the *skin friction coefficient*

$$c_f \equiv \frac{\tau_w}{\frac{1}{2}\rho U_e^2} = 2\left(\frac{u_\tau}{U_e}\right)^2 \tag{14.10}$$

as a parameter (see Fig. 14.1). Velocity measurements, normalized as $\overline{U}/U_e$, are plotted vs. $\log \frac{yU_e}{\nu}$ in the same chart and the value of c_f for which the corresponding f-curve is closest to the measurements is determined. To reduce the subjectiveness of the Clauser chart method, one may compute u_τ directly by solving the law of the wall, rewritten in the form

$$\log \frac{y\overline{U}}{\nu} = \frac{1}{A}\left(\frac{\overline{U}}{u_\tau} - B\right) + \log \frac{\overline{U}}{u_\tau} \tag{14.11}$$

for a selected pair of measurements $(y, \overline{U})$ within the logarithmic sublayer. Because Eq. (14.11) is implicit, its solution can be only graphical [10] or numerical (see Fig. 14.2). The random uncertainty of u_τ can be reduced by the averaging of estimates determined from several pairs of points within the logarithmic sublayer.

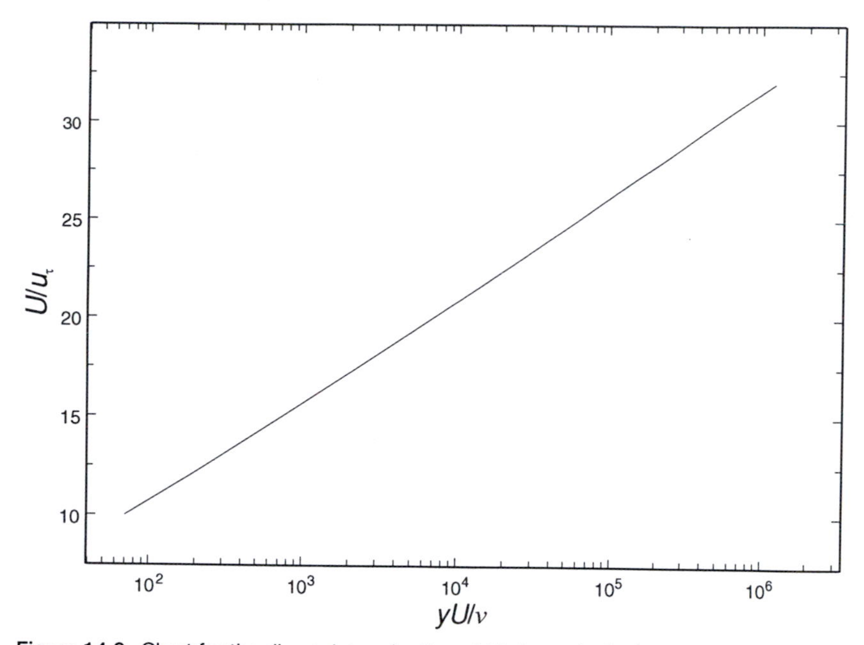

Figure 14.2. Chart for the direct determination of friction velocity from a single measurement point in the logarithmic sublayer according to Eq. (14.11) with $A = 5.9$ and $B = 5.0$.

An alternative approach allows for a partly optimal fitting of the law of the wall to a specific available set of measurements by assuming a value for the constant A only, rather than for both A and B. For this approach, it is more convenient to rewrite the law of the wall as

$$\frac{\overline{U}}{U_e} = \sqrt{\frac{c_f}{2}} A \log \frac{yU_e}{\nu} + \sqrt{\frac{c_f}{2}} \left(\log \sqrt{\frac{c_f}{2}} + B \right). \tag{14.12}$$

Given a set of measured pairs of values $(y, \overline{U})$, plots of $\overline{U}/U_e$ vs. $\log(yU_e/\nu)$ can be fitted by a straight line, extending only in the data range in which linear fitting seems appropriate. The optimal value of c_f can be determined from the slope $\sqrt{c_f/2}A$ of the fitted line, if one assumes a value for A, for instance the value 5.9, corresponding to $\kappa \approx 0.39$. The optimal value of the coefficient B can be determined by extrapolation of the fitted line to $\overline{U}/U_e = 0$. Repeating the procedure with minor adjustments in the boundaries of the 'linear' range and the chosen value of A would provide a measure of uncertainty in the determination of c_f.

Generalized expressions that contain the law of the wall but that are also valid in the buffer, viscous, and/or outer sublayers have also been used for the calculation of wall shear stress [1, 3]. All such methods depend on the applicability of the assumed universal expression to the particular set of measurements at hand. Universal laws of the wall describe well fully developed, two-dimensional, high-Reynolds-number boundary-layer profiles, but their applicability to low-Reynolds-number boundary layers, and flows in their early stages of development, over highly curved and three-dimensional surfaces, or near separated conditions, cannot be taken for granted.

14.2 Estimates from pressure differences

The wall shear stress is intimately related to static-pressure losses in pipes and channels and to pressure differences generated by obstructions at or near the wall. In general, however, the wall shear stress also depends on other factors, so that it cannot be determined from a single measured pressure difference and a simple theoretical expression. The notable exception is fully developed flow in horizontal circular pipes, in which streamwise momentum balance gives the (uniform) wall shear stress in terms of the pipe diameter D and the streamwise wall-pressure gradient, as

$$\tau_w = \frac{D}{4}\left(-\frac{\mathrm{d}p}{\mathrm{d}x}\right), \tag{14.13}$$

an expression valid for both laminar and turbulent flows. The entrance length L for flow development can be estimated as $L/D \approx 0.06\mathrm{Re}$ for laminar flows and $L/D \approx 25$–40 for turbulent ones. Although this configuration is rarely encountered in most applications of interest, it is suitable for the calibration of other wall shear stress measurement methods. Early techniques utilizing empirical correlations between wall shear stress and pressure differences include the use of two wall taps of different sizes, inclined taps [11], taps distorted to resemble forward- or backward-facing steps, and others. Such techniques have been of limited application because of their low sensitivity and the difficulty in being reproduced accurately [1]. In the following subsections, three related techniques, which are relatively easy to apply and have enjoyed some popularity, are discussed.

Preston tubes: The *Preston tube* is essentially a Pitot tube (see Section 8.4) resting on the wall [Fig. 14.3(a)]. This technique presumes universality of an inner boundary-layer scaling law, although not any specific theoretical relationship. It measures the shear stress τ_w, through calibration, from the difference Δp between the total pressure measured by the tube and the local static pressure measured by a nearby wall tap, with the two parameters represented in dimensionless forms as $\tau_w/(\rho\nu^2/d^2)$ and $\Delta p/(\rho\nu^2/d^2)$ (d is the external tube diameter, ρ is the density, and ν is the kinematic viscosity). Most researchers use rather than Preston's [12] original calibration expression, the expression that is due to Patel [13], which correlates the parameters

$$x^* = \log\frac{\Delta p}{4\rho\nu^2/d^2} \quad \text{and} \quad y^* = \log\frac{\tau_w}{4\rho\nu^2/d^2} \tag{14.14}$$

as (the latter of the three expressions has been inverted from its original implicit form)

$$y^* = 0.037 + 0.50x^*,$$

for

$$\begin{aligned} 0 &< x^* < 2.9, \\ 0 &< y^* < 1.5, \\ 0 &< d^+ < 11.2; \end{aligned}$$

$$y^* = 0.8287 - 0.1381x^* + 0.1437x^{*2} - 0.0060x^{*3}, \tag{14.15}$$

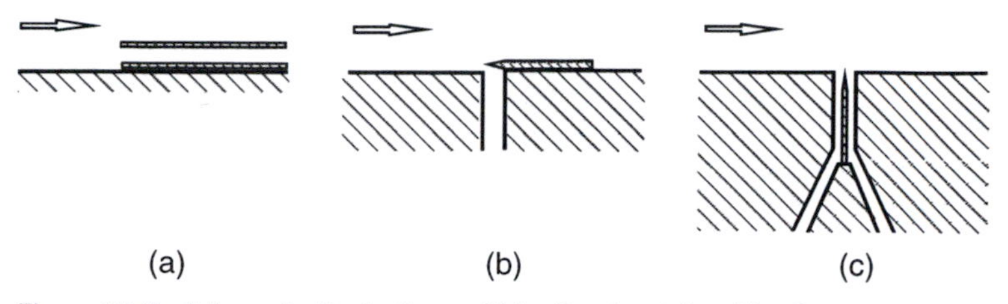

Figure 14.3. Schematic illustrations of (a) a Preston tube, (b) a Stanton gauge, and (c) a sublayer fence.

for

$$2.9 < x^* < 5.6,$$
$$1.5 < y^* < 3.5,$$
$$11.2 < d^+ < 110;$$

$$y^* = -0.9654 + 0.718x^* + 0.0175x^{*2} - 0.0005x^{*3},$$

for

$$5.6 < x^* < 7.6,$$
$$3.5 < y^* < 5.3,$$
$$110 < d^+ < 1600.$$

Notice that the parameter y^* is related to the dimensionless external tube diameter $d^+ = du_\tau/\nu$ (which may also be viewed as the tube Reynolds number), as

$$y^* = 2\log\left(d^+/2\right). \tag{14.16}$$

The three ranges in expressions (14.15) roughly correspond to the viscous, buffer, and logarithmic sublayers. Uncertainties in the expressions for the latter two layers have been cited as ±1% and ±1.5%, respectively. For higher Reynolds numbers, one may use the following alternative expression (also inverted from the original implicit form) [14]:

$$y^* = -1.1649 + 0.784x^* + 0.0104x^{*2} - 0.000235x^{*3}, \tag{14.17}$$

for

$$6.4 < x^* < 11.3,$$
$$4.3 < y^* < 8.7,$$
$$280 < d^+ < 45000.$$

A compound plot of $\tau_w/\left(4\rho\nu^2/d^2\right)$ vs. $\Delta p/\left(4\rho\nu^2/d^2\right)$, based on all the preceding expressions and also showing the variation of d^+, is shown in Fig. 14.4.

The preceding expressions apply to flows with a zero or negligible streamwise pressure gradient, whereas Preston tubes would generally overestimate the wall shear stress when inserted in flows with either a favourable ($\mathrm{d}p/\mathrm{d}x < 0$) or an adverse ($\mathrm{d}p/\mathrm{d}x > 0$) pressure gradient [1, 13]. An error bound of 3% is to be expected for flows in which the pressure gradient parameter $\Delta = \left(\nu/\rho u_\tau^3\right)(\mathrm{d}p/\mathrm{d}x)$ is in the range $-0.005 < \Delta < 0.01$, provided that $d^+ \leq 200$ and the flow is not in the process of relaminarization ($\mathrm{d}\Delta/\mathrm{d}x < 0$, when $\Delta < 0$) [13]. In addition to applying to smooth walls,

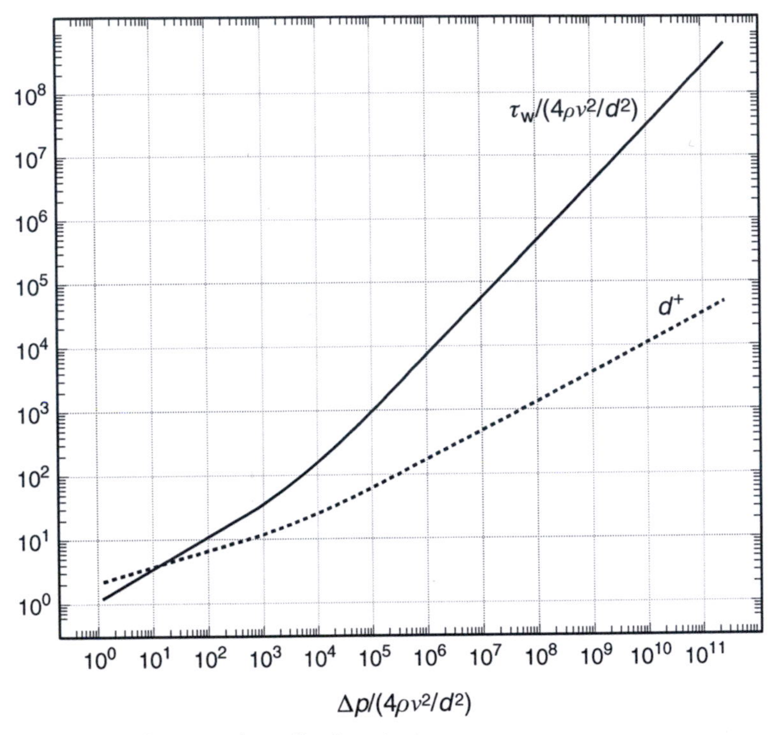

Figure 14.4. Preston tube calibration chart.

the preceding expressions also apply to *hydraulically smooth* walls, namely those having roughness elements whose height k is lower than the laminar sublayer thickness, i.e., $ku_\tau/\nu < 5$. Correction factors to Patel's equations for rough walls have been presented in the literature but carry considerable uncertainty [15]. Additional corrections for effects of turbulence and compressibility are also available [1, 2].

The accuracy of Preston tube measurements also depends on practical details. To minimize flow disturbance, it is preferable to insert the tube through the wall or fasten it on the wall, rather than introduce it, like a Pitot tube, through the flow. To maintain the tube within the inner boundary layer while also ensuring a fairly fast response, one may combine a smaller section as a tip, connected to progressively larger sections downstream. The use of flattened tubes has also been suggested, although this is expected to increase measurement uncertainty, unless a specific calibration is performed. Static-pressure taps should be of good quality and small size and at the same streamwise location as the tube tip. They should be close to the tube side, but far enough to avoid blockage effects (one can test this by comparing measured static pressures with and without the tube). Averaging the readings of two static taps on either side of the tube might also be of some advantage. Preston tube readings do not appear to be particularly sensitive to the inner-to-outer diameter ratio, but it seems prudent to keep this near the main value of 0.6 used by Patel. A few variants of the basic configuration, involving tubes with slanted tips and multiple tubes, have also been employed to a limited extent [1].

Razor blade technique: This approach originated with an arrangement known as the *Stanton tube* [16], consisting of a rectangular tube resting on the wall and sufficiently thin to be immersed within the viscous sublayer. The most widely used variant, sometimes referred to as the *Stanton gauge* [Fig. 14.3(b)], employs a thin blade firmly attached to the wall (e.g., magnetically or by an adhesive) and partly blocking the opening of a static-pressure tap. The blade thickness must also be smaller than the viscous sublayer thickness, and the wall shear stress is measured, through calibration, from the difference between the pressure in the blocked tap and the local static pressure. Empirical expressions, similar to those for Preston tubes, have been proposed by different authors [17, 18], but they are very sensitive to geometrical particulars, thus necessitating the individual calibration of each device. Uncertainty estimates for calibrated Stanton gauges are typically $\pm 3\%$ and increase dramatically in flows with considerable pressure gradients.

Sublayer fence: This device consists of a wall tap with a thin blade mounted inside it such that it partitions the tap into upstream and downstream halves [Fig. 14.3(c)] [4, 19, 20]. The tip of the blade must extend slightly into the flow but must remain immersed within the viscous sublayer, as the name *sublayer fence* indicates. The difference in pressures in the two halves of the tap is linearly related to the wall shear stress, with the proportionality coefficient determined through calibration. Sublayer fences have a higher sensitivity than that of Stanton tubes.

14.3 Floating-element balances

The principle of this method is simple: Wall shear stress is measured as the ratio of the shear force applied on the surface of a sensor mounted flush with the wall and the sensing area. Thus such a sensor is essentially a force balance. Its sensing component is a *floating element* [Fig. 14.5(a)] whose surface is exposed to the flow and is mounted inside a wall cavity with some small clearance around it so that it may move laterally under the influence of wall shear stress. A great variety of designs has been proposed; these designs determine force either by measuring the displacement of the floating element or by measuring the feedback required for restoring the element to its null position. Conventional transducers that have been used to measure the force include LVDTs, strain gauges, and piezoelectric transducers. A detailed description of such devices and a discussion of their limitations have been presented in previous reviews [1–3]. Limitations include relatively low spatial and temporal resolutions and sensitivity to floating-element misalignment with the surface, surrounding clearance size, temperature variation, acceleration, vibrations, and pressure gradients. Although their output is directly proportional to wall shear stress and their response does not require the use of any analogy or theory, wall shear stress balances require calibration under conditions comparable with those in the experiment.

In recent decades, floating-element sensors have been constructed by use of silicon micromachining technology (MEMS devices) [5,21]. Besides the fact that they are much smaller than conventional sensors, they also have the advantage of lower misalignment

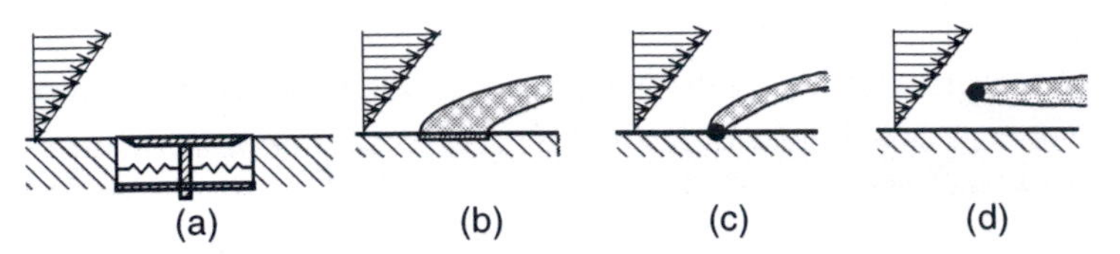

Figure 14.5. Schematic illustrations of (a) a floating-element balance, (b) a flush-mounted hot-film, (c) a groove-mounted hot-wire, and (d) a wall-hot wire.

errors, because of their integrated fabrication. In electrical MEMS sensors, the sensing element is tethered to its support by semiconductor links acting as elastic springs, and its displacement is detected by capacitive or piezoelectric components. In an optical-detection design [22], the floating element is mounted over two photodiodes, which measure displacement by collecting the light of a laser beam, blocked partially by the element. This device is suitable for measuring shear stress fluctuations up to a frequency of 10 kHz, with a resolution of 0.003 Pa.

14.4 Thermal techniques

Thermal sensors for wall shear stress measurement are very similar to those used in hot-wire–hot-film and pulsed-wire anemometry (see Section 11.1). Their main difference is that the thermal fields of the former must be immersed in the viscous sublayer, in which the local flow velocity is proportional to the wall shear stress, according to Eq. (14.5). Like the corresponding velocity thermal sensors, they are divided into two categories: those measuring wall shear stress through its analogy to heat flux from the heated element and those that measure wall shear stress from the time of flight of heated spots. Details on the different sensor designs and their operation have been included in previous reviews [1, 4, 23], so the following discussion focusses on fundamental concepts. The section closes with a recent suggestion for a different thermal technique with some potential for development.

Flush-mounted hot-films: These are metallic (commonly golden) film sensors, mounted flush on the wall and heated electrically by a constant-temperature anemometer circuit. They must be sufficiently small for their thermal layer thickness to be smaller than the viscous sublayer thickness [see Fig. 14.5(b)]. Theoretical analysis predicts that, under such conditions, the heat transfer rate from the heated element would be proportional to the $1/3$ power of the local wall shear stress. To account for small differences among individual sensors, their response is described by an empirical expression analogous to the modified King's law in HWA [Eq. (11.8)], as

$$\frac{E^2}{T_w - T_f} = A + B\tau_w^{1/3}, \tag{14.18}$$

where E is the supplied voltage, T_w is the film temperature, T_f is the bulk fluid temperature, and the coefficients A and B are determined by calibration vs. a Preston tube or other standard. Because of their small size, hot-film sensors are sensitive to wall shear stress fluctuations and may capture unsteady phenomena and turbulent fluctuations, although not necessarily the fine structure, at least in high-speed air flows. Like all thermal sensors, flush-mounted films cannot discriminate between forward and reverse flows. Nevertheless, if mounted on a turntable, they can identify the near-wall mean flow direction, which will be parallel to their long axes at an orientation such that the heat transfer rate is at a minimum. Besides hot-films, hot-wires have also been used under similar conditions, by being partially embedded in a groove on the wall [see Fig. 14.5(c)]. Their sensitivity is generally lower than that of hot-films and they have not found wide application.

Hot-film arrays: Arrays of co-planar, rectangular, thin metallic films are available commercially and may be also fabricated by use of integrated circuit technology. They are epoxied on the wall surface with the film's long axes normal to the flow direction. Each film must be supplied by a separate anemometer circuit in the constant-temperature mode. In principle, they can be calibrated so that each sensor would measure the local wall shear stress; however, the responses of individual sensors would be complicated by interference from the thermal fields of other sensors. In practice, they are only used for qualitative purposes, mainly to identify the locations of flow separation, transition to turbulence, shock waves, and other discontinuities near the wall. For this reason, they may be powered by relatively simple and inexpensive circuitry, which is an essential requirement when considering that such arrays often contain a large number of elements. Their frequency response may be sufficiently high to detect transient and oscillatory phenomena. Micro-hot-film arrays with large numbers of active elements (in the hundreds or even thousands) can be manufactured by use of MEMS technology [24].

Wall hot-wires: These are conventional single-wire sensors, which are mounted on prongs that protrude slightly through the wall while remaining within the viscous sublayer, and measure wall shear stress through its proportionality to local flow velocity [see Fig. 14.5(d)]. Their voltage output is related to wall shear stress by a modified King's law, including a correction for wall effects on the sensor heat transfer [25]. The sensitivity of wall hot-wires improves with increasing distance from the wall, and their frequency response is higher than that of other wall shear stress sensors.

Pulsed films and wires: These transducers consist of a central heated element, which is supplied with a current pulsed at some appropriate frequency, and two temperature-sensing elements on either side of the central one, which detect the convection of heated parcels of fluid. Both flush-mounted pulsed-film arrangements and near-wall-mounted pulsed-wire arrangements have been used. The main advantage of these sensors over other techniques is that they can identify the instantaneous flow direction and measure wall shear stress fluctuations in separated and reverse flows.

Infrared thermography: In this approach, a laser beam is used to heat a spot on the surface of an immersed object, which is made of a material with a low thermal conductivity and a high emissivity [26]. The spot temperature is continuously monitored with an infrared camera. When a steady temperature is reached, the laser is turned off and the spot-temperature time history is recorded. This time history is used as input to a numerical algorithm that calculates heat transfer rates and, through analogy, wall shear stress. The uncertainty of the method is still quite high, about 10%.

14.5 Electrochemical method

The electrochemical method measures wall shear stress from the current that must be supplied to an electrode flush with the wall in order to sustain a chemical reaction in the flowing fluid. It has an advantage over previously discussed methods in that its response is known theoretically, thus eliminating the need for calibration. Its main limitation is that it can be used only in flows of special liquids, being unsuitable for air and water flows. The analytical prediction of the mass transfer from the electrode, which must occur within the viscous sublayer, is analogous to heat transfer analysis from a heated element, both mass flux and heat flux being proportional to the $1/3$ power of the wall shear stress [see Eq. (14.18)]. The most commonly used active substances are ferricyanide, $Fe(CN)_6^{3-}$, and ferrocyanide, $Fe(CN)_6^{4-}$, mixed in sodium hydroxide, NaOH (a neutral electrolyte), as the working fluid. The sensor is a small, thin nickel film, flush with the wall [similar to the hot-film shown in Fig. 14.5(b)], serving as the cathode, whereas the anode is a second, much larger, nickel electrode, positioned downstream. Free electrons released at the cathode reduce ferricyanide ions into ferrocyanide ions and the reaction is reversed at the anode, as

$$Fe(CN)_6^{3-} + e^- \underset{\text{cathode}}{\overset{\text{anode}}{\leftrightarrows}} Fe(CN)_6^{4-} \tag{14.19}$$

The ion flux from a rectangular cathode having an exposed area A and a length L in the streamwise direction is expressed, in dimensionless form, by the *Sherwood number*

$$\mathrm{Sh} = \frac{LI}{c_0 \gamma_c A \mathcal{F}}, \tag{14.20}$$

where I is the electric current, c_0 is the bulk concentration of ferricyanide ions in the mixture, γ_c is the molecular diffusivity of the same ions, and $\mathcal{F} = 94484.6\ \mathrm{C\,mol^{-1}}$ is Faraday's constant. Under appropriate experimental conditions, the diffusion equation can be solved to provide an 'exact' relationship between electric current and wall shear stress as

$$\mathrm{Sh} = 0.807 \left(\frac{L^2}{\mu \gamma_c}\right)^{1/3} \tau_w^{1/3}. \tag{14.21}$$

The same result applies to circular electrodes, provided that the length L is replaced with the effective length $0.8139d$, where d is the electrode diameter. The uncertainty of electrochemical sensor measurements in steady flows is about 3%–4%. Their frequency response is sufficient to resolve turbulent fluctuations under certain conditions, although corrections might be necessary. Various aspects of the response of such probes have been discussed by a number of authors [2, 27].

14.6 Optical techniques

Optical techniques are generally attractive for being non-intrusive and having a response based on theoretical relationships, thus eliminating the need for calibration. On the other hand, they are subjected to other limitations, such as the need for flow seeding or the use of special coatings. In the following subsections, three recently developed optical techniques that have found increasing application are discussed.

Laser Doppler technique: This technique is a variation of LDV, adapted to the measurement of wall shear stress [28, 29]. A laser beam is passed through a diffractive lens and enters the test section through two narrow, closely spaced, parallel slits, etched on a chromium coating of a glass window, as shown in Fig. 14.6. Interference of the light diffracted by the two slits creates interference fringes (see Section 5.3) in the wall region, with a spacing δ that increases linearly with distance y from the wall, as

$$\delta = \frac{\lambda}{S} y, \tag{14.22}$$

where λ is the laser-light wavelength and S is the distance between the slits. Particles seeded into the flow will scatter light as they cross the fringes at the Doppler frequency (see Section 11.2):

$$f_D = \frac{U}{\delta}. \tag{14.23}$$

A receiving lens collects the light from a measuring volume and projects it onto a photodetector. If the measuring volume is in the viscous sublayer, where the velocity varies linearly with y, then all particles will scatter light at the same Doppler frequency and the wall shear stress can be determined as

$$\tau_w = \frac{\mu \lambda}{S} f_D. \tag{14.24}$$

This technique has the advantages of not requiring calibration and providing the instantaneous value of τ_w, from which statistical properties can be determined. With frequency shifting of light transmitted through one of the slits, it can also determine flow reversal. Its disadvantage is that it requires seeding, and it is subjected to the limitations of particle response near the wall.

Besides the original version of this technique, which utilized an Ar-ion laser and conventional optical components [28], a miniaturized version has also been developed by use of a small laser diode, a photodiode, optical fibres, and MOEMS (micro-optical–electrical–mechanical systems) technology [29].

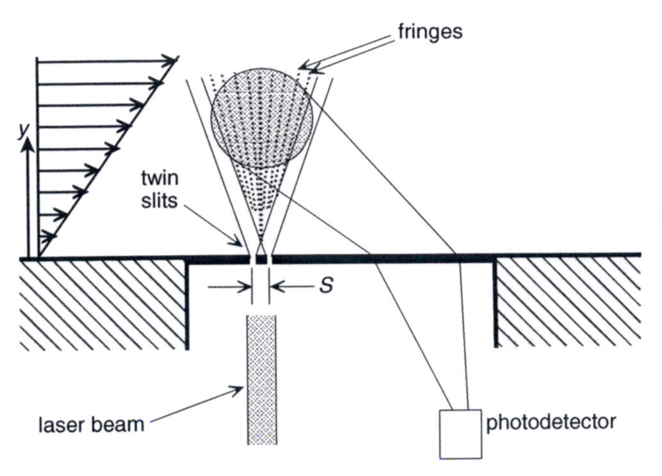

Figure 14.6. Sketch of a laser Doppler wall shear stress measurement setup.

Oil-film interferometry: This method, which is an extension of the oil streak technique used for flow visualization, has found considerable application in wind-tunnel flows over models. It measures wall shear stress through its relationship to the thickness of an oil film in contact with the wall. To demonstrate this principle, consider a small oil droplet applied on a smooth wall. When exposed to a flow, the droplet will deform into a wedge-like film, whose local thickness will continually diminish as time passes (Fig. 14.7).

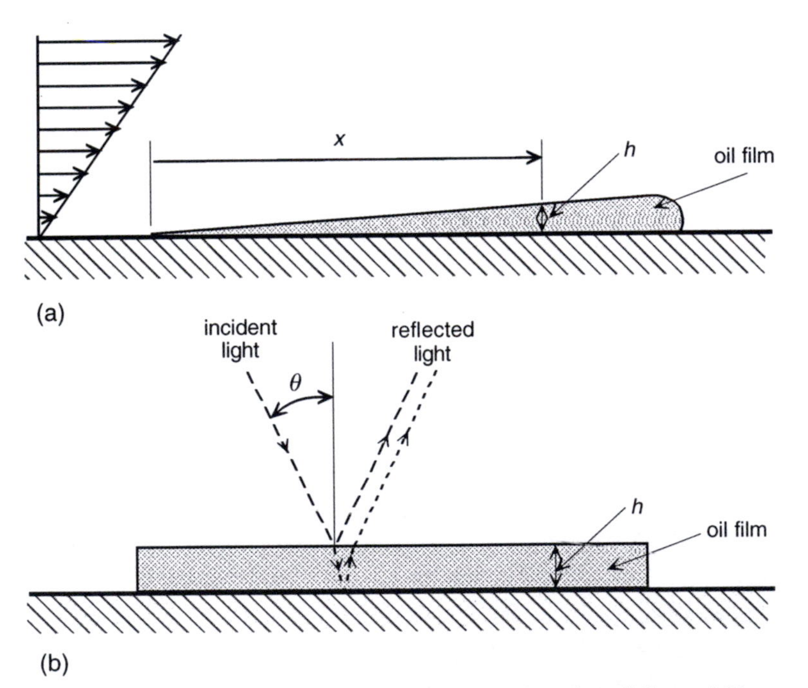

Figure 14.7. Sketches of (a) a streamwise cross-section of an oil film and (b) a spanwise cross-section of the same film at a location x, also showing the incident and reflected light beams.

For simplicity, assume that the oil flow is two dimensional and that the shear stress is uniform. For small film thicknesses, one may neglect the effects of gravity, pressure gradient, and surface tension and use lubrication theory to derive the following relationship among wall shear stress, local thickness h, and time t:

$$\tau_w = \frac{\mu_{oil} x}{h(x,t)t}. \tag{14.25}$$

In practice, the oil thickness is of the order of tens of micrometres and is most conveniently measured by an interferometric setup. Monochromatic light from a sodium or mercury lamp is directed on the film at an angle of incidence θ. Part of this light is reflected off the oil surface, while another part enters the oil. Part of the latter will be reflected off the wall material (commonly used materials include glass, polished steel, and Mylar) and pass for a second time through the film to enter the air stream again. The two reflected beams will interfere with each other, producing dark fringes on the oil surface when their phase shift ϕ is an odd multiple of 2π and bright fringes when the phase shift is an even multiple of 2π. The local film thickness is related to the phase shift as

$$h(x,t) = \frac{\lambda \phi}{4\pi \sqrt{n_{oil}^2 - n_{air}^2 \sin^2 \theta}}, \tag{14.26}$$

where n_{oil} and n_{air} are, respectively, the refractive indices of oil and air and λ is the wavelength of the incident light. Obviously the fringes will shift downstream with time, as the film thickness diminishes. Images of the film are used to identify the locations and orders of the various fringes, from which the wall shear stress can be determined with Eqs. (14.25) and (14.26). Theoretical treatments and practical details on the method can be found in the literature [4–6, 30–33].

Liquid-crystal technique: LC coatings are frequently used for flow visualization and surface temperature mapping, but have also found application for the measurement of wall shear stress [5, 34–36]. Among the different variants of this approach, one that has quantitative capabilities is as follows. A thin coating of LCs is applied on a surface and illuminated by white light, in a direction normal to the surface, within $\pm 15°$. The choice of crystals among the available commercial products is such that their appearance in the absence of shear stress is in the red-orange range. When viewed by an observer positioned upstream, the crystal colour changes towards the blue as the shear stress increases. In contrast, an observer positioned downstream would see no colour change. This happens because the birefringence of LCs depends on the applied shear stress, resulting in changes of the spectral content of light reflected by the crystal. The perceived colour depends both on the magnitude and the orientation of the shear stress. To resolve both, several images of the area of interest are taken by digital cameras from different orientations. The coating has to be replaced after each use. This technique is still quite cumbersome and requires attention to many details, so that it remains the task of specialists.

QUESTIONS AND PROBLEMS

1. LDV measurements in a low-Reynolds-number turbulent boundary layer, generated in a water tunnel at a temperature of 20 °C, consist of the following pairs of distance y from the wall (in millimetres) and mean streamwise velocity $\overline{U}$ (in millimetres per second) [37]:

0.2, 42	1.8, 196	8, 256	26, 307	55, 340
0.3, 65	2, 205	9, 258	28, 312	60, 341
0.4, 85	2.5, 208	10, 264	30, 315	70, 342
0.5, 106	3, 220	12, 273	32, 318	80, 342
0.6, 121	3.5, 225	14, 278	34, 320	90, 342
0.8, 143	4, 232	16, 282	36, 323	100, 341
1, 162	4.5, 234	18, 289	38, 324	110, 341
1.2, 174	5, 239	20, 295	40, 329	120, 341
1.4, 182	6, 245	22, 300	45, 334	130, 342
1.6, 191	7, 249	24, 303	50, 337	150, 341

   a. Determine the wall shear stress by assuming that a number of the measurements closest to the wall are in the viscous sublayer; verify the validity of this assumption.
   b. Determine the wall shear stress by using a Clauser chart.
   c. Determine the wall shear stress by solving Eq. (14.11) for several pairs of measurements within the logarithmic sublayer; compare these values and repeat the procedure for different combinations of values of the coefficients A and B.
   d. Determine the wall shear stress by utilizing Eq. (14.12) and the associated procedure; determine the optimal value of B and compare it with those used in parts b and c; repeat the procedure for different values of A.
   e. Compare the results of the preceding calculations and discuss the accuracy of these procedures.
2. Assume that a Preston tube is used to measure wall shear stress in the boundary layer presented in Problem 1. Estimate the pressure difference between the Preston tube and a nearby wall tap if the tube diameter is 0.5, 1, 3, or 10 mm.
3. A Preston tube with a diameter of 1.2 mm reads a pressure of 3.9 kPa above static in air flow in a duct. Determine the local wall shear stress.
4. Assume that a flush-mounted hot-film that has been calibrated in steady flow is used in a pulsatile flow in which the wall shear stress varies sinusoidally as

$$\tau_w = \overline{\tau}_w(1 + a \sin \omega t),$$

and assume that the frequency response of the sensor is sufficient to capture the preceding fluctuations without appreciable distortion; derive an expression for the ratio of the estimate of $\overline{\tau}_w$, obtained under the assumption that Eq. (14.18) relates it to the mean voltage $\overline{E}$ and the actual mean wall shear stress $\overline{\tau}_w$; assume that the amplitude a is small compared with 1 and use binomial series expansions to produce a first-order estimate; estimate the limits of validity of this approach, if the acceptable uncertainty must be maintained below 5%.

5. Assume that the laser Doppler technique is used to measure the wall shear stress in the boundary layer presented in Problem 1. If a He–Ne laser is used, determine the Doppler frequency f_D that would be measured if the distance S between slits were 10 μm.

REFERENCES

[1] K. G. Winter. An outline of the techniques available for the measurement of skin friction in turbulent boundary layers. *Prog. Aerosp. Sci.*, 18:1–57, 1977.

[2] T. J. Hanratty and J. A. Campbell. Measurement of wall shear stress. In R. J. Goldstein, editor, *Fluid Mechanics Measurements* (2nd Ed.), pp. 575–648. Taylor & Francis, Washington, DC, 1996.

[3] J. H. Haritonidis. The measurement of wall shear stress. In M. Gad-el-Hak, editor, *Advances in Fluid Mechanics Measurements*, pp. 229–261. Springer-Verlag, Berlin, 1989.

[4] H. H. Fernholz, G. Janke, M. Schober, P. M. Wagner, and D. Warnack. New developments and applications of skin-friction measuring techniques. *Meas. Sci. Technol.*, 7:1396–1409, 1996.

[5] J. W. Naughton and M. Sheplak. Modern developments in shear-stress measurement. *Prog. Aerospace Sci.*, 38:515–570, 2002.

[6] C. Mercer (Editor). *Optical Metrology for Fluids, Combustion and Solids*. Kluwer Academic, Dordrecht, The Netherlands, 2003.

[7] F. Durst, R. Mueller, and J. Jovanovic. Determination of the measuring position in laser-doppler anemometry. *Exp. Fluids*, 6:105–110, 1998.

[8] N. Hutchins and K.-S. Choi. Accurate measurements of local skin friction coefficient using hot-wire anemometry. *Prog. Aerospace Sci.*, 38:421–446, 2002.

[9] F. H. Clauser. Turbulent boundary layers in adverse pressure gradients. *J. Aeronaut. Sci.*, 21:91–108, 1954.

[10] V. Ozarapoglu. *Measurements in Incompressible Turbulent Flows*. Ph.D. thesis, Laval University, Quebec, Canada, 1973.

[11] G. Onsrud, L. N. Persen, and L. R. Saetran. On the measurement of wall shear stress. *Exp. Fluids*, 5:11–16, 1987.

[12] J. H. Preston. The determination of turbulent skin friction by means of pitot tubes. *J. R. Aeronaut. Soc.*, 58:109–121, 1954.

[13] V. C. Patel. Calibration of the Preston tube and limitations on its use in pressure gradients. *J. Fluid Mech.*, 23:185–208, 1965.

[14] M. V. Zagarola, D. R. Williams, and A. J. Smits. Calibration of the Preston probe for high Reynolds number flows. *Meas. Sci. Technol.*, 12:495–501, 2001.

[15] A. B. Hollingshead and N. Rajaratnam. A calibration chart for the Preston tube. *J. Hydraual. Res.*, 18:313–326, 1980.

[16] T. E. Stanton, D. Marshall, and C. N. Bryant. On the conditions at the boundary of a fluid in turbulent motion. *Proc. R. Soc. London Ser. A*, 97:413–434, 1920.

[17] L. F. East. Measurement of skin friction at low subsonic speeds by the razor blade technique. Tech. Rep. 3525, Aeronautic Research Council, London, 1967.

[18] B. R. Pai and J. H. Whitelaw. Simplification of the razor blade technique and its application to the measurement of wall-shear stress in wall-jet flows. *Aeronaut. Q.*, 20:355–364, 1969.

[19] M. R. Head and I. Rechenberg. The Preston tube as a means of measuring skin friction. *J. Fluid Mech.*, 14:1–17, 1962.

[20] J. B. Vagt and H. Fernholz. Use of surface fences to measure wall shear stress in three-dimensional boundary layers. *Aeronaut. Q.*, 2:87–91, 1973.

[21] L. Löfdahl and M. Gad-el-Hak. MEMS-based pressure and shear stress sensors for turbulent flows. *Meas. Sci. Technol.*, 10:665–686, 1999.

[22] A. Padmanabhan, H. D. Goldberd, M. A. Schmidt, and K. S . Breuer. A wafer-bonded floating-element shear-stress micro-sensor with optical position sensing by photodiodes. *IEEE J. Microelectricalmechanical Syst.* , 5:307–315, 1996.

[23] H. H. Bruun. *Hot-Wire Anemometry*. Oxford University Press, Oxford, UK, 1995.

[24] C. M. Ho and Y. C. Tai. Micro-electro-mechanical-systems and fluid flows. *Annu. Rev. Fluid Mech.*, 30:579–612, 1998.

[25] J. D. Ruedi, H. Nagib, J. Oesterlund, and P. A. Monkewitz. Evaluation of three techniques for wall-shear measurements in three-dimensional flows. *Exp. Fluids*, 35:389–396, 2003.

[26] R. Mayer, R. A. W. M. Henkes, and J. L. Van Ingen. Wall-shear stress measurement with quantitative IR-thermography. AGARD CP-601, Paper 22. Advisory Group for Aerospace Research and Development, Paris, 1997.

[27] A. A. Van Steenhoven and F. J. H. M. Van De Beucken. Dynamical analysis of electrochemical wall shear rate measurements. *J. Fluid Mech.*, 231:599–614, 1991.

[28] A. A. Naqwi and W. C. Reynolds. Measurement of turbulent wall velocity gradients using cylindrical waves of laser light. *Exp. Fluids*, 10:257–266, 1991.

[29] D. Fourguette, D. Modarress, F. Taugwalder, D. Wilson, M. Koochesfahani, and M. Gharib. Miniature and MOEMS flow sensors. In *Proceedings of the 31st AIAA Fluid Dynamics Conference and Exhibit*, AIAA Paper 2001-2982. American Institute of Aeronautics and Astronautics, Reston, VA, 2001.

[30] L. H. Tanner and L. G. Blows. A study of the motion of oil films on surfaces in air flow, with application to the measurement of skin friction. *J. Phys. E*, 9:194–202, 1976.

[31] J. D. Murphy and R. V. Westphal. The laser interferometer skin-friction meter: A numerical and experimental study. *J. Phys. E*, 19:744–751, 1986.

[32] R. V. Westphal, W. D. Bachalo, and M. H. Houser. Improved skin friction interferometer. NASA Tech. Memo. 88216. NASA Ances Research Center, Moffett Field, CA, 1986.

[33] D. M. Driver. Application of oil film interferometry skin-friction to large wind-tunnels. AGARD CP-601, Paper 25. Advisory Group for Aerospace Research and Development, Paris, 1997.

[34] D. C. Reda and J. J. Muratore. Measurement of surface shear stress vectors using liquid crystal coatings. *AIAA J.*, 32:1576–1582, 1994.

[35] D. C. Reda, M. C. Wilder, R. Mehta, and G. Zilliac. Measurement of continuous pressure and shear distributions using coatings and imaging techniques. *AIAA J.*, 36:895–899, 1998.

[36] A. J. Smits and T. T. Lim (Editors). *Flow Visualization Techniques and Examples*. Imperial College Press, London, 2000.

[37] E. Menu. Turbulence structure near a moving wall. Tech. Rep., Department of Mechanical Engineering, University of Ottawa, 2004.

15 Outlook

Previous chapters have discussed both classical experimental techniques and some state-of-the-art methods in fluids measurement. In closing, it seems worthwhile to identify areas of experimental fluid mechanics in which further development is needed as well as to speculate on the chances of such development occurring in the near future.

The evolution of modern fluid mechanics as an important branch of physical sciences and a vital engineering discipline has been based on both theoretical formulation and empirical observation and documentation. Theoretical and experimental tools alike have grown in sophistication and complexity in response to ever-increasing technological needs. On the theoretical side, significant advances in many fluid mechanical applications have followed the development of analytical methods, such as perturbation analysis, whereas the maturing of computational fluid dynamics during the past few decades has enabled the numerical solution of many problems that could not have been tackled by other means. As earlier euphoria on the potency of computational approaches is being replaced with realism, it has become evident that experimentation retains an indispensable role in solving engineering problems and designing and testing new engineering systems. Moreover, the discovery of new concepts, the understanding of complex physical phenomena and mechanisms, and the development and verification of theories, models, and algorithms still rely heavily on the imaginative conception and meticulous execution of scientific experiments. Recent experiences and current needs point to no risk of experimental fluid mechanics becoming obsolete. But what will the fluid mechanics laboratory look like in 10, 20, or 30 years?

In forecasting forthcoming developments, one must consider that fluid mechanics is a mature and multifaceted field. Thus its development will probably continue to be steady, arduous, and gradual, rather than explosive. Experimental fluid mechanicists have long been exploiting a multitude of mechanical, thermal, acoustic, optical, electrical, and electromagnetic principles and phenomena to measure fluid properties. One may plausibly anticipate that new experimental approaches will be devised, based on hitherto unutilized physical concepts. Nevertheless, a more predictable short-term development is that existing techniques will continue to be improved, as some of their current limitations are overcome and their ranges of application are expanded.

As an example, let us consider particle image velocimetry (PIV), which is currently the most popular velocity measurement method. PIV is already very powerful, in that it provides instantaneous velocity maps, and time-resolved PIV versions have begun to emerge and improve rapidly. Yet any development in PIV technology seems to increase the demand and expectation for further improvement. Desirable future developments include the enhancement of its spatial resolution by at least 1 order of magnitude and its temporal resolution by several orders of magnitude. In addition, a substantial enhancement of its ability to map three-dimensional flow fields in three-dimensional domains would also be welcome. Such improvements would be possible only by the development of pulsed lasers with a higher energy per pulse and a higher repetition rate than currently available. It would also require enhancement of image recording and processing hardware and software. The radical consequence of having a high-resolution, fully three-dimensional PIV system would be that, for the first time, we would be able to measure the turbulence fine structure without making compromises, thus obtaining the kind of statistical information (such as velocity derivatives in arbitrary directions) that is currently provided only by direct numerical simulations.

An optimistic development in sensor technology is the adaptation of new materials and manufacturing methods to replace conventional ones. Microfabrication techniques have already permitted the production of affordable large arrays of pressure transducers, photodetectors, and other sensors. As this approach is still at a relatively early stage, one may only speculate of future advances. As in all pioneering endeavours, success is often accompanied by setbacks. For instance, the effective application of semiconductors as miniature thermal velocity sensors has not yet materialized, despite early promise. The trend, however, is clear: The experimental fluid mechanics laboratory of the future will have at its disposal smaller, faster, less expensive, mass-produced multi-sensors of pressure, velocity, wall shear stress, temperature, radiation, and concentration. This will allow the mapping of these properties in more complex flow configurations, as well as permit the use of distributed feedback to control adverse phenomena, such as flow separation, drag, and transition to turbulence.

A current area of growth, which is expected to continue, is the adaptation of existing measurement methods to new applications, which pose new challenges as the measured properties are beyond the conventional ranges of operation of the method. Examples include measurements in transonic and hypersonic flows, measurements at extremely low Reynolds numbers, measurements in flow systems with length scales in the microrange and the nanorange, measurements in living organisms and biological systems, measurements at critical conditions, measurements in opaque fluids, measurements in multiphase flows, and flows with phase change and measurements at extremely high and low pressures and temperatures. These extensions will necessitate much background support, in terms of establishing the reliability of the method and the associated measurement uncertainty and sensitivity to various influences.

Like any other endeavour into the future, the development of new instrumentation and experimental methods would greatly benefit from the experience of the past. The inventory of any fluid mechanics laboratory will more than likely contain Pitot tubes and liquid manometers, both instruments that have remained essentially unchanged for

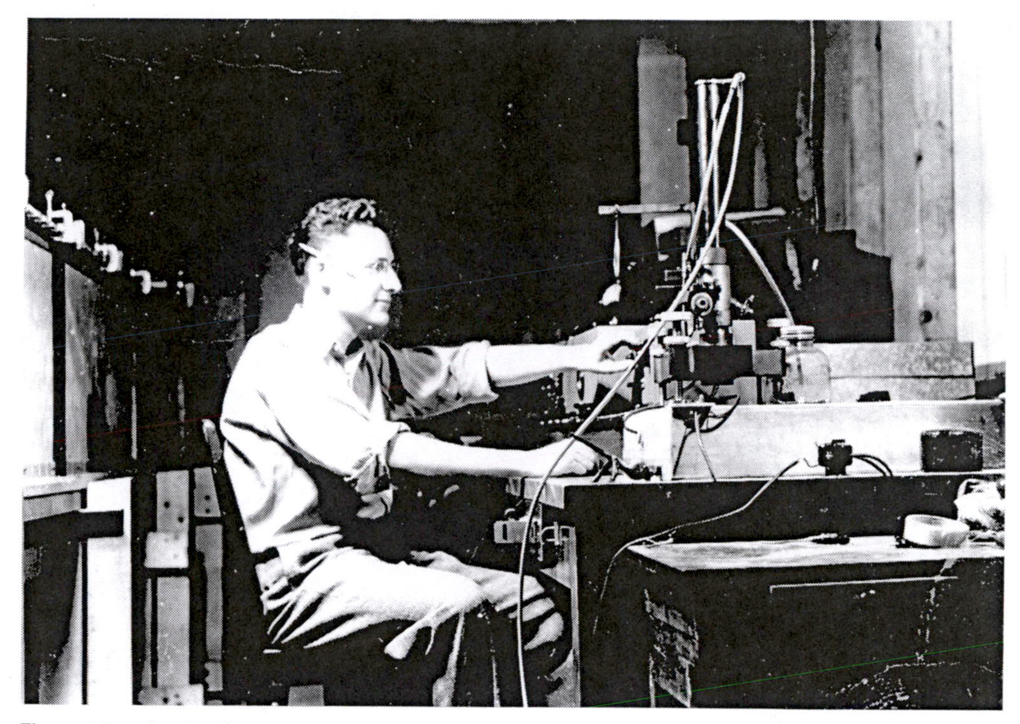

Figure 15.1. Stanley Corrsin operating a hot-wire anemometer at the Graduate Aeronautical Engineering Laboratories of the California Institute of Technology ca. 1943 (with permission of the California Institute of Technology).

hundreds of years. What makes such tools persist, whereas state-of-the-art expensive apparatus has a useful lifetime of just a few years? It is their simplicity of construction and operation, a response based on fundamental laws of mechanics, rather than calibration, and a relative insensitivity to most, although not all, sources of interference. These are qualities that any instrumentation of the future must have if it is to be of lasting service. As a final note, let us once more reconsider the objectives of a 'good' fluid mechanics experiment. It is not to impress with the cost and complexity of apparatus but with the quality of the results. Obviously the availability of suitable and functioning equipment is a prerequisite for a good experiment. That, however, comes with no guarantee of success. The experimeter is the one to decide how to set up the experiment, what to measure, how to use the equipment properly, and how to present the results. Good experimental results can have a lasting significance, much longer than the lifetime of the instruments, and even the persons, that generated them. Figure 15.1 shows the late Stanley Corrsin in action more than 60 years ago. The equipment he used includes an early hot-wire anemometer, but certainly no data acquisition systems, no computers, not even transistors of any sort. Yet Corrsin's heated jet measurements of that era are still cited and have not been superseded by more recent ones.

Index

LaVergne, TN USA
09 June 2010
185449LV00001B/9/P